Fracture Mechanics

The Experimental Method of Caustics and the Det.-Criterion of Fracture

W0044198

George A. Papadopoulos

Fracture Mechanics
The Experimental Method of Caustics
and the Det.-Criterion of Fracture

With 196 Figures

Springer-Verlag
London Berlin Heidelberg New York
Paris Tokyo Hong Kong
Barcelona Budapest

George A. Papadopoulos
Department of Engineering Science, Section of Mechanics, The National
Technical University of Athens, 5, Heroes of Polytechnion Avenue,
GR-157 73, Zographou, Athens, Greece

Cover illustration: Ch.3, Fig.71. A series of photographs of an arrested crack in a polycarbonate plate where the shear stress waves generated at the crack tip travel along the specimen. (Adaptation.)

ISBN-13: 978-1-4471-1994-4   e-ISBN-13: 978-1-4471-1992-0
DOI: 10.1007/978-1-4471-1992-0

British Library Cataloguing in Publication Data
Papadopoulos, George A.
  Fracture Mechanics: Experimental Method
  of Caustics and the Det-criterion of
  Fracture
  I. Title
  620.1126

Library of Congress Cataloging-in-Publication Data
Papadopoulos, George A., 1945-
  Fracture mechanics: the experimental method of caustics and the
det.-criterion of fracture / George A. Papadopoulos.
      p.   cm.
  Includes bibliographical references (p. 273) and indexes.

  1. Fracture mechanics. 2. Stress corrosion. I. Title.
TA409.P365   1992   92-27308
620.1'126 – dc20   CIP

Typeset by Thomson Press (India) Limited, New Delhi
69/3830-543210   Printed on acid-free paper

*To my wife Ioanna
and our daughter Elina*

Preface

This book is concerned with the experimental method of static and dynamic caustics and the determinant-criterion of fracture. Basic theoretical considerations, experimental techniques and applications are presented.

The study of the stress field created in the vicinity of a crack tip is of special importance in understanding the mechanism of fracture in most bodies. The most important processes to be evaluated are the opening mode stress intensity (K_I) factor and the sliding mode factor (K_{II}). The optical method of reflected caustics (static and dynamic), which has been developed during recent years, has been applied to many crack problems and permits the study of the stress state very close to the crack tip, in contrast to other experimental techniques, interferometric and photo-elasticity methods. By the method of dynamic caustics, dynamic crack problems can also be solved, and the path of a propagating crack observed. At each step of crack propagation the caustic gives experimental information about the direction of the next step and the values of the stress intensity factors K_I and K_{II}.

The distribution of the determinant (Det.) of the stress tensor around the crack tip gives information on the expected angle of crack propagation. According to the Det.-criterion, which is based on the determinant of the stress tensor, introduced by the author, a crack propagates in the direction of the maximum of the determinant (Det.(σ_{ij})) distribution around the crack tip. This Det.-criterion of fracture is a new and simple criterion, which has been experimentally validated.

The object of this book is to assist the teaching of experimental fracture mechanics and stress analysis in undergraduate and postgraduate courses.

Chapter 1 is concerned with the fundamentals of the stress field at the crack tip, covering complex stress functions, Westergaard stress functions and modes of deformation.

Chapter 2 contains basic considerations of static caustics. The parametric equations of the caustics are given. The stress-optical constants, which are needed for the caustics, are evaluated by the stress-optical law of geometrical optics. This method converts the stress singularity in the vicinity of the crack tip into an optical singularity, the caustic, which carries all the information concerning the stress field. The generative curve of the caustic is a mathematical curve around the crack tip on the cracked plate and coincides with the boundary of the singular region around it. The generative curve (called the initial curve of the caustic) is, in general, for the case of optically dummy and isotropic materials, a circle with its centre on the crack tip and radius of the order of 0.001 m. Relations for the calculation of stress intensity in terms of the dimensions of the caustic are also given.

Chapter 3 considers the basics of dynamic caustics. The distortion of the form of the corresponding reflected caustic from the lateral faces of a dynamically loaded transparent and optically inert specimen containing a transverse crack running under constant velocity is studied on the basis of complex potential elasticity theory and the influence of its form on the dynamic stress intensity factor is given. The method is applied to the study of a propagating crack under various propagation velocities and the corresponding dynamic stress intensity factors K_I^d and K_{II}^d evaluated. Also, dynamic caustics are applied to various crack problems of composite materials.

Chapter 4 is concerned with the elastic strain energy per unit volume of material which is stored in terms of components of the stress field at the crack tip.

Chapter 5 is concerned with the Det.-criterion of fracture. The stationary value of the second stress invariant as a local fracture parameter is studied. The second stress invariant is calculated along a circle around the crack tip. This circle, which defines the core-region around the crack tip, is the initial curve of the caustics. The distribution of the second stress invariant around the crack tip presents a positive maximum. The crack propagates in the direction of the maximum value of the stress tensor distribution and fracture will initiate when the Det.(σ_{ij}) on the core-region boundary reaches a critical maximum value which is a material property. This condition for initiation of the crack is proposed as the Det.-criterion of fracture. For various crack problems, this criterion gives theoretical results which are in agreement with experimental results. The interest in this study is that the distribution of the determinant of the stress tensor is calculated experimentally from isochromatic and isopachic fringe patterns and so the prediction of the direction of crack initiation can be experimentally calculated. Experimental stress analysis near the crack tip and remote from it, is accomplished by the combination of photoelastic, interferometric and caustics methods. So, for many crack problems, which are not easily solved theoretically, the direction of crack extension can be predicted by this experimental method.

Chapter 6 is concerned with the application of the Det.-criterion to various crack problems, such as plates with cracks or blunt notches under uniaxial or biaxial loading.

Chapter 7 is concerned with the experimental Det.-criterion of fracture which is based on isopachic and isochromatic fringe patterns.

The results presented in this book originate to a great part from the scientific effort of the author who began his investigations in the field of experimental mechanics in 1973 at the National Technical University of Athens, Greece. The author wishes to express his gratitude to many collaborators and co-authors of the papers cited in the bibliography. Wholehearted thanks are also addressed to his collaborator Dr. E. Sideridis for his continuous help.

Athens, Greece   G. A. Papadopoulos
1991

Contents

Symbols

Notation	Name
a, a_0	crack length
c	crack propagation velocity
c_r, c_t, c_f	stress-optical constants
c_1	longitudinal wave velocity
c_2	shear wave velocity
d	thickness of specimen
dW/dV	elastic strain energy density
$\text{Det.}(\sigma_{ij})$	determinant of the stress tensor
D_t^{max}, D_l^{max}	maximum diameters of the caustic
E	modulus of elasticity, Young's modulus
f_p	isopachic fringe value
f_σ	photoelastic material fringe value
f_I, f_{II}, g_I, g_{II}	correction factors of K_I and K_{II} which depend on geometry of the specimen
$F_1(x, y), F(x, y)$	Airy stress functions
G	shear modulus
G_I, G_{II}, G_{III}	elastic energy release rate
$\mathbf{i}, \mathbf{j}, \mathbf{k}$	unit vectors of Cartesian system $Oxyz$
I	light intensity
Im	imaginary part of complex function
k	biaxiality factor
K_I, K_{II}, K_{III}	stress intensity factors for mode I, II, III, respectively
K_I^d, K_{II}^d	dynamic stress intensity factors
l	length of specimen
n	hardening exponent
$N_{1,2}$	number of fringes
r	polar coordinate
r_0	radius of initial curve of caustic
Re	real part of complex function
t	time
T_D	distortional strain energy density
T_V	dilatational strain energy density
u, v, w	displacements

w	width of specimen
$\mathbf{w}$	deviation vector
x, y	Cartesian coordinates
z, z_1, z_2	complex variables
z_i	distance between the specimen plane and the focus of the light beam
z_0	distance between the specimen and the reference screen
$Z_I(z), Z_{II}(z), Z_{III}(z), Z(z)$	analytic functions, Westergaard stress functions
α, β	ratios of the light intensity reduction
β	angle of loading
γ	fracture surface energy
$\gamma_{xy}, \gamma_{xz}, \gamma_{yz}$	components of strain
$\delta_t^{max}, \delta_l^{max}$	correction factors of the diameters of the caustic
$\dot{\varepsilon}$	strain rate
ε_{ij}	strain tensor
$\varepsilon_{xx}, \varepsilon_{yy}, \varepsilon_{zz}, \varepsilon_{xy}, \varepsilon_{xz}, \varepsilon_{yz}$	components of strain
$\varepsilon_1, \varepsilon_2, \varepsilon_3$	principal strains
ε_0	tensile yield strain
ζ	complex variable
η, η_0	refractive index
θ	polar coordinate
κ	ratio of the stress intensity factors
$\varkappa$	ratio of dynamic stress intensity factors
λ	Lamé constant
λ	wavelength
λ_m	magnification ratio
$\lambda_{1,2}$	order of singularity
μ	Lamé constant
μ	shear modulus
ν	Poisson's ratio
ξ, η	Cartesian coordinates
$\xi_{r,t}$	optical anisotropy coefficient
ρ	mass density
ρ	notch tip radius
$\sigma, \sigma_\infty, \tau$	stresses at infinity
σ_{ij}	stress tensor
$\sigma_r, \sigma_\theta, \tau_{r\theta}$	components of stress in polar coordinates
$\sigma_{xx}, \sigma_{yy}, \sigma_{zz}$	components of stress
σ_0	tensile yield stress
$\sigma_1, \sigma_2, \sigma_3$	principal stresses
$\tau_{xy}, \tau_{xz}, \tau_{yz}$	components of stress
ϕ	angular displacement of the caustic
$\varphi(z), x(z)$	complex stress functions
$\Phi(z), \Psi(z)$	complex stress functions
$\Phi(z), \Omega(z)$	Muskhelishvili complex stress functions
ω	angle of crack inclination
$\Omega_j(z_j), \Omega(z_j)$	complex stress functions
∇^2	$\dfrac{\partial^2}{\partial x^2} + \dfrac{\partial^2}{\partial y^2}$
I_1, I_2, I_3	stress invariants

Introduction

At the beginning of the 20th century, the first significant theory of cracks appeared; Inglis [1] solved the problem of stress distribution around an elliptical hole in an elastic body with infinite dimensions which was loaded with a uniform stress field. Later, Muskhelishvili [2] solved the problem for an infinite elastic body with holes which was loaded with any system of loads. In the above works, where a crack is defined as the limit of an elliptical hole as the small half axis tends to zero, the characteristic crack dimensions are not dependent upon load; this is contrary to experimental results. Griffith [3, 4] gave the dependence of the characteristic elements of the crack with load and also evaluated the critical load for crack propagation from the strain energy variation of the body. On the other hand Wolf [5] gave an accurate and simple method of evaluation and Obreimov [6] correlated the crack characteristic parameters with the surface stress. Westergaard [7], Sneddon [8, 9, 10], Elliot [11] and Williams [12] gave the distribution of the stress and strain fields at the discontinuity surfaces of the displacement vector.

Considerable progress in fracture mechanics, mainly in crack problems of brittle and consequently of ductile materials was attained by Irwin [13, 14] and Orowan [15]. Irwin studied the stress field around the crack tip and observed that three independent mechanisms of crack deformation exist, namely: a) Opening mode (mode-I) with stress intensity factor K_I. b) Sliding or shear mode (mode-II) with stress intensity factor K_{II}, and c) Tearing mode (mode-III) with stress intensity factor K_{III}.

The definition of the crack and the application of the classical theory of elasticity in the study of crack problems resulted in the conclusion that the stresses at the crack tip tend to infinity. In 1955, Zheltov and Khristianovitch [16] studying cracked rocks, formulated the idea that at the crack tip the stresses are finite and the crack lips have the same tangent. Zheltov and Khristianovitch's ideas were later used by Barenblatt [17] who presented a complete crack model.

Studies of crack problems in materials which present considerable plastic behaviour were done by Hult and McClintock [18], McClintock [19, 20, 21], Rice [22, 23, 24], Smith [25, 26] and Bilby et al. [27].

In 1960 Dugdale [28] reported a substantial study on mode-I cracking. He formulated the idea that the crack extends by a length equal to the length of the plastic zone. The stress, which is applied on the extension crack, equals the yield stress, so that the stresses at the supposed crack tip do not tend to infinity. Dugdale's proposals for the solution of the plastic crack problem complement those of

Barenblatt, and so this is referred to as the Dugdale–Barenblatt model. The Dugdale–Barenblatt model was used by Goodier and Field [29] for the evaluation of displacements at the crack tip and by Theocaris and Gdoutos [30].

The experimental method of transmitted caustics was developed first by Manogg [31], while the experimental method of reflected caustics was developed by Theocaris [32–37]. The experimental method of caustics, which is based on the laws of geometrical optics, transforms the stress singularity into an optical singularity. This optical singularity gives much information for the evaluation of the stress field.

According to the method of caustics, a coherent light beam from a laser impinges normally on the specimen in the vicinity of the crack tip, and the reflected rays are received on a reference screen at some distance from the specimen. When a certain load is applied to the specimen the reflected light rays in the vicinity of the crack tip, where there is an abrupt thickness variation due to the existence of a singularity, are scattered and, when projected on a reference screen placed at some distance from the specimen, are concentrated along a curve, the so-called caustic.

The optical method of reflected caustics, as it has been developed during the last twenty years, was extensively applied to various elastic problems containing singularities and especially to problems with cracked plates which were made of isotropic or birefringent materials [38–41]. While in all these problems, cracked plates under any combination of the three modes of deformation were studied, in the case of uniaxial loading of the plate, the problem of the influence of the biaxiality of loading was strangely always omitted. The influence of the component of stress parallel to the crack tip, which was added to the singular expression of stresses, was certainly taken into account but this was valid only for the case when the crack-axis was normal to the applied tensile load at infinity.

It was only in 1977 that Liebowitz and his co-workers [42–45] considered problems of infinite plates containing slant cracks where the influence of biaxiality of loading to the plate was taken into account for stationary cracks. Thus, Eftis, Subramonian and Liebowitz [42] have shown that the one-parameter representation of the stress field in the vicinity of the crack tip is a satisfactory approximation only for simple cases of transverse cracks and uniaxial loading normal to the crack-axis. When a biaxial load at infinity is applied to the cracked plate and the crack is oblique to the principal stresses applied to the plate, this representation may lead to erroneous results. It was also shown that the second term in the series representation for the stresses contributes significantly and is independent of the distance from the crack tip. The effect of higher terms is best indicated in the problem of a biaxially loaded infinite plate containing a transverse central crack [43] under biaxial loading at infinity. Liebowitz and co-workers continued their study on the influence of biaxiality and have shown in a third paper [44] that the elastic strain energy density also depends on the biaxiality of the applied load. Finally, in a fourth paper Liebowitz, Lee and Eftis [45] have shown that the elastic stress intensity factor as well as the J-integral are not sensitive to the presence of the biaxial load. They also extended their studies to internal cracks in finite plates and to cases of small scale yielding.

Simultaneously, some extensive experimental studies with photoelasticity were undertaken, where crack tip stress patterns were analysed on the basis of two and multiple parameter characterization of the stress components in the vicinity of the crack tip. It was shown that at least a two-parameter representation of the stress field around the crack tip is necessary and in many cases is sufficient to yield results in agreement with experiment [46–50]. For a critical review of these two-parameter methods see Ref. [50].

A three-parameter method taking into account the third term in the Taylor series expansion of the Westergaard complex stress function was introduced in Ref. [51] for the evaluation of the mode-I stress intensity factor at a crack tip, where the idea of an appropriate selection of the polar direction for determining K_1 was introduced which depended on the distance of the point of measurement of isochromatics from the crack tip. Three and several-term approximations in the series expansions of $Z(z)$ were also introduced by Sanford and Dally [52, 53] in order to liberate the measurements from the requirement of being made in the vicinity of the crack tip, but these methods are rather complicated and necessitate considerable computer work with doubtful results. Finally, Rossmanith [54] gave an extensive analysis of the mixed-mode isochromatic patterns in the vicinity of the crack tip in a plate containing an oblique crack.

On the other hand, Cotterell [55] studied the influence that the coefficient of the second term of the Williams asymptotic expansion has on the shape and orientation of the isostatic loops. Similarly, Williams and Ewing [56] have studied the influence of the applied biaxial stress at infinity in a plate containing an internal slant crack and discussed the influence of the constant term in the Taylor series expansion of the stress function on the position of the critical angle of fracture. The procedure adopted in the experiments indicates empirically the significance of inclusion of the constant and higher terms of the series expansion of the Westergaard complex stress function. However it fails to disclose the fundamental influence of these terms on the stress distribution around the crack tip and consequently on the form of the isochromatics.

With caustics the situation is different. While the isochromatics are proportional to $|\bar{z}\Phi'(z) + \Psi(z)|$ (where $\Phi(z)$, and $\Psi(z) = -2\Phi(z)/z$, are the Muskhelishvili complex stress functions) and the isopachics are proportional to Re $\Phi(z)$, in caustics the initial curve and the corresponding caustic depend on $\Phi''(z)$ and $\overline{\Phi'(z)}$ respectively and for the K_1 mode of deformation of the crack these curves are independent of the constant term, at least in the series expansion of $\Phi(z)$. An extensive study of the influence of the biaxiality factor k on the dynamic crack propagation in slant cracks of any obliqueness was undertaken by Theocaris [57] and Theocaris and Papadopoulos [58], whereas closed-form solutions for the stress field in cracked plates under biaxial load were given in Ref. [59].

The influence of birefringence on the shape and form of the caustic was first mentioned by Manogg [31] who introduced the coefficient of anisotropy, ξ, and it was studied by Theocaris and his co-workers [38, 40, 41].

While the method of transmitted or reflected caustics was extensively used in problems of stationary cracks for defining the mixed-mode stress intensity factor, $K^* = K_1 - iK_{II}$, at the crack tip, only recently has the method been extended to dynamic problems.

In the beginning, dynamic cases were dealt with by introducing the corresponding dynamic values for the mechanical and ultimately the optical constants into the corresponding static analysis of the problem. A basic improvement of the method was the introduction not only of some standard dynamic values for the moduli of the material used in the dynamic tests, but more accurately the exact values of the characteristic properties of the material under the same dynamic conditions as each particular test [60, 61].

The error introduced by using the elastic static analysis in a dynamic problem of crack propagation was empirically corrected by Schirrer [62] who, based on the theoretical analyses of Broberg [63] and Freund and Clifton [64] introduced a correction factor to the static analysis taking care of the dynamic phenomenon.

Furthermore, Rosakis [65], in a recent report, mentions that Kalthoff *et al.* used such a correction factor to take care of the error introduced by the use of a static analysis in problems for dynamically propagating cracks.

However, for the complete and accurate analysis of propagating cracks by the method of caustics (or any other experimental method) it is necessary to consider the dynamic expressions for the stress components about the propagating crack. Whereas for the more general cases associated with dynamic crack propagation it is very difficult to derive the stress and displacement fields theoretically [66], there are approximate solutions based on asymptotic analysis [67], where it is assumed that the total internal energy of the cracked body is finite and therefore the internal energy density may be integrated. In this case the near crack-tip stress field of a running crack in a plane elastic problem admits solutions of a universal spatial dependence in a local coordinate system which is associated with the moving crack tip.

While Rosakis [65] used asymptotic expressions based on approximations for the stress components at the running crack tip under constant velocity c to derive a simplified expression for the equation of the caustic for the case of a running crack under mode-I deformation, Theocaris and Ioakimidis [68] gave accurate expressions for the form of distortion of the caustic at the tip of a running crack, under similar conditions of constant velocity c, based on complex stress function theory.

Various criteria have been introduced for the description of mode of propagation of a crack subjected to combined mode I and II in-plane deformations.

The prediction of the direction of crack growth in fracture mechanics is actually based on different conceptions, the principal of which considers a fracture configuration by assuming that the crack extends by infinitesimal segments, which may be of variable length and orientation. In this case the energy-release rate increases as the branch of crack extension tends to zero, thus generating two singularities, one at the crack tip and the other at the re-entrant corner at the web–branch junction.

Local criteria are intended to circumvent the above difficulties by predicting the near-tip behaviour existing before the onset of crack propagation. These criteria are the maximum stress [69], and the minimum elastic strain-energy density [72–76]. The maximum tangential stress criterion assumes that fracture occurs when this stress on an element at a critical distance from the crack tip reaches a critical value. A more satisfactory criterion, which depends on a combination of all components of stresses, and which may be considered as a parallel to Mises yield criterion, is the criterion based on the minimum value of the strain-energy density. This criterion was advanced by Sih [77, 78].

It was shown that both local criteria cease to be valid as values of the polar distance r from the crack tip tend to zero. Hence the necessity of introducing a critical distance r_0 defining a circular core region around the crack tip, along the boundary of which these criteria apply. Sih [75] has defined the size r_0 of the core region by assuming the crack to be a sharp ellipse with a small, but finite, radius of curvature ρ along its major axis. Stresses derived from the predictions of the minimum strain-energy density criterion at $\rho = r_0$ then coincide with those derived from perfectly sharp cracks by classical elastic theory.

Experiments, executed by Sih and Kipp [79] and by Williams and Ewing [56], showed the necessity of testing all local fracture criteria outside the core region. While Sih and Kipp [79] indicated the validity of the S-criterion for the whole range of orientation of crack direction relative to the loading axis, Williams and Ewing [56] improved the results of the maximum tangential stress criterion, by applying it along the boundary of the core region. In particular, Williams and Ewing considered additional terms (and especially the constant term, which corresponds to an axial load applied along the crack axis) in the series asymptotic expansion of the complex-stress function, describing the stress field at the cracked plate.

Riedmüller [80] has determined experimentally the distribution of the energy density in the neighbourhood of a crack tip for cracks subjected to modes I and II deformation, by applying classical photoelasticity and interferometry. These experimental methods are very convenient for classical whole-field elasticity problems. However, they lose their potentialities and accuracy when applied to singular stress fields, since they necessitate extrapolation of data far away from the singularity in order to evaluate the singular stress field.

In a series of papers it was shown that the core region, as defined by Sih [79], Williams [56], and others, coincides with the initial curve, on the specimen, generating the caustic, since this curve marks the transition from linear to nonlinear behaviour of the material [81, 82]. Furthermore, the distribution of the elastic strain-energy density was studied at stationary cracks for complex fracture modes (modes I and II) by using the S-criterion and the initial curve of the caustic defining the core region [83]. In this paper, [83], a model was introduced explaining the zig-zag propagation of slant cracks. It was shown that the application of the S-criterion in cases of stationary cracks yielded satisfactory results. On the other hand, it has been established that in the case of propagating cracks, the dynamic effect alters, sometimes significantly, the results. For this reason, the study of the influence of the dynamic state of stress and strain in a running slant crack was undertaken in this paper [83]. The whole study was based on the validity of the minimum elastic strain-energy density criterion.

Recently, new criteria of fracture have been developed. Theocaris and Andrianopoulos [84] have developed the T-criterion, while Papadopoulos (the author) [85, 86] has proposed the experimental Det.-criterion of fracture. The Det.-criterion of fracture is based on the second stress invariant and for plane-conditions, on the determinant of the stress tensor Det.(σ_{ij}). The development of the Det.-criterion will be covered in Part II of this book.

Part I
The Experimental Method of Caustics

Chapter 1
Theory of Cracks

1.1 General Aspects

The determination of stress and strain field according to classical elasticity, for conditions that do not correspond to fracture or cracking of the materials, is realized by conversion to a boundary problem which is characterized by either an elastic or a plastic law of deformation. The boundaries and the strain behaviour of the body are considered as known and the applied loads cause small-scale deformations of the boundary regions.

However, above a certain critical value of the load, cracks which correspond to surface discontinuities of the displacement vector appear. Now, the pre-existent stress state, at the positions where cracks have appeared, has vanished and the boundaries of the body have changed. Thus, the definition of the stress and strain field is very difficult, because, relative to the problem, more conditions for the determination of the body boundaries are needed. For this reason, in the theory of cracks the problem which is solved is: "In a body which is subjected to a system of external loads the magnitudes of which are continuously increased, a system of cracks is given. It is required to determine the crack surfaces and the stress–strain distribution for each step of loading. It is suggested that the increase of loads takes place smoothly so that the influence of dynamic loading is neglected". According to the theory of elasticity this problem is formulated as follows: "It is required to solve the differential equations of elasticity in a region, where there are cracks, which is defined by the boundary Σ, so that the boundary conditions are satisfied". The difference of this problem from ordinary problems of elasticity is fundamental, because it is a problem with unknown boundaries which must be defined during the solution of the problem.

Another difficulty in crack problems is the choice of an appropriate model on which to base the solution of the problem. There are two models which are used. a) The particle model which forms the basis of Newtonian dynamics, and b) The continuum model on which the theory of elasticity is based. Solution of the crack problem, mainly in the area around the crack tip, is based on the particle principle because the development and propagation of a crack consists in the separation of pairs of neighbouring atoms. But solution of elastic problems according to the particle model is very difficult, involving complex mathematical analysis. Far from the crack tip where the micromolecular structure of the material does not appreciably influence the stress field, the problem can be solved through the continuous medium model.

To a first approximation, a crack can be considered as a cavity in the body with the distance between its lips small relative to its length. The first relatively complete consideration of crack structure at the tip is due to Griffith [3,4] who, using the classical theory of elasticity and the Inglis solution, arrived at infinite stresses at the crack tip which was circular, with a radius of curvature of the order of the molecular distance, a fact not admissible by continuum mechanics. It is observed that the stresses at the crack tip may be finite if the opposite lips of the crack have a common tangent at the point under consideration and are not free of stresses. In this type of crack boundary, it can be proved that, for infinite variation of the crack boundaries, the released energy is equal to zero. Thus, only such curves can form the boundaries of cracks in equilibrium. If the crack surfaces are considered to be free of stresses, then for any shape of crack boundary, the stresses become infinite and cracks in equilibrium cannot exist. Thus in order to arrive at admissible solutions, molecular forces of connection must be considered at the crack tip.

1.2 Physical Law

Linear and elastic materials are those materials in which the developing strains are elastic and there is a linear relationship between the stresses and the corresponding strains. This linear relationship expresses the physical law, termed as Hooke's law [87, 88].

The relationship between stresses and strains for an isotropic and homogeneous polycrystalline material for the three-dimensional case is given by:

$$\varepsilon_{xx} = \frac{1}{E}[\sigma_{xx} - \nu(\sigma_{yy} + \sigma_{zz})]$$

$$\varepsilon_{yy} = \frac{1}{E}[\sigma_{yy} - \nu(\sigma_{zz} + \sigma_{xx})]$$

$$\varepsilon_{zz} = \frac{1}{E}[\sigma_{zz} - \nu(\sigma_{xx} + \sigma_{yy})]$$

$$\gamma_{xy} = \frac{\tau_{xy}}{G}, \quad \gamma_{yz} = \frac{\tau_{yz}}{G}, \quad \gamma_{zx} = \frac{\tau_{zx}}{G}, \quad G = \frac{E}{2(1+\nu)} \tag{1}$$

where E is Young's modulus or the modulus of elasticity, ν is Poisson's ratio, G is the shear modulus and σ_{ij}, ε_{ij} are the elements of the stress and strain tensors respectively. The stress tensor is:

$$\sigma_{ij} = \begin{pmatrix} \sigma_{xx} & \tau_{xy} & \tau_{xz} \\ \tau_{yx} & \sigma_{yy} & \tau_{yz} \\ \tau_{zx} & \tau_{zy} & \sigma_{zz} \end{pmatrix} \tag{2}$$

with:

$$\tau_{xy} = \tau_{yx}, \quad \tau_{xz} = \tau_{zx}, \quad \tau_{yz} = \tau_{zy}$$

and the strain tensor is:

$$\varepsilon_{ij} = \begin{pmatrix} \varepsilon_{xx} & \varepsilon_{xy} = \gamma_{xy}/2 & \varepsilon_{xz} = \gamma_{xz}/2 \\ \varepsilon_{yx} = \gamma_{yx}/2 & \varepsilon_{yy} & \varepsilon_{yz} = \gamma_{yz}/2 \\ \varepsilon_{zx} = \gamma_{zx}/2 & \varepsilon_{zy} = \gamma_{zy}/2 & \varepsilon_{zz} \end{pmatrix} \tag{3}$$

with:

$$\varepsilon_{xy} = \varepsilon_{yx}, \quad \varepsilon_{xz} = \varepsilon_{zx}, \quad \varepsilon_{yz} = \varepsilon_{zy}$$

The strains are defined by the following relations:

$$\varepsilon_{xx} = \frac{\partial u}{\partial x}, \quad \varepsilon_{yy} = \frac{\partial v}{\partial y}, \quad \varepsilon_{zz} = \frac{\partial w}{\partial z}$$

$$\gamma_{xy} = 2\varepsilon_{xy} = \frac{\partial u}{\partial y} + \frac{\partial v}{\partial x}$$

$$\gamma_{xz} = 2\varepsilon_{xz} = \frac{\partial u}{\partial z} + \frac{\partial w}{\partial x}$$

$$\gamma_{yz} = 2\varepsilon_{yz} = \frac{\partial v}{\partial z} + \frac{\partial w}{\partial y} \tag{4}$$

where u, v, w are the displacements in x, y, z directions, respectively.

By inverting Eqs (1), we obtain:

$$\sigma_{xx} = \frac{E}{(1+v)(1-2v)}[(1-v)\varepsilon_{xx} + v(\varepsilon_{yy} + \varepsilon_{zz})]$$

$$\sigma_{yy} = \frac{E}{(1+v)(1-2v)}[(1-v)\varepsilon_{yy} + v(\varepsilon_{zz} + \varepsilon_{xx})]$$

$$\sigma_{zz} = \frac{E}{(1+v)(1-2v)}[(1-v)\varepsilon_{zz} + v(\varepsilon_{xx} + \varepsilon_{yy})]$$

$$\tau_{xy} = G\gamma_{xy}, \quad \tau_{yz} = G\gamma_{yz}, \quad \tau_{xz} = G\gamma_{xz} \tag{5}$$

1.3 Plane-Stress Deformation

If the applied stresses on the body are parallel to a plane, then the problems of elasticity simplify considerably. Most engineering applications involve plane deformation, that is, the distribution of the stresses is plane. For a thick plate subjected to a system of loads acting at its boundary and parallel to the plane of the plate, and uniformly distributed through the thickness, if the Oxy-plane of the $Oxyz$-system is taken to coincide with the mean plane of the plate, then any stress having a z suffix may be set to zero, yielding:

$$\sigma_{zz} = \tau_{xz} = \tau_{yz} = 0 \tag{6}$$

on both surfaces of the plate and without major error, throughout the thickness.

This state of stress is called "plane-stress deformation". When the thickness of the plate tends to a very small value, that is, the plate becomes a thin membrane, then an ideal state of plane-stress deformation is obtained, and the stresses σ_{xx}, σ_{yy} and τ_{xy} can be considered constant through the thickness of the plate.

In this case, the stress–strain relationships are given as:

$$\varepsilon_{xx} = \frac{1}{E}(\sigma_{xx} - \nu\sigma_{yy}), \quad \varepsilon_{yy} = \frac{1}{E}(\sigma_{yy} - \nu\sigma_{xx}), \quad \varepsilon_{zz} = -\frac{\nu}{E}(\sigma_{xx} + \sigma_{yy})$$

$$\gamma_{xy} = \frac{\tau_{xy}}{G}, \quad \gamma_{yz} = \gamma_{zx} = 0 \tag{7}$$

By inverting Eqs (7), we obtain:

$$\sigma_{xx} = \frac{E}{1 - \nu^2}(\varepsilon_{xx} + \nu\varepsilon_{yy}), \quad \sigma_{yy} = \frac{E}{1 - \nu^2}(\varepsilon_{yy} + \nu\varepsilon_{xx}) \tag{8}$$

1.4 Plane-Strain Deformation

When the plate is sufficiently thick, through-thickness strains are prevented. The strain deformation in this case, is plane as in the previous case, and therefore we have "plane-strain deformation". Thus, it can be assumed that there is no change in the distribution of strain over the (x, y) plane, in the z-direction. Then, any strain having a z suffix may be set to zero, while the displacements u and v are functions of x and y and the w displacement is zero. Thus, we have:

$$\varepsilon_{xx} = \varepsilon_{xx}(x, y), \qquad \varepsilon_{yy} = \varepsilon_{yy}(x, y), \qquad \gamma_{xy} = \gamma_{xy}(x, y)$$

$$\gamma_{yz} = \frac{\partial v}{\partial z} + \frac{\partial w}{\partial y} = 0, \quad \gamma_{xz} = \frac{\partial u}{\partial z} + \frac{\partial w}{\partial x} = 0, \quad \varepsilon_{zz} = \frac{\partial w}{\partial z} = 0 \tag{9}$$

The stress–strain Eqs (1) and (2) with the aid of Eqs (9) after some manipulation, are reduced to:

$$\varepsilon_{xx} = \frac{1 + \nu}{E}[(1 - \nu)\sigma_{xx} - \nu\sigma_{yy}] = \frac{1 - \nu^2}{E}\left[\sigma_{xx} - \frac{\nu}{1 - \nu}\sigma_{yy}\right]$$

$$\varepsilon_{yy} = \frac{1 + \nu}{E}[(1 - \nu)\sigma_{yy} - \nu\sigma_{xx}] = \frac{1 - \nu^2}{E}\left[\sigma_{yy} - \frac{\nu}{1 - \nu}\sigma_{xx}\right]$$

$$\gamma_{xy} = \frac{\tau_{xy}}{G} = 2\frac{1 - \nu^2}{E}\left(1 + \frac{\nu}{1 - \nu}\right)\tau_{xy} \tag{10}$$

and:

$$\sigma_{xx} = \frac{E}{(1 + \nu)(1 - 2\nu)}[(1 - \nu)\varepsilon_{xx} + \nu\varepsilon_{yy}]$$

$$\sigma_{yy} = \frac{E}{(1 + \nu)(1 - 2\nu)}[(1 - \nu)\varepsilon_{yy} + \nu\varepsilon_{xx}]$$

$$\sigma_{zz} = \frac{\nu E}{(1 + \nu)(1 - 2\nu)}[\varepsilon_{xx} + \varepsilon_{yy}]$$

$$\tau_{xy} = G\gamma_{xy}, \quad \tau_{yz} = \tau_{zx} = 0 \tag{11}$$

Equations (10) are of exactly the same form as those for plane stress (Eqs (7)) if E is replaced by $E' = E/(1 - v^2)$ and v is replaced by $v' = v/(1 - v)$. It may thus be concluded that the solution of the plane-strain deformation problem is restricted to the evaluation of the stresses $\sigma_{xx}, \sigma_{yy}, \tau_{xy}$, as in the case of plane-stress deformation.

1.5 Generalized Plane-Stress Deformation

For generalized plane-stress deformation, the same assumptions relative to the stresses are valid, as in the case of plane-stress deformation. The only difference is that the stresses are not functions of the variable z, giving:

$$\sigma_{xx} = \sigma_{xx}(x, y), \quad \sigma_{yy} = \sigma_{yy}(x, y), \quad \tau_{xy} = \tau_{xy}(x, y)$$

$$\sigma_{zz} = \tau_{zx} = \tau_{zy} = 0 \tag{12}$$

The stresses $\sigma_{xx}, \sigma_{yy}, \tau_{xy}$ may be derived from the stress function of x and y, $F_1 = F_1(x, y)$, normally termed the Airy stress function, as follows:

$$\sigma_{xx} = \frac{\partial^2 F_1}{\partial y^2} + V, \quad \sigma_{yy} = \frac{\partial^2 F_1}{\partial x^2} + V, \quad \tau_{xy} = -\frac{\partial^2 F_1}{\partial x \partial y} \tag{13}$$

the stress function F_1 satisfies the equation:

$$\nabla^4 F_1 = -(1 - v)\nabla^2 V \tag{14}$$

where:

$$\nabla^2 V = \text{constant} \tag{15}$$

Since the stresses $\sigma_{xx}, \sigma_{yy}, \tau_{xy}$ are functions of the variables x and y only, all the boundary conditions of the plane problem are satisfied.

1.6 Complex-Stress Function

In order to solve the plane-stress problem, definitions of the stresses σ_{xx}, σ_{yy} and τ_{xy}, which satisfy the equilibrium equations, are needed:

$$\frac{\partial \sigma_{xx}}{\partial x} + \frac{\partial \tau_{xy}}{\partial y} = 0 \tag{16}$$

$$\frac{\partial \tau_{xy}}{\partial x} + \frac{\partial \sigma_{yy}}{\partial y} = 0 \tag{17}$$

Also, the compatibility equation expressed in terms of stresses:

$$\nabla^2 (\sigma_{xx} + \sigma_{yy}) = \left(\frac{\partial^2}{\partial x^2} + \frac{\partial^2}{\partial y^2} \right)(\sigma_{xx} + \sigma_{yy}) = 0 \tag{18}$$

Equations (16) and (17) are satisfied by the stresses σ_{xx}, σ_{yy} and τ_{xy} as long as the stresses are derived from the function F, as follows:

$$\sigma_{xx} = \frac{\partial^2 F}{\partial y^2}, \quad \sigma_{yy} = \frac{\partial^2 F}{\partial x^2}, \quad \tau_{xy} = -\frac{\partial^2 F}{\partial x \partial y} \tag{19}$$

From Eqs (19), we obtain:

$$\sigma_{xx} + \sigma_{yy} = \left(\frac{\partial^2}{\partial x^2} + \frac{\partial^2}{\partial y^2} \right) F = \nabla^2 F \tag{20}$$

By substituting Eq. (20) into Eq. (18), we obtain:

$$\nabla^2 \nabla^2 F = \left(\frac{\partial^2}{\partial x^2} + \frac{\partial^2}{\partial y^2} \right) \left(\frac{\partial^2}{\partial x^2} + \frac{\partial^2}{\partial y^2} \right) F = 0 \tag{21}$$

or:

$$\frac{\partial^4 F}{\partial x^4} + 2 \frac{\partial^4 F}{\partial x^2 \partial y^2} + \frac{\partial^4 F}{\partial y^4} = 0 \tag{22}$$

Equation (22) must be satisfied by the function $F = F(x, y)$, which is a biharmonic function, termed the "Airy function" or "stress function".

By introducing the stress function $F = F(x, y)$, the solution of Eqs (16–18) corresponds to the solution of Eq. (22), in which its constants are defined so that the boundary conditions are satisfied. Every stress function $F(x, y)$ which satisfies Eq. (22), may be represented, with the aid of two functions $\varphi(z)$ and $x(z)$ of the complex variable $z = x + iy$, by the relation [89]:

$$F = \mathrm{Re} \, [\bar{z} \varphi(z) + x(z)] \tag{23}$$

or:

$$2F = \bar{z} \varphi(z) + z \overline{\varphi(z)} + x(z) + \overline{x(z)} \tag{24}$$

From Eqs (19) and (24) it is concluded that the plane-stress and plane-strain fields are derived from the equations:

$$\sigma_{xx} + \sigma_{yy} = 2[\varphi'(z) + \overline{\varphi'(z)}] = 4 \, \mathrm{Re} \, \varphi'(z) = 4 \, \mathrm{Re} \, \Phi(z)$$

$$= 2[\Phi(z) + \overline{\Phi(z)}] \tag{25}$$

$$\sigma_{yy} - \sigma_{xx} + 2\tau_{xy} i = 2[\bar{z} \varphi''(z) + \psi'(z)] = 2[\bar{z} \Phi'(z) + \Psi(z)] \tag{26}$$

$$2\mu(u + iv) = \kappa \varphi(z) - z \overline{\varphi'(z)} - \overline{\psi(z)} \tag{27}$$

where:

$$\psi(z) = \frac{dx}{dz}, \quad \Phi(z) = \varphi'(z), \quad \Psi(z) = \psi'(z) \tag{28}$$

and:

$$\kappa = \frac{\lambda + 3\mu}{\lambda + \mu}, \qquad \text{for plane strain} \tag{29}$$

$$\kappa = \kappa' = \frac{\lambda' + 3\mu}{\lambda' + \mu}, \quad \text{for plane stress} \tag{30}$$

with:

$$\lambda' = \frac{2\lambda\mu}{\lambda + 2\mu} \tag{31}$$

where λ, μ are the Lamé constants.

The functions $\varphi(z)$ and $\psi(z)$ are arbitrary. They can be defined as follows:

a) When the stresses are given, it must be satisfied that:

$$\varphi(0) = 0, \quad \text{Im}\, \varphi'(0) = 0, \quad \psi(0) = 0 \tag{32}$$

b) When the strains are given, it must be satisfied that:

$$\varphi(0) = 0 \tag{33}$$

1.7 Westergaard Stress Function

According to Westergaard, it is possible to solve two-dimensional crack problems as long as the stress function is defined by the relation:

$$F = \text{Re}\, \bar{\bar{Z}} + y\, \text{Im}\, \bar{Z} \tag{34}$$

where Z is a analytic function of the complex variable $z = x + iy$, Re and Im are real and imaginary parts respectively, and:

$$\bar{Z} = \frac{\mathrm{d}\bar{\bar{Z}}}{\mathrm{d}z} \tag{35}$$

From the Cauchy–Riemann conditions, we have:

$$\frac{\partial}{\partial x}(\text{Re}\, \bar{Z}) = \frac{\partial}{\partial y}(\text{Im}\, \bar{Z}) = \text{Re}\, Z \tag{36}$$

$$\frac{\partial}{\partial x}(\text{Im}\, \bar{Z}) = -\frac{\partial}{\partial y}(\text{Re}\, \bar{Z}) = \text{Im}\, Z \tag{37}$$

$$\nabla^2(\text{Re}\, \bar{Z}) = \nabla^2(\text{Im}\, \bar{Z}) = 0 \tag{38}$$

where:

$$Z = \frac{\mathrm{d}\bar{Z}}{\mathrm{d}z} \tag{39}$$

From the above relations, we have:

$$\frac{\partial F}{\partial x} = \text{Re}\, \bar{Z} + y\, \text{Im}\, Z, \quad \frac{\partial F}{\partial y} = y\, \text{Re}\, Z \tag{40}$$

$$\frac{\partial^2 F}{\partial x^2} = \text{Re}\, Z + y\, \text{Im}\, Z', \quad \frac{\partial^2 F}{\partial y^2} = \text{Re}\, Z - y\, \text{Im}\, Z' \tag{41}$$

$$\frac{\partial^2 F}{\partial x\, \partial y} = \frac{\partial^2 F}{\partial y\, \partial x} = y\, \text{Re}\, Z' \tag{42}$$

where:

$$Z' = \frac{\mathrm{d}Z}{\mathrm{d}z} \tag{43}$$

According to Eqs (19), it is now possible to write expressions for the stresses, σ_{xx},

σ_{yy}, τ_{xy}, in terms of Z:

$$\sigma_{xx} = \frac{\partial^2 F}{\partial y^2} = \operatorname{Re} Z - y \operatorname{Im} Z' \tag{44}$$

$$\sigma_{yy} = \frac{\partial^2 F}{\partial x^2} = \operatorname{Re} Z + y \operatorname{Im} Z' \tag{45}$$

$$\tau_{xy} = -\frac{\partial^2 F}{\partial x \partial y} = -y \operatorname{Re} Z' \tag{46}$$

Also, it is observed that Eqs (16), (17) of the equilibrium, and Eq. (18) of the compatibility, are satisfied by Eqs (44), (45) and (46). Thus, we have:

$$\frac{\partial \sigma_{xx}}{\partial x} = \operatorname{Re} Z' - y \operatorname{Im} Z'', \quad \frac{\partial \tau_{xy}}{\partial y} = -\operatorname{Re} Z' + y \operatorname{Im} Z'' \tag{47}$$

$$\frac{\partial \tau_{xy}}{\partial x} = -y \operatorname{Re} Z'', \quad \frac{\partial \sigma_{yy}}{\partial y} = y \operatorname{Re} Z'' \tag{48}$$

$$\sigma_{xx} + \sigma_{yy} = 2 \operatorname{Re} Z \tag{49}$$

and:

$$\nabla^2 (\sigma_{xx} + \sigma_{yy}) = 2 \nabla^2 (\operatorname{Re} Z) = 0 \tag{50}$$

It is possible to get the same solution by using the method of complex functions and defining the functions $\Phi(z)$ and $\Psi(z)$ by the relations:

$$\Phi(z) = \frac{Z}{2}, \quad \Psi(z) = -\frac{zZ'}{2} \tag{51}$$

By substituting Eqs (51) into Eqs (25) and (26), we obtain:

$$\sigma_{xx} + \sigma_{yy} = Z + \bar{Z} = 2 \operatorname{Re} Z \tag{52}$$

$$\sigma_{yy} - \sigma_{xx} + 2\tau_{xy}i = \bar{z}Z' - zZ' = -2iyZ' = -2iy(\operatorname{Re} Z' + i \operatorname{Im} Z') \tag{53}$$

and:

$$\sigma_{yy} - \sigma_{xx} = 2y \operatorname{Im} Z' \tag{54}$$

$$\tau_{xy} = -y \operatorname{Re} Z' \tag{55}$$

From Eqs (52) and (54), we obtain:

$$\sigma_{xx} = \operatorname{Re} Z - y \operatorname{Im} Z' \tag{56}$$

$$\sigma_{yy} = \operatorname{Re} Z + y \operatorname{Im} Z' \tag{57}$$

Equations (55), (56) and (57) are the same as those which were found using the Westergaard Z function.

In order to derive normal and shear stress components in an (x', y') coordinate system, from those in the (x, y) coordinate system, where the new axes are found by rotating the original axes through a positive angle θ, it is necessary to consider the equilibrium of the element. The resulting equations are well known [90] namely:

$$\sigma_{x'x'} = \frac{\sigma_{xx} + \sigma_{yy}}{2} + \frac{\sigma_{xx} - \sigma_{yy}}{2} \cos 2\theta + \tau_{xy} \sin 2\theta$$

$$\sigma_{y'y'} = \frac{\sigma_{xx} + \sigma_{yy}}{2} - \frac{\sigma_{xx} - \sigma_{yy}}{2} \cos 2\theta - \tau_{xy} \sin 2\theta$$

$$\tau_{x'y'} = -\frac{\sigma_{xx} - \sigma_{yy}}{2} \sin 2\theta + \tau_{xy} \cos 2\theta \tag{58}$$

These may be written in the alternative form [88]:

$$\sigma_{x'x'} + \sigma_{y'y'} = \sigma_{xx} + \sigma_{yy}$$

$$\sigma_{y'y'} - \sigma_{x'x'} + 2i\tau_{x'y'} = (\sigma_{yy} - \sigma_{xx} + 2i\tau_{xy})e^{2i\theta} \tag{59}$$

Equations (58) or (59) form the basis of Mohr's stress circle construction [88]. The principal stresses, σ_1 and σ_2, may be obtained from the relationship:

$$\left. \begin{matrix} \sigma_1 \\ \sigma_2 \end{matrix} \right\} = \frac{\sigma_{xx} + \sigma_{yy}}{2} \pm \frac{1}{2}\sqrt{(\sigma_{xx} - \sigma_{yy})^2 + 4\tau_{xy}^2} \tag{60}$$

Equations (52) and (53) in the (x', y') coordinate system may be written:

$$\sigma_{x'x'} + \sigma_{y'y'} = 2\,\mathrm{Re}\,Z$$

$$\sigma_{y'y'} - \sigma_{x'x'} + 2i\tau_{x'y'} = -2iy(\mathrm{Re}\,Z' + i\,\mathrm{Im}\,Z')e^{2i\theta} \tag{61}$$

Alternatively, in polar coordinates:

$$\sigma_r + \sigma_\theta = 2\,\mathrm{Re}\,Z$$

$$\sigma_\theta - \sigma_r + 2i\tau_{r\theta} = -2iy(\mathrm{Re}\,Z' + i\,\mathrm{Im}\,Z')e^{2i\theta} \tag{62}$$

It is possible to write expressions for the stresses in polar coordinates:

$$\sigma_r = \mathrm{Re}\,Z - y\,\mathrm{Im}\,Z'e^{2i\theta}$$

$$\sigma_\theta = \mathrm{Re}\,Z + y\,\mathrm{Im}\,Z'e^{2i\theta}$$

$$\tau_{r\theta} = -y\,\mathrm{Re}\,Z'e^{2i\theta} \tag{63}$$

So, according to the Westergaard method, an analytic stress function is demanded, with the exception of the region which is defined by the crack, so that the boundary conditions are satisfied. Such an analytic stress function is:

$$Z = \frac{\sigma z}{(z^2 - a^2)^{\frac{1}{2}}} \tag{64}$$

which is a Westergaard function for the crack:

$$-a < x < a, \quad y = 0 \tag{65}$$

in an infinite body, which is subjected to a biaxial loading σ at infinity, with crack length $2a$.

Other forms of Airy stress functions are:

$$F = -y\,\mathrm{Re}\,\bar{Z} \tag{66}$$

which represents mode-II deformation, with expressions for the stresses:

$$\sigma_{xx} = 2\,\mathrm{Im}\,Z + y\,\mathrm{Re}\,Z' \tag{67}$$

$$\sigma_{yy} = -y\,\mathrm{Re}\,Z' \tag{68}$$

$$\tau_{xy} = \mathrm{Re}\,Z - y\,\mathrm{Im}\,Z' \tag{69}$$

Similarly, for mode-III deformation, the expressions for the stresses are:

$$\tau_{xz} = \operatorname{Im} Z' \tag{70}$$

$$\tau_{yz} = \operatorname{Re} Z' \tag{71}$$

Finally, the Westergaard stress function by any mode of deformation can be obtained by combination of the stress functions of the above three modes of deformation.

1.8 Modes of Deformation

By supposing that, in a body, it is possible to consider cracks as discontinuous surfaces of the displacement vector, Irwin [14, 91, 92] observed that there exist three independent modes of crack lip deformation, as shown in Fig. 1. Any other mode of deformation results from the superposition of these three modes. The three modes of crack deformation are:

Tensile Mode-I or Opening Mode. The crack lips tend to separate symmetrically relative to the crack plane before deformation. So there is symmetry about the (x, y) and (x, z) planes. For mode-I crack deformation, the Westergaard function is given by Eq. (64). By referring this function to the Cartesian coordinate system, (x, y, z)

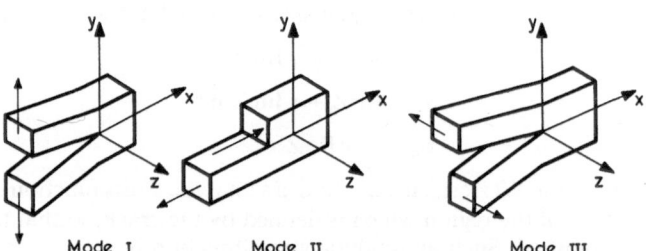

Fig. 1. The three modes of crack deformation.

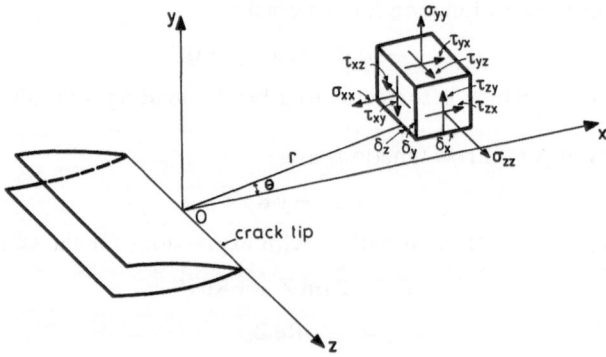

Fig. 2. Cartesian coordinate system $Oxyz$ with origin at the crack tip; and stress system.

with origin at the crack tip (Fig. 2) and applying the transformation:

$$\zeta = z - a \tag{72}$$

we obtain:

$$Z_1 = \frac{\sigma(\zeta + a)}{\sqrt{\zeta}\sqrt{\zeta + 2a}} \tag{73}$$

The quantity $1/\sqrt{\zeta + 2a}$ can be represented by a series expansion in powers of ζ:

$$\frac{1}{\sqrt{\zeta + 2a}} = \frac{1}{\sqrt{2a}}\left[1 - \frac{1}{2}\left(\frac{\zeta}{2a}\right) + \frac{1 \cdot 3}{2 \cdot 4}\left(\frac{\zeta}{2a}\right)^2 - \cdots\right] \tag{74}$$

then Eq. (73) becomes:

$$Z_1 = \frac{\sigma(\zeta + a)}{\sqrt{\zeta}}\frac{1}{\sqrt{2a}}\left[1 - \frac{1}{2}\left(\frac{\zeta}{2a}\right) + \frac{1 \cdot 3}{2 \cdot 4}\left(\frac{\zeta}{2a}\right)^2 - \cdots\right] \tag{75}$$

By omitting the powers of (ζ/a), since ζ is very small relative to a, we have:

$$Z_1 = \frac{\sigma\sqrt{\pi a}}{\sqrt{2\pi\zeta}} \tag{76}$$

and, by setting:

$$K_1 = \sigma\sqrt{\pi a} \tag{77}$$

then Eq. (76) becomes:

$$Z_1 = \frac{K_1}{\sqrt{2\pi\zeta}} \tag{78}$$

K_1 defines the magnitude of the crack tip stress field singularity, and is termed the mode-I stress intensity factor [14], or tensile stress intensity factor. K_1 can also be defined by the relation:

$$K_1 = \lim_{z \to a}(2\pi)^{\frac{1}{2}}(z - a)^{\frac{1}{2}}\sigma_{yy} \tag{79}$$

and:

$$K_1 = \lim_{z \to a}(2\pi)^{\frac{1}{2}}(z - a)^{\frac{1}{2}}Z_1(z) \tag{80}$$

By setting:

$$\zeta = re^{i\theta} \tag{81}$$

and by substituting into Eqs (44), (45) and (46) of Westergaard, we obtain the expressions of the stresses in terms of polar coordinates (r, θ):

$$\sigma_{xx} = \frac{K_1}{\sqrt{2\pi r}}\cos\frac{\theta}{2}\left(1 - \sin\frac{\theta}{2}\sin\frac{3\theta}{2}\right) \tag{82}$$

$$\sigma_{yy} = \frac{K_1}{\sqrt{2\pi r}}\cos\frac{\theta}{2}\left(1 + \sin\frac{\theta}{2}\sin\frac{3\theta}{2}\right) \tag{83}$$

$$\tau_{xy} = \frac{K_{\mathrm{I}}}{\sqrt{2\pi r}} \sin\frac{\theta}{2} \cos\frac{\theta}{2} \cos\frac{3\theta}{2} \tag{84}$$

$$\tau_{xz} = \tau_{yz} = 0 \tag{85}$$

According to Hooke's law the strains are:

$$\varepsilon_{xx} = \frac{\partial u}{\partial x} = \frac{1}{E}[\sigma_{xx} - \nu(\sigma_{yy} + \sigma_{zz})] \tag{86}$$

$$\varepsilon_{yy} = \frac{\partial v}{\partial y} = \frac{1}{E}[\sigma_{yy} - \nu(\sigma_{zz} + \sigma_{xx})] \tag{87}$$

and by substituting the stresses σ_{xx}, σ_{yy}, τ_{xy} for the *plane strain* condition ($\varepsilon_{zz} = 0$), we obtain:

$$\sigma_{zz} = \nu(\sigma_{xx} + \sigma_{yy}) \tag{88}$$

$$u = \frac{K_{\mathrm{I}}}{G}\sqrt{\frac{r}{2\pi}}\cos\frac{\theta}{2}\left(1 - 2\nu + \sin^2\frac{\theta}{2}\right) \tag{89}$$

$$v = \frac{K_{\mathrm{I}}}{G}\sqrt{\frac{r}{2\pi}}\sin\frac{\theta}{2}\left(2 - 2\nu - \cos^2\frac{\theta}{2}\right) \tag{90}$$

$$w = 0 \tag{91}$$

while for the *plane stress* condition ($\sigma_{zz} = 0$), we obtain:

$$u = \frac{K_{\mathrm{I}}}{G}\sqrt{\frac{r}{2\pi}}\cos\frac{\theta}{2}\left(\frac{1 - \nu}{1 + \nu} + \sin^2\frac{\theta}{2}\right) \tag{92}$$

$$v = \frac{K_{\mathrm{I}}}{G}\sqrt{\frac{r}{2\pi}}\sin\frac{\theta}{2}\left(\frac{2}{1 + \nu} - \cos^2\frac{\theta}{2}\right) \tag{93}$$

$$w = -\frac{\nu}{E}\int(\sigma_{xx} + \sigma_{yy})\mathrm{d}z \tag{94}$$

where G is the shear modulus of elasticity and:

$$K_{\mathrm{I}} = (2\pi r)^{\frac{1}{2}}\lim_{r \to 0}\sigma_{yy} \tag{95}$$

Sliding, or Inplane Shear Mode (Mode-II). This mode has anti-symmetry about the (x, z) plane, and symmetry about the (x, y) plane. The crack lips tend to separate in the opposite directions but in the same plane (Fig. 1). The Westergaard stress function for mode-II, is given by:

$$Z_{\mathrm{II}} = \frac{K_{\mathrm{II}}}{\sqrt{2\pi\zeta}}, \quad K_{\mathrm{II}} = \tau\sqrt{\pi a} \tag{96}$$

K_{II} defines the magnitude of the crack tip stress field singularity, and is termed the mode-II stress intensity factor or inplane shear stress intensity factor. By substituting into Eqs (67), (68) and (69) of Westergaard, we obtain expressions for the stresses in terms of polar coordinates (r, θ):

$$\sigma_{xx} = -\frac{K_{\mathrm{II}}}{\sqrt{2\pi r}} \sin \frac{\theta}{2} \left(2 + \cos \frac{\theta}{2} \cos \frac{3\theta}{2} \right) \tag{97}$$

$$\sigma_{yy} = \frac{K_{\mathrm{II}}}{\sqrt{2\pi r}} \cos \frac{\theta}{2} \sin \frac{\theta}{2} \cos \frac{3\theta}{2} \tag{98}$$

$$\tau_{xy} = \frac{K_{\mathrm{II}}}{\sqrt{2\pi r}} \cos \frac{\theta}{2} \left(1 - \sin \frac{\theta}{2} \sin \frac{3\theta}{2} \right) \tag{99}$$

$$\tau_{xz} = \tau_{yz} = 0 \tag{100}$$

The displacement field for the *plane strain* condition $(\varepsilon_{zz} = 0)$ is:

$$\sigma_{zz} = v(\sigma_{xx} + \sigma_{yy}) \tag{101}$$

$$u = \frac{K_{\mathrm{II}}}{G} \sqrt{\frac{r}{2\pi}} \sin \frac{\theta}{2} \left(2 - 2v + \cos^2 \frac{\theta}{2} \right) \tag{102}$$

$$v = \frac{K_{\mathrm{II}}}{G} \sqrt{\frac{r}{2\pi}} \cos \frac{\theta}{2} \left(2v - 1 + \sin^2 \frac{\theta}{2} \right) \tag{103}$$

$$w = 0 \tag{104}$$

while for the *plane stress* condition $(\sigma_{zz} = 0)$, we obtain:

$$u = \frac{K_{\mathrm{II}}}{G} \sqrt{\frac{r}{2\pi}} \sin \frac{\theta}{2} \left(\frac{2}{1+v} + \cos^2 \frac{\theta}{2} \right) \tag{105}$$

$$v = \frac{K_{\mathrm{II}}}{G} \sqrt{\frac{r}{2\pi}} \cos \frac{\theta}{2} \left(\frac{v-1}{v+1} + \sin^2 \frac{\theta}{2} \right) \tag{106}$$

$$w = -\frac{v}{E} \int (\sigma_{xx} + \sigma_{yy}) \, dz \tag{107}$$

where:

$$K_{\mathrm{II}} = (2\pi r)^{\frac{1}{2}} \lim_{r \to 0} \tau_{xy} \tag{108}$$

Tearing Mode (Mode-III) or Anti-plane Shear Mode. This mode has anti-symmetry about the (x, y) and (x, z) planes. The crack lips tend to separate in the opposite transverse directions, under equal and opposite forces which are perpendicular to the body plane (Fig. 1). The Westergaard stress function for mode-III, is given by:

$$Z_{\mathrm{III}} = \frac{K_{\mathrm{III}}}{\sqrt{2\pi \zeta}} \tag{109}$$

K_{III} defines the magnitude of the crack tip stress field singularity, and is termed the mode-III stress intensity factor or anti-plane shear stress intensity factor. By substituting into Eqs (70) and (71) of Westergaard, we obtain the expressions of shear stresses in terms of polar coordinates (r, θ):

$$\tau_{xz} = -\frac{K_{\text{III}}}{\sqrt{2\pi r}} \sin \frac{\theta}{2} \qquad (110)$$

$$\tau_{yz} = \frac{K_{\text{III}}}{\sqrt{2\pi r}} \cos \frac{\theta}{2} \qquad (111)$$

$$\sigma_{xx} = \sigma_{yy} = \sigma_{zz} = \tau_{xy} = 0 \qquad (112)$$

The displacement field for mode-III, is:

$$u = 0, \quad v = 0, \quad w = w(x, y) \qquad (113)$$

and from Hooke's law, we obtain:

$$\gamma_{xz} = \frac{\tau_{xz}}{G} = \frac{\partial w}{\partial x} \qquad (114)$$

$$\gamma_{yz} = \frac{\tau_{yz}}{G} = \frac{\partial w}{\partial y} \qquad (115)$$

The equations of equilibrium, in x and y directions are satisfied by identity, while in the z direction we have:

$$\frac{\partial \tau_{xz}}{\partial x} + \frac{\partial \tau_{yz}}{\partial y} = 0 \qquad (116)$$

By substituting Eqs (114) and (115) into Eq. (116), we obtain:

$$\Delta w = 0 \qquad (117)$$

By setting:

$$w = \frac{1}{G} \operatorname{Im} Z_{\text{III}} \qquad (118)$$

and by comparing Eqs (70) and (71), we obtain for the displacement w:

$$w = \frac{K_{\text{III}}}{G} \sqrt{\frac{2r}{\pi}} \sin \frac{\theta}{2} \qquad (119)$$

where:

$$K_{\text{III}} = (2\pi r)^{\frac{1}{2}} \lim_{r \to 0} \tau_{yz} \qquad (120)$$

The equations of the stress and displacement fields of the above three modes are exact only at the limit as r approaches zero, but will represent a good approximation in the region where $r \ll a$ and $r \ll L$, where L represents the smallest planar dimension of the body.

The strain energy release rate G, or energy per unit crack extension, for mode-I deformation, is given by:

$$G_{\text{I}} = \frac{K_{\text{I}}^2}{E}, \qquad \text{for plane stress} \qquad (121)$$

$$G_{\text{I}} = (1 - v^2)\frac{K_{\text{I}}^2}{E}, \quad \text{for plane strain.} \qquad (122)$$

For mode-II deformation, G is given by:

$$G_{II} = \frac{K_{II}^2}{E}, \qquad \text{for plane stress} \tag{123}$$

$$G_{II} = (1 - v^2)\frac{K_{II}^2}{E}, \quad \text{for plane strain} \tag{124}$$

and for mode-III deformation, is given by:

$$G_{III} = (1 + v)\frac{K_{III}^2}{E} \tag{125}$$

For crack extension under combined-mode deformation, the total energy release rate is given by:

$$G = G_I + G_{II} + G_{III} \tag{126}$$

Chapter 2
The Optical Method of Static Caustics

2.1 General Aspects

In crack problems, the most interesting region is that around the crack tip because in this region the variation of the stresses and strains is steep. So, while solution of the problem becomes difficult by classical experimental methods of stress and strain analysis, in contrast, mathematical analysis of the problem shows significant progress.

The experimental method of two- and three-dimensional photoelasticity uses information resulting from the interference fringes which appear. The isochromatic fringe patterns have high density at the crack tip, and so the calculation of the intensities is very difficult in this region. The isoclinic fringe patterns pass through the crack tip. This means that a stress singularity exists at this point.

The experimental method based on Moiré fringes which are formed by the use of grids is unable to follow the steep variation of the stresses at the crack tip.

The measurement of strains by any type of strain gauge and in particular the measurement of the rapid change in strain at the crack tip, is difficult. Also, the experimental methods of optical interferometry and holography present disadvantages in the study of the thickness variation at the crack tip.

The optical method of caustics, introduced by Manogg [31] and Theocaris [32], has been shown to be very effective for the determination of the characteristic parameters of singular elastic fields, where the corresponding stress parameters in these particular regions are governed by singularities which render the solution of the problem difficult, if not impossible, by conventional stress analysis methods. The difficulty arises mainly from the fact that the highly-strained region at the singularity is very small and therefore the information gathered by classical experimental methods is rather vague and inaccurate.

According to the method of reflected caustics, a light beam impinges in the immediate vicinity of the singularity and is reflected from the front and rear faces of the specimen (for the case of transparent materials). The reflected rays are collected along a singular surface, which is strongly illuminated. A reference screen, placed some distance from the specimen, intersects this surface and yields a singular curve, the caustic, which is, for all cases studied up to now, a generalized epicycloid. In this way, and by applying simple laws of geometric optics – that is, the reflection laws – the singular stress field is transformed into an optical singularity (i.e. the caustic), which yields all the necessary information for evaluation of the stress singularity.

The method of reflected caustics was used for the solution of several elasticity problems of particular interest in engineering applications. The method was applied primarily to cases with singularities, such as those appearing in cracked plates under plane-stress conditions; to contact problems; and to problems of multiwedges (composite materials). However, the method works equally satisfactorily in elastic problems with any type of stress concentration, not necessarily including singularities. In such cases the caustic is generated from a deformed boundary instead of from a singular curve in the interior of the specimen (initial curve). For a review of all these applications the reader is referred to a review paper by Theocaris [37].

2.2 The Stress-Optical Law

Assume a flat, transparent specimen, with parallel surfaces, and a monochromatic light ray in the plane of the diagram impinging at a point P (Fig. 3). The light ray is either reflected from the front face of the specimen or refracted and transmitted through the specimen to the rear face, where it is again reflected or refracted. The reflected light ray from the rear face is once again refracted or reflected at the front face of the specimen. These successive reflections and refractions of the light ray at the specimen surface continue for infinite time, so that a series of rays emerge from the front and rear surfaces of the specimen.

The initial intensity I of the impinging light ray is systematically attenuated by the successive reflections and refractions.

Let β be the ratio of the intensity reduction during reflection, α the ratio of the intensity reduction during refraction into the specimen and α' the ratio of the intensity reduction during refraction out of the specimen. If I is the intensity of the impinging ray, then the intensity of the first reflected ray will be βI and the intensity of the first refracted ray into the specimen will be αI. The first refracted ray is again refracted out of the specimen with intensity $\alpha' \alpha I$ and is reflected from the rear face

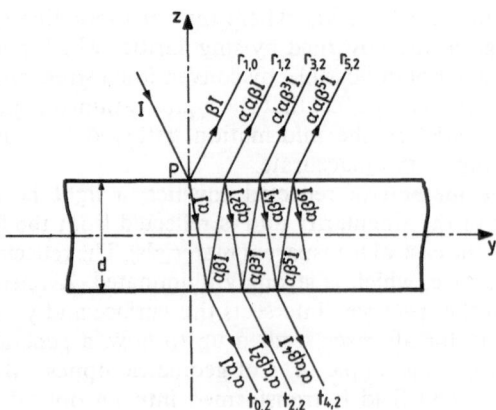

Fig. 3. Variation of light beam intensity transmitting through a transparent material.

with intensity $\alpha\beta I$. This reflected ray impinging on the front face of the specimen is refracted out of the specimen with intensity $\alpha'\alpha\beta I$ and is reflected with intensity $\alpha\beta^2 I$. The intensity variation law for subsequent rays emerging from the front or rear face of the specimen is obvious. So, the total intensity of the rays emerging from the front face of the specimen is:

$$\beta I + \alpha'\alpha\beta I + \alpha'\alpha\beta^3 I + \alpha'\alpha\beta^5 I + \cdots = \beta I + \alpha'\alpha\beta I (1 + \beta^2 + \beta^4 + \cdots)$$

$$= \beta I + \alpha'\alpha\beta I \frac{1}{1-\beta^2} = \beta I \left(1 + \frac{\alpha'\alpha}{1-\beta^2} \right), \quad \beta < 1 \tag{127}$$

and the total intensity of the rays emerging from the rear face of the specimen is:

$$\alpha'\alpha I + \alpha'\alpha\beta^2 I + \alpha'\alpha\beta^4 I + \cdots = \alpha'\alpha I (1 + \beta^2 + \beta^4 + \cdots) = \frac{\alpha'\alpha I}{1-\beta^2} \tag{128}$$

By applying the principle of luminous energy conservation and by omitting the radiation absorbed by the specimen, we obtain:

$$I = \beta I \left(1 + \frac{\alpha'\alpha}{1-\beta^2} \right) + \frac{\alpha'\alpha I}{1-\beta^2} \tag{129}$$

from which it follows that:

$$\alpha'\alpha = (1-\beta)^2 \tag{130}$$

If $r_{m,l}$ denotes the rays emergent from the front face of the specimen and $t_{m,l}$ denotes the rays emergent from the rear face of the specimen, where subscript m represents the number of reflections and subscript l represents the number of refractions, then the intensities of the rays are:

$$I_{r_{1,0}} = \beta I, \quad I_{r_{1,2}} = \beta(1-\beta)^2 I, \quad I_{r_{3,2}} = \beta^3 (1-\beta)^2 I, \ldots \tag{131}$$

and:

$$I_{t_{0,2}} = (1-\beta)^2 I, \quad I_{t_{2,2}} = \beta^2 (1-\beta)^2 I, \quad I_{t_{4,2}} = \beta^4 (1-\beta)^2 I, \ldots \tag{132}$$

or, in general:

$$I_{r,t_{m,l}} = \beta^m (1-\beta)^l I \tag{133}$$

where the reduction ratio β is given by the relation [93]:

$$\beta = \left(\frac{\eta - 1}{\eta + 1} \right)^2 \tag{134}$$

where η is the index of refraction of the material.

The refractive index of glass and plexiglas is $\eta = 1.5$. Then the ratio $\beta = 0.04$. So, from Rels (131) and (132) we obtain:

$$I_{r_{1,0}} = 0.04000 I, \quad I_{r_{1,2}} = 0.03686 I, \quad I_{r_{3,2}} = 0.00006 I, \ldots \tag{135}$$

and:

$$I_{t_{0,2}} = 0.92160 I, \quad I_{t_{2,2}} = 0.00147 I, \quad I_{t_{4,2}} = 2.4 \times 10^{-6} I, \ldots \tag{136}$$

From the above relationships we observe that the rays $r_{1,0}$, $r_{1,2}$ and $t_{0,2}$ have significant intensity, whereas the remaining rays have negligible intensities.

We assume light ray ABDE (Fig. 4) polarized to the direction of the principal stress σ_1 or σ_2 and perpendicularly impinging on the surface of the specimen which is in a plane-stress condition. For the light ray reflected from the rear face of the

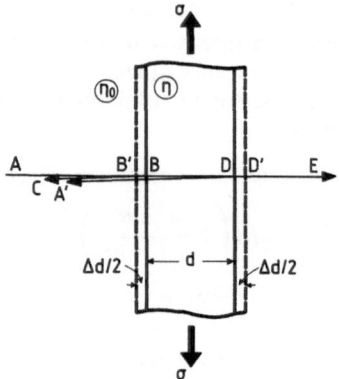

Fig. 4. Geometry of a perpendicularly impinging light ray on the specimen.

specimen, the optical path between points A and A' of the ray ABDA' (AB = A'B), for the unloaded specimen, is:

$$s_A = 2(AB)\eta_0 + 2d\eta \tag{137}$$

where d is the thickness of the specimen, η_0 is the refractive index of the environment and η is the refractive index of the unloaded specimen material.

For a loaded specimen, the optical path of the light ray between points A and A', is:

$$s_{L_{1,2}} = 2\left((AB) - \frac{\Delta d}{2}\right)\eta_0 + 2(d + \Delta d)\eta_{1,2} \tag{138}$$

where Δd is the variation of the specimen thickness and $\eta_{1,2}$ are the refractive indices of the loaded specimen material in the directions of the principal stresses $\sigma_{1,2}$, respectively.

By setting:

$$\eta_{1,2} = \eta + \Delta\eta_{1,2} \tag{139}$$

where $\Delta\eta_{1,2}$ are the variations of the refractive indices, the variation of the optical path of the ray ABDA', caused by the loading, is:

$$\Delta s_{r_{1,2}} = s_{L_{1,2}} - s_A = 2\left[(\eta_{1,2} - \eta)d + \left(\eta - \frac{\eta_0}{2}\right)\Delta d\right] \tag{140}$$

Likewise, for light ray ABDE transmitted through the specimen, the variation of the optical path caused by the loading is:

$$\Delta s_t = (\eta_{1,2} - \eta)d + (\eta - \eta_0)\Delta d \tag{141}$$

and for the light ray ABC reflected from the front face of the specimen, the absolute variation of the optical path is:

$$\Delta s_f = -\eta_0\Delta d \tag{142}$$

The light rays DA' and BC reflected from the rear and front faces of the specimen respectively, produce interference fringes, the number of which is related to the difference in optical path $\delta_{1,2}$ given by:

$$\delta_{1,2} = (\Delta s_{r_{1,2}} - \Delta s_f) = 2[(\eta_{1,2} - \eta)d + \eta\Delta d] \tag{143}$$

The number of the fringes is given by:

$$N_{1,2} = \delta_{1,2}/\lambda \tag{144}$$

where λ is the wavelength of the monochromatic illumination.

For the plane-stress problem, the variations of the refractive index η in the directions of the principal stresses $\sigma_{1,2}$, are given by the relations [94]:

$$\Delta\eta_{1,2} = \eta_{1,2} - \eta = b_1\varepsilon_{1,2} + b_2(\varepsilon_{2,1} + \varepsilon_3) \tag{145}$$

where b_1, b_2 are constants and ε_1, ε_2, ε_3 are the principal strains. The Rels (145) express the *stress optical law* of Neumann–Maxwell [95, 96]. By substituting Rels (145) into Rels (143) and (144), we obtain:

$$N_{1,2} = 2d(a^*\sigma_{1,2} + \beta^*\sigma_{2,1}) \tag{146}$$

where:

$$a^* = \frac{1}{E\lambda}(b_1 - 2vb_2 - v\eta) \tag{147}$$

$$\beta^* = \frac{1}{E\lambda}[b_2 - v(b_1 + b_2) - v\eta] \tag{148}$$

Variation of the optical paths ABDA' and ABDE, caused by the loading is obtained similarly. By introducing the generalized Hooke's law into Rels (145), we obtain from Rels (140) and (141) the conditions of Favre [97]:

$$\Delta s_{r_{1,2}} = 2c_r[(\sigma_1 + \sigma_2) \pm \xi_r(\sigma_1 - \sigma_2)]d \tag{149}$$

and:

$$\Delta s_{t_{1,2}} = c_t[(\sigma_1 + \sigma_2) \pm \xi_t(\sigma_1 - \sigma_2)]d \tag{150}$$

where:

$$c_r = \frac{a_r + \beta_r}{2}, \quad c_t = \frac{a_t + \beta_t}{2}, \quad \xi_{r,t} = \frac{a_{r,t} - \beta_{r,t}}{a_{r,t} + \beta_{r,t}} \tag{151}$$

where $\xi_{r,t}$ is the coefficient of optical anisotropy of the material, c_r and c_t are the stress-optical constants and a_r, a_t, β_r, β_t are given by the relations:

$$a_r = \frac{1}{E}\left[b_1 - 2vb_2 - v\left(\eta - \frac{\eta_0}{2}\right)\right] \tag{152}$$

$$\beta_r = \frac{1}{E}\left[b_2 - v(b_1 + b_2) - v\left(\eta - \frac{\eta_0}{2}\right)\right] \tag{153}$$

$$a_t = \frac{1}{E}[b_1 - 2vb_2 - v(\eta - \eta_0)] \tag{154}$$

$$\beta_t = \frac{1}{E}[b_2 - v(b_1 + b_2) - v(\eta - \eta_0)] \tag{155}$$

From the above relations it can be concluded that the variation of optical path during loading of a birefringent material consists of isotropic and anisotropic parts. The isotropic part depends on the sum of the principal stresses $(\sigma_1 + \sigma_2)$, whereas the anisotropic part depends on the difference of the principal stresses $(\sigma_1 - \sigma_2)$.

For isotropic material, we have:

$$b_1 = b_2 = b \tag{156}$$

and:

$$\Delta s_{r_1} = \Delta s_{r_2} = 2c_r(\sigma_1 + \sigma_2)d \tag{157}$$

$$\Delta s_{t_1} = \Delta s_{t_2} = c_t(\sigma_1 + \sigma_2)d \tag{158}$$

with:

$$a_r = \beta_r = c_r = \frac{1}{E}\left[(1-2v)b - v\left(\eta - \frac{\eta_0}{2}\right)\right] \tag{159}$$

$$a_t = \beta_t = c_t = \frac{1}{E}[(1-2v)b - v(\eta - \eta_0)] \tag{160}$$

From Rels (159), (160) and (151), we obtain:

$$c_r = c_t - \frac{v\eta_0}{2E}, \quad \frac{\xi_r}{\xi_t} = \frac{c_r}{c_t} \tag{161}$$

2.3 Theory of Caustics

The light rays reflected from the rear face of the specimen form wave fronts:

$$s(x, y, z) = \text{const} \tag{162}$$

If $s(x, y)$ expresses the optical path of the light ray between two planes parallel to the middle plane of the loaded plate and lying in the faces of the plate, then the following relation (162) is valid:

$$s(x, y, z) = z - s(x, y) = \text{const} \tag{163}$$

and:

$$\text{grad } s(x, y, z) = \mathbf{k} - \frac{\partial s}{\partial x}\mathbf{i} - \frac{\partial s}{\partial y}\mathbf{j} \tag{164}$$

where $\mathbf{i}, \mathbf{j}, \mathbf{k}$ are the unit vectors of the (x, y, z) Cartesian coordinate system whose origin is at the crack tip. The Ox-axis coincides with the crack-axis, whereas the Oz-axis is perpendicular to the specimen, and $s(x, y)$ is the optical path through the specimen at every point P of it (Fig. 5).

The deviation of the reflected light rays at a distance z_0 from the middle plane of the specimen, is expressed by the vector $\mathbf{w}$ on the plane $z = z_0$ and according to the theory of Econal, is given by [93]:

$$\mathbf{w} = z_0 \text{ grad } s(x, y, z) \tag{165}$$

or:

$$\mathbf{w} = -z_0 \text{ grad } s(x, y) = -z_0\left(\frac{\partial s}{\partial x}\mathbf{i} + \frac{\partial s}{\partial y}\mathbf{j}\right) \tag{166}$$

Given that the incident wave fronts of the light rays on the middle plane of the plate are parallel to this plane, $s(x, y)$ can be written as:

$$s(x, y) = s_0 + \Delta s_{r,t}(x, y) \tag{167}$$

where s_0 is a constant of the wave front and $\Delta s_{r,t}$ expresses the variation of the optical path through the specimen and is given by Rels (149) and (150). Then, by

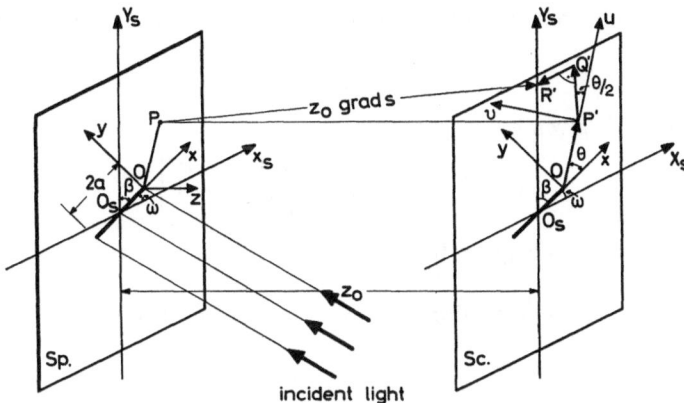

Fig. 5. Geometry of a cracked specimen and the relative position of specimen and viewing screen.

substituting Rel. (167) into Rel. (166), we obtain:

$$\mathbf{w} = -z_0 \, \mathrm{grad} \, \Delta s_{r,t}(x, y) \tag{168}$$

The tip of the vector $\mathbf{w}$ on the plane $z = z_0$ prescribes a surrounding line (caustic), which is closely related to the mode of deformation of the plate regions from which the light rays emerged.

By substituting Rels (149), (150), (157) and (158) into Rel. (168), we obtain the following.

1. *For optically anisotropic materials* with a coefficient of optical anisotropy $\xi_{r,t}$, we have:
For the first light ray reflected from the rear face of the specimen:

$$\mathbf{w} = -2z_0 dc_r \, \mathrm{grad} \left[(\sigma_1 + \sigma_2) \pm \xi_r(\sigma_1 - \sigma_2) \right] \tag{169}$$

For the first light ray transmitted through the specimen:

$$\mathbf{w} = -z_0 dc_t \, \mathrm{grad} \left[(\sigma_1 + \sigma_2) \pm \xi_t(\sigma_1 - \sigma_2) \right] \tag{170}$$

Relations (169) and (170) can be written as:

$$\mathbf{w} = -\varepsilon z_0 dc_{r,t} \, \mathrm{grad} \left[(\sigma_1 + \sigma_2) \pm \xi_{r,t}(\sigma_1 - \sigma_2) \right] \tag{171}$$

where the constant ε takes values $\varepsilon = 2$ for the first light ray reflected from the rear face of the specimen and denoted by subscript r and $\varepsilon = 1$ for the first light ray transmitted through the specimen and denoted by subscript t.

2. *For optically isotropic materials*, we have:
For the first light ray reflected from the rear face of the specimen:

$$\mathbf{w} = -2z_0 dc_r \, \mathrm{grad}(\sigma_1 + \sigma_2) \tag{172}$$

For the first light ray transmitted through the specimen:

$$\mathbf{w} = -z_0 dc_t \, \mathrm{grad}(\sigma_1 + \sigma_2) \tag{173}$$

Relations (172) and (173) can be written as:

$$\mathbf{w} = -\varepsilon z_0 dc_{r,t} \, \mathrm{grad}(\sigma_1 + \sigma_2) \tag{174}$$

If α is the angle of distortion caused by deformation, then for the first light ray reflected from the front face of the specimen, the angle of deviation ϕ is given by Snell's law:

$$\phi = 2\alpha \tag{175}$$

and the angle α is given by:

$$\alpha = \operatorname{grad} \frac{\Delta d}{2} = -\operatorname{grad}\left[\frac{vd}{2E}(\sigma_1 + \sigma_2)\right] \tag{176}$$

Therefore, the angle ϕ is given by:

$$\phi = -\frac{vd}{E}\operatorname{grad}(\sigma_1 + \sigma_2) = -dc_f\operatorname{grad}(\sigma_1 + \sigma_2) \tag{177}$$

where:

$$c_f = v/E \tag{178}$$

is the stress-optical constant for the first light ray reflected from the front face of the plate. The deviation vector $\mathbf{w}$ of the light on the screen at a distance z_0, for this case is:

$$\mathbf{w} = -z_0 dc_f \operatorname{grad}(\sigma_1 + \sigma_2) \tag{179}$$

Generally, for an isotropic material, Rels (174) and (179) can be written as:

$$\mathbf{w} = -\varepsilon z_0 dc_{r,t,f} \operatorname{grad}(\sigma_1 + \sigma_2) \tag{180}$$

where $\varepsilon = 2$ for the light ray (r) reflected from the rear face of the plate and $\varepsilon = 1$ for the light ray (t) transmitted through the plate and for the light ray (f) reflected from the front face of the plate. The stress optical constants $c_{r,t}$ are calculated by the interferometric method using monochromatic polarized light (laser) [98, 99], whereas the stress-optical constant c_f is given by Rel. (178).

For divergent $(+)$ or convergent $(-)$ light beams the magnification ratio λ_m is:

$$\lambda_m = \frac{z_0 \pm z_i}{z_i} \tag{181}$$

where z_0 is the distance of the reference screen from the specimen (Fig. 5) and z_i is the distance of the light beam focus from the specimen.

By setting [32]:

$$C^*_{r,t,f} = -\frac{\varepsilon z_0 dc_{r,t,f}}{\lambda_m} \tag{182}$$

relation (180) for isotropic materials, can be more generally written:

$$\mathbf{w} = C^*_{r,t,f} \operatorname{grad}(\sigma_1 + \sigma_2) \tag{183}$$

2.4 Application of the Method of Caustics in Plane Crack Problems

2.4.1 For Optically Isotropic Materials

According to the solution of the plane-stress problem by the complex-stress function method, the sum of stresses σ_{xx} and σ_{yy} is given by Rel. (25). Let us define $\Phi(z)$

as follows:

$$\Phi(z) = u(x, y) + iv(x, y) \tag{184}$$

where $u(x, y)$, $v(x, y)$ are real functions of the Cauchy–Riemann relations:

$$\frac{\partial u}{\partial x} = \frac{\partial v}{\partial y}, \quad \frac{\partial u}{\partial y} = -\frac{\partial v}{\partial x} \tag{185}$$

From Rels (185) it follows that the functions u and v satisfy Laplace's biharmonic equation:

$$\frac{\partial^2 u}{\partial x^2} + \frac{\partial^2 u}{\partial y^2} = 0, \quad \frac{\partial^2 v}{\partial x^2} + \frac{\partial^2 v}{\partial y^2} = 0 \tag{186}$$

The deviation vector of light for isotropic materials, Rel. (183), can be written as:

$$\mathbf{w} = C_{r,t,f}^* \operatorname{grad}(\sigma_1 + \sigma_2) = 4C_{r,t,f}^* \operatorname{grad}[\operatorname{Re}\Phi(z)]$$

$$= 4C_{r,t,f}^* \left(\frac{\partial u}{\partial x}\mathbf{i} + \frac{\partial u}{\partial y}\mathbf{j} \right) \tag{187}$$

or, on the complex plane:

$$w = 4C_{r,t,f}^* \left(\frac{\partial u}{\partial x} + i\frac{\partial u}{\partial y} \right) \tag{188}$$

The vector $\mathbf{w}$ expresses the relative deviation of light between every point P of the specimen (plate) and its image on the reference screen. By referencing this deviation to the (x', y') coordinate system on the reference screen (Fig. 5), we have:

$$W = z + w \tag{189}$$

and from Rel. (188), we obtain:

$$W = z + 4C_{r,t,f}^* \left(\frac{\partial u}{\partial x} + i\frac{\partial u}{\partial y} \right) \tag{190}$$

By setting:

$$z = x + iy \tag{191}$$

relation (190) becomes:

$$W = \left(x + 4C_{r,t,f}^* \frac{\partial u}{\partial x} \right) + i\left(y + 4C_{r,t,f}^* \frac{\partial u}{\partial y} \right) \tag{192}$$

or:

$$W = x' + iy' \tag{193}$$

where:

$$x' = x + 4C_{r,t,f}^* \frac{\partial u}{\partial x} \tag{194}$$

$$y' = y + 4C_{r,t,f}^* \frac{\partial u}{\partial y} \tag{195}$$

The quantity W expresses the projection of the deviations of the light rays on the reference screen. These deviations are different in the plastic region around the crack tip, being dependent on variation in the thickness and the refractive index of the materials. The concentration of these light rays defines the caustic on the reference screen. The caustic is a strongly illuminated, generally singular, curve. The condition of existence of the singularity is the zeroing of the functional determinant. Jacobian J:

$$J = \frac{\partial(x', y')}{\partial(x, y)} = 0 \tag{196}$$

or:

$$\begin{vmatrix} \dfrac{\partial x'}{\partial x} & \dfrac{\partial x'}{\partial y} \\[2mm] \dfrac{\partial y'}{\partial x} & \dfrac{\partial y'}{\partial y} \end{vmatrix} = 0 \tag{197}$$

where:

$$\frac{\partial x'}{\partial x} = 1 + 4C^*_{r,t,f}\frac{\partial^2 u}{\partial x^2}, \quad \frac{\partial x'}{\partial y} = 4C^*_{r,t,f}\frac{\partial^2 u}{\partial x \partial y} \tag{198}$$

$$\frac{\partial y'}{\partial x} = 4C^*_{r,t,f}\frac{\partial^2 u}{\partial x \partial y}, \quad \frac{\partial y'}{\partial y} = 1 + 4C^*_{r,t,f}\frac{\partial^2 u}{\partial y^2} \tag{199}$$

Then, Rel. (197) can be written:

$$\begin{vmatrix} 1 + 4C^*_{r,t,f}\dfrac{\partial^2 u}{\partial x^2} & 4C^*_{r,t,f}\dfrac{\partial^2 u}{\partial x \partial y} \\[3mm] 4C^*_{r,t,f}\dfrac{\partial^2 u}{\partial x \partial y} & 1 + 4C^*_{r,t,f}\dfrac{\partial^2 u}{\partial y^2} \end{vmatrix} = 0 \tag{200}$$

or:

$$1 + 4C^*_{r,t,f}\left(\frac{\partial^2 u}{\partial x^2} + \frac{\partial^2 u}{\partial y^2}\right) + 16C^{*2}_{r,t,f}\left[\frac{\partial^2 u}{\partial x^2}\frac{\partial^2 u}{\partial y^2} - \left(\frac{\partial^2 u}{\partial x \partial y}\right)^2\right] = 0 \tag{201}$$

and from the first of Rels (186), we obtain:

$$1 + 16C^{*2}_{r,t,f}\left[\frac{\partial^2 u}{\partial x^2}\frac{\partial^2 u}{\partial y^2} - \left(\frac{\partial^2 u}{\partial x \partial y}\right)^2\right] = 0 \tag{202}$$

The second derivative of the analytic function $\Phi(z) = u(x, y) + iv(x, y)$ is:

$$\Phi''(z) = \frac{\partial^2 u}{\partial x^2} + i\frac{\partial^2 v}{\partial x^2} \tag{203}$$

and according to Rels (185), we have:

$$\Phi''(z) = \frac{\partial^2 u}{\partial x^2} - i\frac{\partial^2 u}{\partial x \partial y} \tag{204}$$

or the modulus of $\Phi''(z)$:

$$|\Phi''(z)|^2 = \left(\frac{\partial^2 u}{\partial x^2}\right)^2 + \left(\frac{\partial^2 u}{\partial x \partial y}\right)^2 \tag{205}$$

and according to the first of Rels (186), we have:

$$|\Phi''(z)|^2 = -\frac{\partial^2 u}{\partial x^2}\frac{\partial^2 u}{\partial y^2} + \left(\frac{\partial^2 u}{\partial x \partial y}\right)^2 \tag{206}$$

By substituting Rel. (206) into Rel. (202), we obtain:

$$|4C_{r,t,f}^* \Phi''(z)|^2 = 1 \tag{207}$$

or:

$$|4C_{r,t,f}^* \Phi''(z)| = 1 \tag{208}$$

which is the equation of the *initial curve* of the caustic.

By using the relation:

$$\overline{\Phi'(z)} = \frac{\partial u}{\partial x} + i\frac{\partial u}{\partial y} \tag{209}$$

the relation (188) gives:

$$W = z + 4C_{r,t,f}^* \overline{\Phi'(z)} \tag{210}$$

which is the equation of the *caustic*.

By using the Westergaard stress function Z, which is related to $\Phi(z)$ by the first of Rel. (51), the equation of the initial curve (208), gives:

$$|2C_{r,t,f}^* Z''(z)| = 1 \tag{211}$$

and the equation of the caustic (210), gives:

$$W = \lambda_m [z + 2C_{r,t,f}^* \overline{Z'(z)}] \tag{212}$$

where λ_m is the magnification ratio which is given by Rel. (181).

2.4.1.1 *Analysis of the Elastic Transverse-Crack Problem*

The stress function of Westergaard for the case of a transverse crack, of length $2a$, in an infinite elastic body, is given by Rel. (78):

$$Z_1 = \frac{K_I}{\sqrt{2\pi\zeta}}, \quad K_I = \sigma\sqrt{\pi a} \tag{213}$$

where σ is the uniform stress at infinity, perpendicular to the crack axis and ζ is a complex variable given by Rel. (81). The first and second derivatives of Rel. (213), are:

$$Z_1'(\zeta) = \frac{dZ_1}{d\zeta} = -\frac{K_I}{2\sqrt{2\pi}}\zeta^{-\frac{3}{2}}$$

$$\overline{Z_1'(\zeta)} = \overline{Z_1'}(\bar{\zeta}) = \frac{K_I}{2\sqrt{2\pi}}\bar{\zeta}^{-\frac{3}{2}}$$

$$Z_1''(\zeta) = \frac{d^2 Z_1}{d\zeta^2} = \frac{3K_I}{4\sqrt{2\pi}}\zeta^{-\frac{5}{2}}$$

Then, the equation of the initial curve of the caustic (211), becomes:

$$\left| 2C_{r,t,f}^* \frac{3K_I}{4\sqrt{2\pi}}\zeta^{-\frac{5}{2}} \right| = 1$$

or:

$$|\zeta| = r_0 = \left(\frac{3}{2}C_{r,t,f}^* \frac{K_I}{\sqrt{2\pi}} \right)^{\frac{2}{5}} \tag{214}$$

and by setting:

$$|C_{r,t,f}| = \frac{|C_{r,t,f}^*|K_I}{\sqrt{2\pi}} = \frac{\varepsilon z_0 dc_{r,t,f} K_I}{\lambda_m \sqrt{2\pi}} \tag{215}$$

equation (214) gives:

$$|\zeta| = r_0 = \left(\frac{3}{2} C_{r,t,f}\right)^{\frac{2}{3}} \tag{216}$$

Relation (216) shows that the radius which defines the envelope of the highly strained zone on the midplane of the specimen is constant. Therefore, the envelope is a circle.

Likewise, for the equation of the caustic (212), we have:

$$\frac{W}{\lambda_m} = \zeta + C_{r,t,f}\bar{\zeta}^{-\frac{3}{2}} \tag{217}$$

and by using Rel. (81), we have:

$$\zeta = re^{i\theta}, \quad \bar{\zeta} = re^{-i\theta} \tag{218}$$

Then relation (217) is written as:

$$\frac{W}{\lambda_m} = re^{i\theta} + C_{r,t,f} r^{-\frac{3}{2}} e^{3i\theta/2} \tag{219}$$

By substituting Rel. (216) into Rel. (219), we obtain:

$$\frac{W}{\lambda_m} = r_0 e^{i\theta} + C_{r,t,f} r_0^{-\frac{3}{2}} e^{3i\theta/2} \tag{220}$$

or:

$$\frac{W}{\lambda_m} = r_0 (e^{i\theta} + \tfrac{2}{3} e^{3i\theta/2}) \tag{221}$$

By setting:

$$e^{i\theta} = \cos\theta + i\sin\theta$$
$$e^{3i\theta/2} = \cos\frac{3\theta}{2} + i\sin\frac{3\theta}{2} \tag{222}$$

relation (221) gives:

$$\frac{W}{\lambda_m} = r_0\left(\cos\theta + \frac{2}{3}\cos\frac{3\theta}{2}\right) + r_0\left(\sin\theta + \frac{2}{3}\sin\frac{3\theta}{2}\right)i \tag{223}$$

From Rel. (223) we have the parametric equations of the caustic:

$$x'_{r,t,f} = \lambda_m r_0\left(\cos\theta + \frac{2}{3}\cos\frac{3\theta}{2}\right) \tag{224}$$

$$y'_{r,t,f} = \lambda_m r_0\left(\sin\theta + \frac{2}{3}\sin\frac{3\theta}{2}\right) \tag{225}$$

In this case, the caustic is a symmetric curve with axis of symmetry coincident with the crack axis. The caustic, expressed by Rels (224) and (225), is an epicycloid, a periodic curve with period 4π. This epicycloid consists of two branches (θ positive or negative) which cross on the negative x'-axis at the mid-distance between the lips of the crack and extend from each edge of the crack to opposite sides of the plate. The extended tails depend on the values of the initial curve. Besides this principal epicycloid, several other epicycloids are given by the interferogram, which correspond to various values of the initial curve.

The radius ρ of the caustic is given by the relation:

$$\rho = (x'^2 + y'^2)^{\frac{1}{2}} = r_0 \left(\frac{13}{9} + \frac{4}{3} \cos \frac{\theta}{2} \right)^{\frac{1}{2}} \tag{226}$$

This radius presents extrema which are calculated by the condition:

$$\frac{\partial \rho}{\partial \theta} = 0 \tag{227}$$

which gives:

$$\theta_I^{\max} = 0, 2\pi, \dots \tag{228}$$

This means that the caustic presents maximum and minimum values on the axis of symmetry (Fig. 6). These values of ρ are:

$$\text{for} \quad \theta_{I(1)}^{\max} = 0, \quad \rho_{\max} = \frac{5r_0}{3} \lambda_m \tag{229}$$

$$\text{for} \quad \theta_{I(2)}^{\max} = 2\pi, \quad \rho_{\min} = \frac{r_0}{3} \lambda_m \tag{230}$$

The axis of symmetry $O'x_1$ (Fig. 6) crosses the caustic at two points. The distance between the two points is:

$$D_I^{\max} = 3r_0 \lambda_m \tag{231}$$

This distance defines the maximum longitudinal diameter of the caustic along the axis of symmetry $O'x_1$.

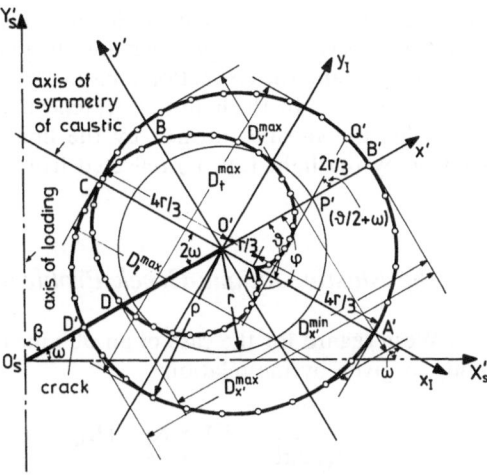

Fig. 6. Form of principal epicycloid around the crack tip for uniaxial tensile loading.

The maximum transverse diameter, $D_t^{\max}$, of the caustic can be derived from the condition:

$$\frac{\partial y'}{\partial \theta} = 0 \tag{232}$$

which gives:

$$\theta_t^{\max} = 72° \tag{233}$$

Relation (225), for $\theta = \theta_t^{\max} = 72°$, gives:

$$D_t^{\max} = 2y' = 3.1702 r_0 \lambda_m \tag{234}$$

Relations (231) and (234) constitute the characteristic invariables of the caustic.

From Rels (224) and (225) the shape of the caustic can be constructed by geometry. Draw the line $O'P' = r_0$ (initial curve) at an angle θ with the axis of symmetry of the caustic. Then, from the point P' draw the vector $P'Q' = 2r_0/3$ which forms an angle $\theta/2$ with the line $O'P'$ (Fig. 6). The point Q' is a point of the principal caustic.

By setting:

$$\delta_l^{\max} = 3.00, \quad \delta_t^{\max} = 3.1702 \tag{235}$$

and defining $\delta_{t,l}^{\max}$ as correction factors of the diameters $D_t^{\max}$ and $D_l^{\max}$, respectively, Rels (231) and (234) can be written:

$$D_l^{\max} = \lambda_m r_0 \delta_l^{\max} \tag{236}$$

$$D_t^{\max} = \lambda_m r_0 \delta_t^{\max} \tag{237}$$

From Rels (215), (216), (236) and (237) the stress intensity factor K_I can be derived as:

$$K_I = \frac{2(2\pi)^{\frac{1}{2}}}{3\varepsilon z_0 d\lambda_m^{\frac{3}{2}} c_{r,t,f}} \left(\frac{D_{t,l}^{\max}}{\delta_{t,l}^{\max}} \right)^{\frac{5}{2}} \tag{238}$$

Relation (238) gives the experimental stress intensity factor K_I from the diameters of the caustics formed at the crack tip. This experimentally calculated stress intensity factor depends on the thickness d of the specimen, the magnification ratio λ_m of the optical set-up, the distance z_0 between the specimen and the reference plane of view, the stress optical constants $c_{r,t,f}$ and the correction factors $\delta_{t,l}^{\max}$. The constant ε takes values $\varepsilon = 2$ for the caustics formed by the light rays (r) reflected from the rear face of the specimen and $\varepsilon = 1$ for the caustics formed by the light rays (t) transmitted through the specimen or by the light rays (f) reflected from the front face of the specimen.

2.4.1.2 Analysis of the Elastic Oblique-Crack Problem

The stress function of Westergaard for the case of an oblique crack of length $2a$, in an infinite elastic body, is given by the relation:

$$Z = \frac{K^*}{\sqrt{2\pi\zeta}}, \quad K^* = K_I - iK_{II} \tag{239}$$

This case is equivalent to those for the transverse crack under both stresses, tensile and shear at infinity. The first and the second derivatives of Rel. (239), are:

$$Z'(\zeta) = \frac{dZ}{d\zeta} = -\frac{K^*}{2\sqrt{2\pi}}\zeta^{-\frac{3}{2}} \tag{240}$$

$$\overline{Z'(\zeta)} = \bar{Z}'(\bar{\zeta}) = \frac{\overline{K^*}}{2\sqrt{2\pi}}\bar{\zeta}^{-\frac{3}{2}} \tag{241}$$

$$Z''(\zeta) = \frac{d^2Z}{d\zeta^2} = \frac{3K^*}{4\sqrt{2\pi}}\zeta^{-\frac{5}{2}} \tag{242}$$

Then the equation of the initial curve of the caustic (211), becomes:

$$\left| \frac{3}{2}C_{r,t,f}\frac{K^*}{K_I}\zeta^{-\frac{3}{2}} \right| = 1 \tag{243}$$

from which we obtain:

$$|\zeta| = r_0 = \left(\frac{3}{2}C_{r,t,f}\right)^{\frac{2}{3}}\left(\frac{|K^*|}{|K_I|}\right)^{\frac{2}{3}} \tag{244}$$

and by setting:

$$\left(\frac{|K^*|}{|K_I|}\right)^{\frac{2}{3}} = \left(\frac{K_I^2 + K_{II}^2}{K_I^2}\right)^{\frac{1}{3}} = (1 + \kappa^2)^{\frac{1}{3}} \tag{245}$$

with:

$$\kappa = \frac{K_{II}}{K_I} \tag{246}$$

equation (244) becomes:

$$|\zeta| = r_0 = (\tfrac{3}{2}C_{r,t,f})^{\frac{2}{3}}(1 + \kappa^2)^{\frac{1}{3}} \tag{247}$$

Relation (247) is the initial curve of the caustic which is a circle of radius r_0.
   Likewise, the equation of the caustic (212), becomes:

$$W = \zeta + C_{r,t,f}\frac{\overline{K^*}}{K_I}\bar{\zeta}^{-\frac{3}{2}} \tag{248}$$

By using Rels (81) for ζ and $\bar{\zeta}$, we obtain:

$$W = re^{i\theta} + C_{r,t,f}\frac{\overline{K^*}}{K_I}r^{-\frac{3}{2}}e^{3i\theta/2} \tag{249}$$

By substituting Rel. (247) into Rel. (249), we obtain:

$$W = r_0 e^{i\theta} + C_{r,t,f}\frac{\overline{K^*}}{K_I}r_0^{-\frac{3}{2}}e^{3i\theta/2} \tag{250}$$

or:

$$W = r_0\left(e^{i\theta} + \frac{2}{3}\frac{\overline{K^*}}{K_I}e^{3i\theta/2}\right) \tag{251}$$

By setting:

$$\frac{\overline{K^*}}{|K^*|} = \frac{K_I}{\sqrt{K_I^2 + K_{II}^2}} + i\frac{K_{II}}{\sqrt{K_I^2 + K_{II}^2}} \tag{252}$$

and:

$$\frac{\overline{K^*}}{|K^*|} = e^{i\omega} \tag{253}$$

we obtain:

$$\tan \omega = \frac{K_{II}}{K_I} = \kappa \tag{254}$$

Rel. (251) then becomes:

$$W = r_0(e^{i\theta} + \tfrac{2}{3}e^{i\omega}e^{3i\theta/2}) \tag{255}$$

or:

$$W = r_0(e^{i\theta} + \tfrac{2}{3}e^{i(\omega + 3\theta/2)}) \tag{256}$$

or:

$$W = r_0\left[\cos\theta + \frac{2}{3}\cos\left(\frac{3\theta}{2} + \omega\right)\right] + r_0\left[\sin\theta + \frac{2}{3}\sin\left(\frac{3\theta}{2} + \omega\right)\right]i \tag{257}$$

From Eq. (257) we get the parametric equation of the caustic relative to the (x', y') coordinate system placed at the crack tip with the $O'x'$-axis coincident with the crack-axis:

$$x'_{r,t,f} = r_0\left[\cos\theta + \frac{2}{3}\cos\left(\frac{3\theta}{2} + \omega\right)\right] \tag{258}$$

$$y'_{r,t,f} = r_0\left[\sin\theta + \frac{2}{3}\sin\left(\frac{3\theta}{2} + \omega\right)\right] \tag{259}$$

Then the parametric equations of the caustic (258) and (259) can be written as:

$$x'_{r,t,f} = r_0\left[\cos\theta + \frac{2}{3}\cos\frac{3\theta}{2}\cos\omega - \frac{2}{3}\sin\frac{3\theta}{2}\sin\omega\right] \tag{260}$$

$$y'_{r,t,f} = r_0\left[\sin\theta + \frac{2}{3}\sin\frac{3\theta}{2}\cos\omega + \frac{2}{3}\cos\frac{3\theta}{2}\sin\omega\right] \tag{261}$$

From Rel. (254), we get the expressions:

$$\sin\omega = \frac{\tan\omega}{\sqrt{1 + \tan^2\omega}} = \frac{\kappa}{\sqrt{1 + \kappa^2}}$$

$$\cos\omega = \frac{1}{\sqrt{1 + \tan^2\omega}} = \frac{1}{\sqrt{1 + \kappa^2}} \tag{262}$$

By substituting the expressions (262) into Eqs (260) and (261), we obtain:

$$x'_{r,t,f} = r_0\left[\cos\theta + \frac{2}{3}\cos\frac{3\theta}{2}\frac{1}{\sqrt{1 + \kappa^2}} - \frac{2}{3}\sin\frac{3\theta}{2}\frac{\kappa}{\sqrt{1 + \kappa^2}}\right] \tag{263}$$

$$y'_{r,t,f} = r_0\left[\sin\theta + \frac{2}{3}\sin\frac{3\theta}{2}\frac{1}{\sqrt{1 + \kappa^2}} + \frac{2}{3}\cos\frac{3\theta}{2}\frac{\kappa}{\sqrt{1 + \kappa^2}}\right] \tag{264}$$

or:

$$x'_{r,t,f} = r_0 \left[\cos\theta + \frac{2}{3}(1+\kappa^2)^{-\frac{1}{2}}\left(\cos\frac{3\theta}{2} - \kappa\sin\frac{3\theta}{2} \right) \right] \tag{265}$$

$$y'_{r,t,f} = r_0 \left[\sin\theta + \frac{2}{3}(1+\kappa^2)^{-\frac{1}{2}}\left(\sin\frac{3\theta}{2} + \kappa\cos\frac{3\theta}{2} \right) \right] \tag{266}$$

The parametric equations of the caustic, relative to the (x_1, y_1) coordinate system placed at the crack tip and forming an angle $2(\pi - \omega)$ with the (x', y') coordinate system (Fig. 6), are obtained by applying the transformation:

$$\begin{pmatrix} x_{1,r,t,f} \\ y_{1,r,t,f} \end{pmatrix} = \begin{pmatrix} \cos 2\omega & -\sin 2\omega \\ \sin 2\omega & \cos 2\omega \end{pmatrix} \begin{pmatrix} x'_{1,r,t,f} \\ y'_{1,r,t,f} \end{pmatrix} \tag{267}$$

Therefore, we get:

$$x_{1,r,t,f} = x'_{1,r,t,f} \cos 2\omega - y'_{1,r,t,f} \sin 2\omega \tag{268}$$

$$y_{1,r,t,f} = x'_{1,r,t,f} \sin 2\omega + y'_{1,r,t,f} \cos 2\omega \tag{269}$$

and by substituting Rels (265) and (266) into Rels (268) and (269), we obtain:

$$x_{1,r,t,f} = r_0 \left[\cos(\theta + 2\omega) + \frac{2}{3}\cos\left(\frac{3\theta}{2} + 3\omega \right) \right] \tag{270}$$

$$y_{1,r,t,f} = r_0 \left[\sin(\theta + 2\omega) + \frac{2}{3}\sin\left(\frac{3\theta}{2} + 3\omega \right) \right] \tag{271}$$

and by setting:

$$\theta + 2\omega = \tau \tag{272}$$

we have:

$$x_{1,r,t,f} = r_0 \left(\cos\tau + \frac{2}{3}\cos\frac{3\tau}{2} \right)$$

$$y_{1,r,t,f} = r_0 \left(\sin\tau + \frac{2}{3}\sin\frac{3\tau}{2} \right) \tag{273}$$

These relations demonstrate that the caustic is a symmetric curve, whose axis of symmetry forms an angle (-2ω) with the $O'x'$-axis along the length of the crack (Fig. 6).

For the caustic relative to the (x_1, y_1) coordinate system, values of the invariants (231) and (234) are obtained and the stress intensity factor K_I is determined by the relation:

$$K_I = \frac{2(2\pi)^{\frac{1}{2}}}{3\varepsilon z \, d\lambda_m^{\frac{3}{2}} c_{r,t,f}} \left(\frac{D_{t,l}^{max}}{\delta_{t,l}^{max}} \right)^{\frac{3}{2}} \cos\frac{\phi}{2} \tag{274}$$

The stress intensity factor K_{II} is determined by Rel. (254) when the stress intensity factor K_I and the angle of rotation of the caustic $(-2\omega) = \phi$ or the angle of loading β are given.

For the caustic relative to the (x', y') coordinate system, the stress intensity factors K_I and K_{II} are determined by Rels (238) and (254) respectively, by using different values of correction factors $\delta_{t,l}^{max}$ because they depend on the ratio $\kappa = K_{II}/K_I$.

From the above it follows that, if the plate contains a transverse crack and it is subjected to pure tension, then $K_{II} = 0$ and therefore $\omega = 0°$ or $\beta = 90°$ (Fig. 6) and the corresponding generalized epicycloid is symmetric to the direction of the crack with its maximum and minimum radii lying on the positive part of the $O'x'$-axis. For the case where $\omega = 90°$ or $\beta = 0°$, when $K_I \to 0$ and $K_{II} \to 0$ and therefore $\kappa = K_{II}/K_I = \tan\omega = \cot\beta \to \infty$, the corresponding caustic is angularly displaced by π and is reduced in size, tending to a single point. For this case, the medium radius of the epicycloid lies on the positive part of the $O'x'$-axis. For any other relationship between K_I and K_{II} the epicycloid rotates around the crack tip O'.

2.4.2 For Optically Anisotropic Materials

The deviation vector of the light, for anisotropic materials with a coefficient of optical anisotropy $\xi_{r,t}$, is given by the Rel. (171) [100]:

$$\mathbf{w} = C_{r,t}^* \operatorname{grad}\left[(\sigma_1 + \sigma_2) \pm \xi_{r,t}(\sigma_1 - \sigma_2)\right] \tag{275}$$

with:

$$C_{r,t}^* = -\frac{\varepsilon z_0 \, dc_{r,t}}{\lambda_m} \tag{276}$$

The sum and the difference of the principal stresses are given by the Rels (52) and (53), respectively:

$$\sigma_1 + \sigma_2 = \sigma_{xx} + \sigma_{yy} = 2\operatorname{Re} Z \tag{277}$$

$$\sigma_1 - \sigma_2 = |\sigma_{yy} - \sigma_{xx} + 2\tau_{xy}\mathrm{i}| = |2y\operatorname{Im} Z' - 2y\operatorname{Re} Z'\mathrm{i}|$$
$$= 2y\left[(\operatorname{Im} Z')^2 + (\operatorname{Re} Z')^2\right]^{\frac{1}{2}} \tag{278}$$

where Z is the stress function of Westergaard.

By substituting Rels (277) and (278) into Rel. (275), we obtain [31, 38, 40, 41]:

$$\mathbf{w} = C_{r,t}^* \operatorname{grad}(\sigma_1 + \sigma_2) \pm C_{r,t}^* \xi_{r,t} \operatorname{grad}(\sigma_1 - \sigma_2)$$
$$= 2C_{r,t}^* \operatorname{grad}(\operatorname{Re} Z) \pm 2C_{r,t}^* \xi_{r,t} \operatorname{grad}\left\{y\left[(\operatorname{Im} Z')^2 + (\operatorname{Re} Z')^2\right]^{\frac{1}{2}}\right\} \tag{279}$$

2.4.2.1 Analysis of the Elastic Transverse-Crack Problem

The stress function of Westergaard for the case of a transverse crack, of length $2a$, in an infinite elastic body, is given by the Rel. (78):

$$Z_{\mathrm{I}} = \frac{K_{\mathrm{I}}}{\sqrt{2\pi\zeta}}, \quad K_{\mathrm{I}} = \sigma\sqrt{\pi a} \tag{280}$$

With $\zeta = r\mathrm{e}^{\mathrm{i}\theta}$, we have:

$$Z_{\mathrm{I}} = \frac{K_{\mathrm{I}}}{\sqrt{2\pi}} r^{-\frac{1}{2}} \mathrm{e}^{-\mathrm{i}\theta/2} = \frac{K_{\mathrm{I}}}{\sqrt{2\pi r}}\left(\cos\frac{\theta}{2} - \mathrm{i}\sin\frac{\theta}{2}\right) \tag{281}$$

$$Z_{\mathrm{I}}' = -\frac{K_{\mathrm{I}}}{2\sqrt{2\pi}} \zeta^{-\frac{3}{2}} = -\frac{K_{\mathrm{I}}}{2\sqrt{2\pi}} r^{-\frac{3}{2}}\left(\cos\frac{3\theta}{2} - \mathrm{i}\sin\frac{3\theta}{2}\right) \tag{282}$$

Then, from these relations, we obtain:

$$\operatorname{Re} Z_1 = \frac{K_1}{\sqrt{2\pi r}} \cos \frac{\theta}{2} \tag{283}$$

$$\operatorname{Im} Z_1' = \frac{K_1}{2\sqrt{2\pi}} r^{-\frac{3}{2}} \sin \frac{3\theta}{2} \tag{284}$$

$$\operatorname{Re} Z_1' = -\frac{K_1}{2\sqrt{2\pi}} r^{-\frac{3}{2}} \sin \frac{3\theta}{2} \tag{285}$$

By substituting Rels (283), (284), (285) and the expression for y:

$$y = 2r \sin \frac{\theta}{2} \cos \frac{\theta}{2} \tag{286}$$

into Rel. (279), we obtain:

$$\mathbf{w} = -2C_{r,t} \operatorname{grad}\left(r^{-\frac{1}{2}} \cos \frac{\theta}{2}\right) \pm C_{r,t} \xi_{r,t} \operatorname{grad}(r^{-\frac{1}{2}} \sin \theta) \tag{287}$$

with:

$$C_{r,t} = \frac{\varepsilon z_0 \, dc_{r,t} \, K_1}{\lambda_m \sqrt{2\pi}} \tag{288}$$

The gradients in the Cartesian coordinates (u, v) shown in Fig. 5, are:

$$\operatorname{grad}_{u,v}\left(r^{-\frac{1}{2}} \cos \frac{\theta}{2}\right) = \frac{\partial}{\partial r}\left(r^{-\frac{1}{2}} \cos \frac{\theta}{2}\right)\mathbf{u} + \frac{1}{r}\frac{\partial}{\partial \theta}\left(r^{-\frac{1}{2}} \cos \frac{\theta}{2}\right)\mathbf{v}$$

$$= \left(-\frac{1}{2} r^{-\frac{3}{2}} \cos \frac{\theta}{2}\right)\mathbf{u} + \left(-\frac{1}{2} r^{-\frac{3}{2}} \sin \frac{\theta}{2}\right)\mathbf{v}$$

and:

$$\operatorname{grad}_{u,v}(r^{-\frac{1}{2}} \sin \theta) = \frac{\partial}{\partial r}(r^{-\frac{1}{2}} \sin \theta)\mathbf{u} + \frac{1}{r}\frac{\partial}{\partial \theta}(r^{-\frac{1}{2}} \cos \theta)\mathbf{v}$$

$$= (-\tfrac{1}{2} r^{-\frac{3}{2}} \sin \theta)\mathbf{u} + (r^{-\frac{3}{2}} \cos \theta)\mathbf{v}$$

where:

$$\mathbf{u} = \cos \theta \, \mathbf{i} + \sin \theta \, \mathbf{j}$$
$$\mathbf{v} = -\sin \theta \, \mathbf{i} + \cos \theta \, \mathbf{j}$$

Therefore, we obtain:

$$\operatorname{grad}_{u,v}\left(r^{-\frac{1}{2}} \cos \frac{\theta}{2}\right) = -\frac{1}{2} r^{-\frac{3}{2}}\left(\cos \frac{3\theta}{2}\mathbf{i} + \sin \frac{3\theta}{2}\mathbf{j}\right)$$

and:

$$\operatorname{grad}_{u,v}(r^{-\frac{1}{2}} \sin \theta) = -\tfrac{1}{4} r^{-\frac{3}{2}}[3 \sin 2\theta \, \mathbf{i} + (1 + 3 \cos 2\theta)\mathbf{j}]$$

Relation (287) relative to the (x', y', z') system with unit vectors $\mathbf{i}, \mathbf{j}, \mathbf{k}$, becomes:

$$\mathbf{w} = C_{r,t} r^{-\frac{3}{2}}\left[\left(\cos \frac{3\theta}{2} \pm \frac{3\xi_{r,t}}{4} \sin 2\theta\right)\mathbf{i} + \left(\sin \frac{3\theta}{2} \pm \frac{\xi_{r,t}}{4}(1 + 3 \cos 2\theta)\right)\mathbf{j}\right] \tag{289}$$

The deflection of light, either reflected from, or passing through a generic point P of the plate in the vicinity of the crack tip (Fig. 5), is given by the deviation vector **W**, which, for a birefringent material, is expressed by [31, 38]:

$$\mathbf{W} = \mathbf{w} + \mathbf{r} = x'_{r,t}\mathbf{i} + y'_{r,t}\mathbf{j} \tag{290}$$

where:

$$\mathbf{r} = r\cos\theta\,\mathbf{i} + r\sin\theta\,\mathbf{j} \tag{291}$$

From Rels (289), (290) and (291) the parametric equations of the caustic are expressed by:

$$x'_{r,t} = r\cos\theta + C_{r,t}r^{-\frac{3}{2}}\left(\cos\frac{3\theta}{2} \pm \frac{3}{4}\xi_{r,t}\sin 2\theta\right) \tag{292}$$

$$y'_{r,t} = r\sin\theta + C_{r,t}r^{-\frac{3}{2}}\left[\sin\frac{3\theta}{2} \mp \frac{1}{4}\xi_{r,t}(1 + 3\cos 2\theta)\right] \tag{293}$$

Since the vector **W** represents the position on the screen Sc of a point Q (Fig. 5) which, in the deflected light, corresponds to the generic point P on the specimen Sp, the zeroing of the Jacobian determinant:

$$J = \frac{\partial(x'_{r,t}, y'_{r,t})}{\partial(r, \theta)} = \begin{vmatrix} \dfrac{\partial x'_{r,t}}{\partial r} & \dfrac{\partial x'_{r,t}}{\partial \theta} \\[2mm] \dfrac{\partial y'_{r,t}}{\partial r} & \dfrac{\partial y'_{r,t}}{\partial \theta} \end{vmatrix} = 0 \tag{294}$$

is the necessary condition for the formation of a caustic on the screen by means of a concentration of light rays. Thus, we have:

$$\frac{\partial x'_{r,t}}{\partial r} = \cos\theta - \frac{3}{2}C_{r,t}r^{-\frac{3}{2}}\left(\cos\frac{3\theta}{2} \pm \frac{3}{4}\xi_{r,t}\sin 2\theta\right)$$

$$\frac{\partial x'_{r,t}}{\partial \theta} = -r\sin\theta - \frac{3}{2}C_{r,t}r^{-\frac{3}{2}}\left(\sin\frac{3\theta}{2} \mp \xi_{r,t}\cos 2\theta\right)$$

$$\frac{\partial y'_{r,t}}{\partial r} = \sin\theta - \frac{3}{2}C_{r,t}r^{-\frac{5}{2}}\left[\sin\frac{3\theta}{2} \mp \frac{1}{4}\xi_{r,t}(1 + 3\cos 2\theta)\right]$$

$$\frac{\partial y'_{r,t}}{\partial \theta} = r\cos\theta + \frac{3}{2}C_{r,t}r^{-\frac{3}{2}}\left(\cos\frac{3\theta}{2} \pm \xi_{r,t}\sin 2\theta\right) \tag{295}$$

and the Rel. (294) yields:

$$r = r_0 = \left\{\frac{3}{2}C_{r,t}\left[\mp\xi_{r,t}\sin\theta + \left(1 \mp \frac{1}{4}\xi_{r,t}\left(7\sin\frac{\theta}{2} - \sin\frac{3\theta}{2}\right)\right.\right.\right.$$

$$\left.\left.\left. + \frac{1}{32}\xi_{r,t}^2(25 + 9\cos 2\theta)\right)^{\frac{1}{2}}\right]\right\}^{\frac{2}{3}} \tag{296}$$

Relation (296), depending on the value of $C_{r,t}$, expresses the equation of the generatrix curve on the specimen and around the singularity, which creates the caustic surface in space. This curve, if it exists, produces the caustic and is termed the *initial curve*.

Introducing Rel. (296) into the parametric Eqs (292) and (293), the parametric equations for the caustic are obtained. These relations, for $\pm \xi_{r,t} = \xi_{r,t}$, are:

$$x'_{r,t} = \left(\frac{3}{2}C_{r,t}\right)^{\frac{2}{3}}\left[A^{\frac{2}{3}}\cos\theta + \frac{2}{3}A^{-\frac{1}{3}}\left(\cos\frac{3\theta}{2} + \frac{3}{4}\xi_{r,t}\sin 2\theta\right)\right] \qquad (297)$$

$$y'_{r,t} = \left(\frac{3}{2}C_{r,t}\right)^{\frac{2}{3}}\left[A^{\frac{2}{3}}\sin\theta + \frac{2}{3}A^{-\frac{1}{3}}\left(\sin\frac{3\theta}{2} - \frac{1}{4}\xi_{r,t}(1 + 3\cos 2\theta)\right)\right] \qquad (298)$$

where A is expressed by:

$$A = -\frac{1}{4}\xi_{r,t}\sin\theta + \left[1 + \frac{1}{4}\xi_{r,t}\left(7\sin\frac{\theta}{2} - \sin\frac{3\theta}{2}\right) + \frac{1}{32}\xi_{r,t}^2(25 + 9\cos 2\theta)\right]^{\frac{1}{2}} \qquad (299)$$

and therefore the radius of the initial curve becomes:

$$r = r_0 = (\tfrac{3}{2}C_{r,t}A)^{\frac{2}{3}} \qquad (300)$$

Relations (297) and (298), together with the expression for A given in Rel. (299), yield the caustic curves formed at a distance z_0 from the specimen on a reference screen Sc. The angle θ varies between:

$$-\pi \leq \theta \leq \pi \qquad (301)$$

For values of θ between $-\pi$ and π and for $\xi_{r,t}$ positive, we obtain one branch of the initial curve and its caustic, and for values of θ between $-\pi$ and π for $\xi_{r,t}$ negative, the other branch. These branches are symmetric with respect to the crack axis and coincide in the case of isotropic materials, where $\xi_{r,t}$ and $A = 1$. For optically anisotropic materials with $|\xi_{r,t}| \neq 0$, the two branches of the above-mentioned curves lie inside and outside of the corresponding curves for $\xi_{r,t} = 0$.

Figure 7 shows the shapes of the caustics, plotted by computer, for values of $|\xi|$ varying between zero and unity. It is clear from this figure that as $|\xi|$ increases, the shapes of the caustics are progressively distorted. These curves are now formed of

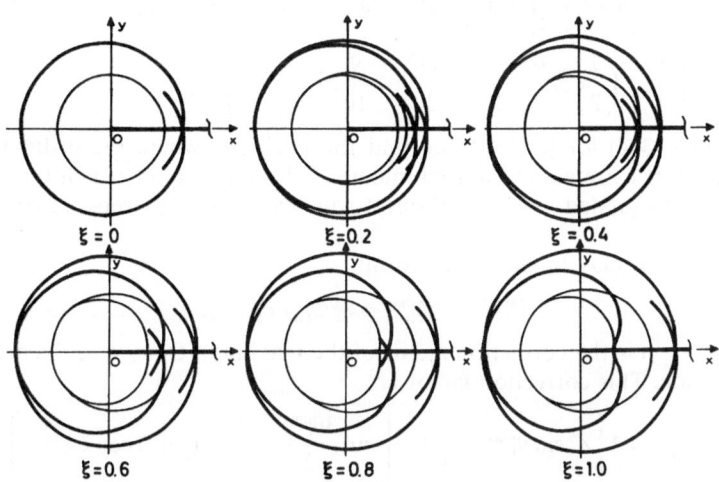

Fig. 7. Initial curves and their respective caustics in a plate with crack in tension made of birefringent materials with $|\xi| = 0$, 0.2, 0.4, 0.6, 0.8 and 1.0, as plotted by computer for $-\pi \leq \theta \leq \pi$.

two branches lying on either side of the curve corresponding to $\xi_{r,t} = 0$. The shapes of either branch are either squeezed or swelled on the one side along the longitudinal $O'x'$-axis, while on the other side they coincide and remain tangent to a line parallel to the $O'y'$-axis. Thus, they cease to resemble circles, as is the case for the caustic and the initial curves for $\xi_{r,t} = 0$. This distortion of the caustics results in a relative displacement of the maximum diameters and of other characteristic points of the caustics and implies the necessity of measuring more than one characteristic quantity from each caustic for the accurate evaluation of K_I.

While for optically isotropic materials the initial curve is a circle and the caustic a generalized epicycloid traced out by the tip of vector $\mathbf{W}$ which is the sum of two other vectors, for optically anisotropic materials the initial curve is not a circle and the caustic is formed by a vector $\mathbf{W}$ which is the sum of four component vectors. We seek, therefore, to find the most suitable points on these curves for the evaluation K_I of the specimen.

For this reason we determine the extrema of the positive branch $(\xi_{r,t} > 0)$ and the negative branch $(\xi_{r,t} < 0)$ of the caustic. The conditions of determination of the extrema are:

$$\frac{\partial y'_{r,t}}{\partial \theta} = 0 \tag{302}$$

$$\frac{\partial^2 y'_{r,t}}{\partial \theta^2} < 0 \tag{303}$$

and for $\xi_{r,t}$ positive or negative, the Rel. (302) gives:

$$\frac{2}{5} A^{-\frac{3}{5}} \dot{A} \sin\theta + A^{\frac{2}{5}} \cos\theta - \frac{2}{5} A^{-\frac{8}{5}} \dot{A} \left[\sin\frac{3\theta}{2} - \frac{1}{4}\xi_{r,t}(1 + 3\cos 2\theta) \right]$$

$$+ A^{-\frac{3}{5}} \left[\cos\frac{3\theta}{2} + \xi_{r,t}\sin 2\theta \right] = 0 \tag{304}$$

where:

$$\dot{A} = -\frac{1}{4}\xi_{r,t}\cos\theta + \frac{1}{2}\left[1 + \frac{1}{4}\xi_{r,t}\left(7\sin\frac{\theta}{2} - \sin\frac{3\theta}{2} \right) + \frac{1}{32}\xi_{r,t}^2(25 + 9\cos\theta) \right]^{-\frac{1}{2}}$$

$$\times \left[\frac{1}{4}\xi_{r,t}\left(\frac{7}{2}\cos\frac{\theta}{2} - \frac{3}{2}\cos\frac{3\theta}{2} \right) - \frac{9}{16}\xi_{r,t}^2\sin^2\theta \right] \tag{305}$$

Solving Rel. (304) for $|\xi_{r,t}| \leq 1$, we find the positions where the ordinates of the caustic present extrema and indeed maxima. The angles θ for which these maxima appear are designated as $\theta_t^{\max}$ (subscript t indicates the transverse direction of the caustic).

Relation (298) may be put in the form:

$$D_t^{\max} = 2y_{r,t}'^{\max} = \left(\tfrac{3}{2} C_{r,t}\right)^{\frac{2}{3}} \delta_t^{\max}(\xi_{r,t}) \tag{306}$$

where $\delta_t^{\max}(\xi_{r,t})$ is the correction factor of the maximum transverse diameter, $D_t^{\max}$, of the caustic. This correction factor is:

$$\delta_t^{\max}(\xi_{r,t}) = 2\left\{ A^{\frac{2}{5}}\sin\theta_t^{\max} + \frac{2}{3} A^{-\frac{3}{5}} \left[\sin\frac{3\theta_t^{\max}}{2} - \frac{1}{4}\xi_{r,t}(1 + 3\cos 2\theta_t^{\max}) \right] \right\} \tag{307}$$

For the position where $\theta = \theta_t^{\max}$, Rel. (307) has a maximum and therefore the transverse diameter of the caustic $D_t^{\max} = 2y_{r,t}'^{\max}$ is also a maximum.

In order to define the longitudinal diameter of the caustic D_l^{max} we must find the angles $\theta_{l(j)}^{max}$, $j = 1, 2$, for which $y'_{r,t}$ (Rel. (298)) is zero. Setting Rel. (298) equal to zero, we obtain two angles $\theta_{l(1)}^{max}$ and $\theta_{l(2)}^{max}$ which when introduced into Rel. (297), yield:

$$D_l^{max} = x'^{max}_{r,t(1)} + x'^{max}_{r,t(2)} = (\tfrac{3}{2} C_{r,t})^{\frac{3}{2}} \delta_l^{max}(\zeta_{r,t}) \tag{308}$$

where $\delta_l^{max}(\zeta_{r,t})$ is the correction factor of the diameter D_l^{max} of the caustic. This correction factor is:

$$\delta_l^{max}(\zeta_{r,t}) = \sum_{j=1}^{2} \left[A_j^{\frac{2}{3}} \cos \theta_{l(j)}^{max} + \frac{2}{3} A_j^{-\frac{1}{3}} \left(\cos \frac{3\theta_{l(j)}^{max}}{2} + \frac{3}{4} \zeta_{r,t} \sin 2\theta_{l(j)}^{max} \right) \right] \tag{309}$$

with:

$$A_j = -\frac{1}{4} \zeta_{r,t} \sin \theta_{l(j)}^{max} + \left[1 + \frac{1}{4} \zeta_{r,t} \left(7 \sin \frac{\theta_{l(j)}^{max}}{2} - \sin \frac{3\theta_{l(j)}^{max}}{2} \right) \right.$$
$$\left. + \frac{1}{32} \zeta_{r,t}^2 (25 + 9 \cos 2\theta_{l(j)}^{max}) \right]^{\frac{1}{2}} \tag{310}$$

Figure 8 presents the variation of θ_t^{max} and $\theta_{l(j)}^{max}$ for different values of the coefficient of optical anisotropy ξ. Due to the unilateral squeezing and swelling of the two branches of each branch of the caustic along the $O'x'$-axis the angles θ corresponding to maximum diameters of the respective branches of each caustic are much different. For the correct evaluation of K_I in such cases more than one measurement on the caustic is necessary, therefore, the exact position of these maxima is imperative for an accurate evaluation of K_I. This figure indicates the differences in the positions

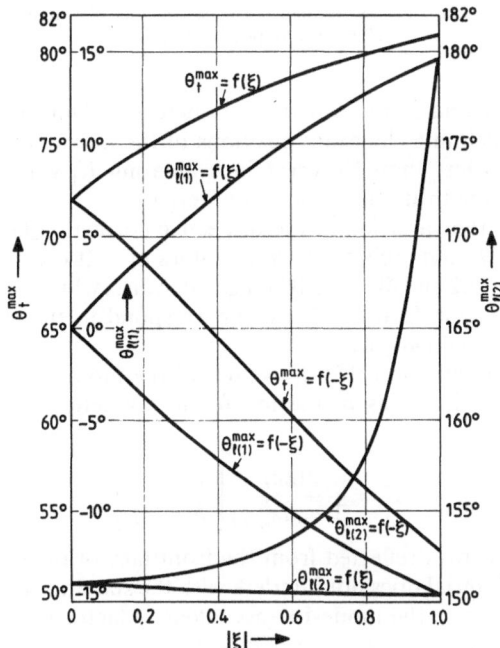

Fig. 8. Variation of the positions of maxima in the parametric equations of caustics for various values of the coefficient of optical anisotropy ξ.

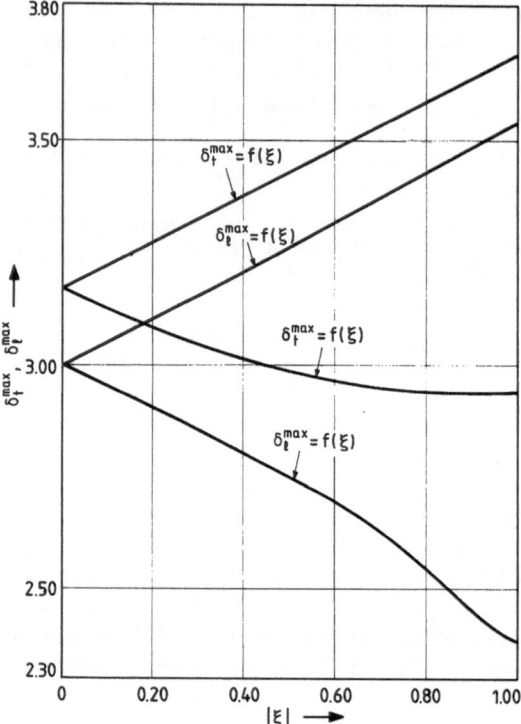

Fig. 9. Variation of transverse and longitudinal correction factors δ_t^{max} and δ_l^{max} plotted against the coefficient of optical anisotropy ξ.

of the respective maxima in the caustic as ξ increases. Thus, measurements made on the caustics by defining the exact positions of the maximum diameters of their branches resulted in less than 5% error in evaluating K_I when this experimental value was compared with its theoretical counterpart.

Figure 9 presents the variation of the respective correction factors δ_t^{max} and δ_l^{max} as given by Rels (307) and (309) for various values of ξ. It can be derived from this figure that $\delta_t^{max} = 3.1702$ and $\delta_l^{max} = 3.00$ for $\xi = 0$, values which are valid for isotropic materials. The correction factors $\delta_{t,l}^{max}(+\xi)$ correspond to the external caustic and $\delta_{t,l}^{max}(-\xi)$ to the internal caustic.

From Rels (307), (309) and (215) the values of the mode-I stress intensity factor K_I may be expressed in terms of the maximum diameters D_t^{max} and D_l^{max} of the caustic (Rel. (238)):

$$K_I = \frac{2(2\pi)^{\frac{1}{2}}}{3\varepsilon z_0 d\lambda_m^{\frac{3}{2}} c_{r,t}} \left[\frac{D_{t,l}^{max}}{\delta_{t,l}^{max}(\xi)} \right]^{\frac{5}{2}} \tag{311}$$

For the case of light rays reflected from the front face of the specimen the optical anisotropy of the material does not interfere with the size and shape of the generated caustic, since in this case the mode-I stress intensity factor is expressed by:

$$K_I = \frac{2(2\pi)^{\frac{1}{2}} E}{3 z_0 d\lambda_m^{\frac{3}{2}} v} \left[\frac{D_{t,l}^{max}}{\delta_{t,l}^{max}} \right]^{\frac{5}{2}} \tag{312}$$

where E and v are the modulus of elasticity and Poisson's ratio of the material and $\delta_t^{max} = 3.1702$, $\delta_l^{max} = 3.00$. The same Rel. (312) applies to opaque and reflecting materials, and this fact forms one of the main advantages of the method of reflected caustics over the method of transmitted caustics [39].

In order to show the potential of the method to evaulate easily and accurately the mode-I stress intensity factors in plates with cracks, typcial tests were executed on thin tension plates of width $w = 49.8 \times 10^{-3}$ m, thickness $d = 3 \times 10^{-3}$ m, and length $l = 250 \times 10^{-3}$ m containing transverse cracks and prepared from large sheets of polycarbonate of bisphenol A (PCBA). This material presents an initial optical anisotropy with $\xi_r = 0.153$ and $\xi_t = 0.223$ [9]. In the mid-length of all specimens a transverse edge-crack was sawn having an initial length $a = 10 \times 10^{-3}$ m. All specimens were subjected to a tension load of the order of $P = 15 \times 10^2$ N in an Instron tester.

The experimental set-up was simple (Fig. 10). A coherent light beam from a He–Ne gas laser impinged on the specimen. Screens in front and behind the specimen were placed at distances z_0 and parallel to specimen mid-plane. On these screens the caustics from reflected and transmitted light rays were formed and recorded by a camera. The magnification factor λ_m of the optical system was given by:

$$\lambda_m = \frac{z_0 + z_i}{z_i} \tag{313}$$

where z_0 was the distance between the mid-plane of the specimen and the screen and z_i the distance between the focus of the bundle and the mid-plane of the specimen. For a positive value of z_i (the focus lying in front of the specimen) a double caustic is formed on the screen due to reflections from the rear face of the specimen. For a negative value of z_i (the focus lying behind the specimen) a simple caustic is formed on the screen due to reflections from the front face of the specimen.

The experimental stress intensity factor K_1 is calculated by the Rel. (311), while the theoretical K_1 is given by:

$$K_I = 1.1215 \, \sigma \sqrt{\pi a} \tag{314}$$

where the factor 1.1215 was evaluated from Ref. [102] to take care of the finite width of the plate. The theoretical value thus calculated was $K_1 = 20.88 \times 10^5$ N/m$^{\frac{3}{2}}$. With this value for K_1, the theoretical shapes of the caustic reflected from the rear

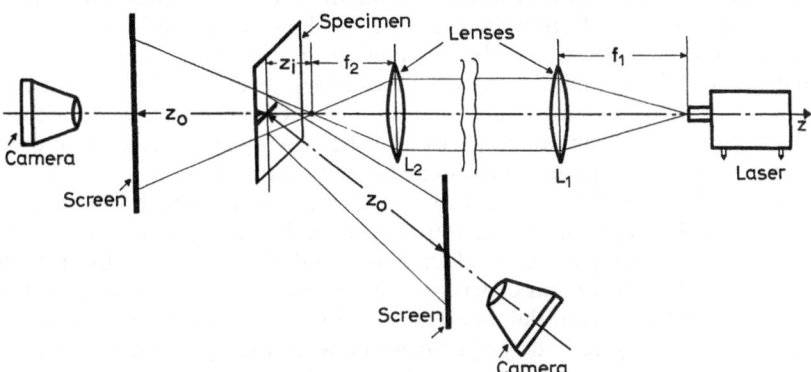

Fig. 10. The optical set-up for reflected and transmitted caustics.

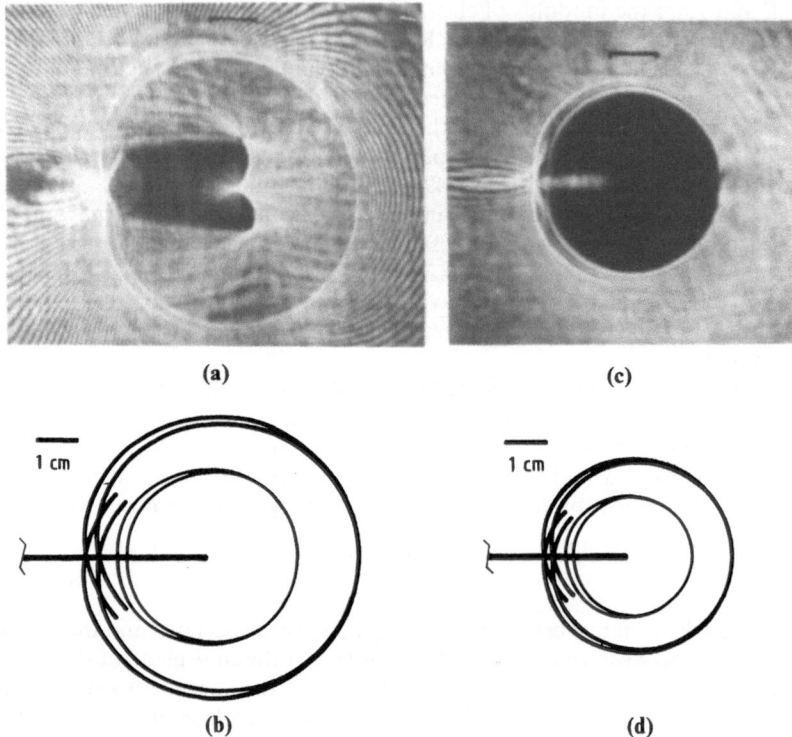

Fig. 11. Experimentally obtained caustics from reflected (a) and transmitted light rays (c) in a plate made of polycarbonate of bisphenol A (PCBA) with $K_1 = 20.88 \times 10^5$ N/m$^{\frac{3}{2}}$ and their respective theoretical curves (b) and (d) as plotted by computer for the same condition and scales.

and of the transmitted caustic were calculated and plotted by computer. Their forms are given in Figs 11(b) and 11(d) respectively, at the same scale as the experimentally obtained caustics shown in Figs 11(a) and 11(c) respectively. The photographs in Figs 11(a) and 11(c) were obtained for a positive value of z_i. The two branches of the caustic in Fig. 11(a) were obtained by reflection from the rear face of the specimen, whereas those in Fig. 11(c) were obtained from the rays transmitted by the specimen. The branches due to reflection are greater than those to transmission since $c_r > c_t$. In Fig. 11(a) a small caustic can be observed inside the main caustic and is due to the reflection of light rays from the front face of the specimen. For a negative value of z_i, the position of the caustics in Fig. 11(a) is interchanged and the same formulae hold but with $c_f = \nu/E$. By a comparison of the two groups of caustics a complete coincidence of their sizes and shapes can be demonstrated.

For the evaluation of K_I we need the values for c_r and c_t which were determined experimentally by applying the interferometric method developed in Refs [98] and [99]. The corresponding correction factors δ_r^{max} and δ_t^{max} were taken from the curves of Fig. 9. The theoretical and experimental values for K_I are tabulated in Table 1.

It can be concluded from the theoretically and experimentally determined values of K_I that there is a satisfactory agreement between theory and experiment.

The sensitivity and accuracy of the method may be immediately estimated, without further calculation, by the coincidence of the experimentally obtained caustics with

Table 1. Experimental determination of stress intensity factor K_1 by caustics in polycarbonate. Properties of material: $E = 2.75 \times 10^9$ N/m², $\nu = 0.36$, $\xi_r = 0.153$, $\xi_t = 0.223$. Theoretical value for stress intensity factor $K_1 = 20.88 \times 10^5$ N/m$^{\frac{3}{2}}$

	z_0 (m)	z_i (m)	λ_m	c_r (m²/N)	c_i (m²/N)	δ_r^{max}	δ_l^{max}	D_r^{max} (m)	D_l^{max} (m)	K_1 (N/m$^{\frac{3}{2}}$)	K_1^* (N/m$^{\frac{3}{2}}$)
External caustic from reflected light from rear face	1.26	0.18	8.00	-2.24×10^{-10}		3.245	3.075	6.1×10^{-2}	5.6×10^{-2}	21.13×10^5 19.52×10^5	20.33×10^5
Internal caustic from reflected light from rear face						3.100	2.926	5.9×10^{-2}	5.3×10^{-2}	21.79×10^5 19.26×10^5	20.53×10^5
External caustic from transmitted light	1.47	0.18	9.17		-1.43×10^{-10}	3.245	3.075	4.54×10^{-2}	4.1×10^{-2}	22.11×10^5 19.60×10^5	20.86×10^5
Internal caustic from transmitted light						3.100	2.926	4.2×10^{-2}	3.95×10^{-2}	20.40×10^5 20.21×10^5	20.31×10^5

the curves traced theoretically by computer and plotter, by introducing into the respective relations the exact values of K_I and K_{II}.

It can be further stated that the values of K_I derived from measurements of the transverse external diameter of the caustic are always greater than the theoretical value, while K_I derived from measurements of the longitudinal diameter is always smaller. However, the maximum discrepancies between theoretical and experimental values are of the order of $\pm 5\%$ for measurements of both the transverse and longitudinal diameters. These differences are acceptable since they include the experimental error in evaluating the diameters of the caustics and since the theoretical value is also approximate, being corrected empirically for the limited dimensions of the plate.

The best approach is to take the average values from transverse and longitudinal diameters. Indeed, the stress intensity factors K_I^* derived from the average values of the caustics are very close to the theoretical ones, the differences being of the order of 1–3%.

2.4.2.2 Analysis of the Elastic Oblique-Crack Problem

A thin elastic plate of a birefringent material containing an internal crack of length $2a$ is submitted to a biaxial state of stress in its plane. A Cartesian reference frame Oxy with its origin at the crack tip and its Ox-axis matching with the crack axis coincides with the middle plane of the plate (Fig. 5). To this system another Cartesian frame $O'x'y'$ is referred with the respective axes parallel to each other and lying on the reference plane placed at distance z_0 from the cracked plate. The origin O' of this system coincides with the image of the crack tip on the reference plane. Finally a system of polar coordinates r and θ is connected with the $O'x'y'$ system as indicated in Fig. 5 and a system of Cartesian coordinates u and v accompanies every point P' which is the projection on the reference plane of a generic point P of the specimen lying in the close vicinity of the singularity due to the crack.

The Westergaard complex stress function Z for the case of a cracked plate subjected to a biaxial state of stress at infinity is expressed by Rel. (239). The stress function of Westergaard for the case of an oblique crack in an infinite elastic body, is given by Rel. (239).

The parametric equations of the caustic, referred to the $O'x'y'$ Cartesian system, are given by Rels (265) and (266), while the initial curve is given by Rel. (247). The parametric equations of the caustic referred to the $O'x_1y_1$ Cartesian system (Fig. 6) are given by Rel. (273).

The parametric equations of the caustic (273), for optically anisotropic materials, become [40, 41]:

$$x_{1,r,t} = \left(\frac{3}{2}C_{r,t}\right)^{\frac{2}{3}}\left\{A^{\frac{2}{3}}\cos\tau + \frac{2}{3}A^{-\frac{2}{3}}\left[\cos\frac{3\tau}{2} + \frac{4}{3}\xi_{r,t}\sin 2\tau\right]\right\} \tag{315}$$

$$y_{1,r,t} = \left(\frac{3}{2}C_{r,t}\right)^{\frac{2}{3}}\left\{A^{\frac{2}{3}}\sin\tau + \frac{2}{3}A^{-\frac{2}{3}}\left[\sin\frac{3\tau}{2} - \frac{1}{4}\xi_{r,t}(1 + 3\cos 2\tau)\right]\right\} \tag{316}$$

with:

$$A = -\frac{1}{4}\xi_{r,t}\sin\tau + \left[1 + \frac{1}{4}\xi_{r,t}\left(7\sin\frac{\tau}{2} - \sin\frac{3\tau}{2}\right) + \frac{1}{32}\xi_{r,t}^2(25 + 9\cos 2\tau)\right]^{\frac{1}{2}} \tag{317}$$

and:

$$\tau = \theta + 2\omega \tag{318}$$

By applying the inverse transformation (267), the parametric equations of the caustic referred to the $O'x'y'$ Cartesian system, are:

$$x'_{r,t} = \left(\frac{3}{2}C_{r,t}\right)^{\frac{2}{3}}(1+\kappa^2)^{\frac{1}{3}}\left\{ A'^{\frac{2}{3}}\cos\theta + \frac{2}{3}A'^{-\frac{1}{3}}(1+\kappa^2)^{-\frac{1}{2}}\left[\cos\frac{3\theta}{2} - \kappa\sin\frac{3\theta}{2} \right.\right.$$
$$\left.\left. - \frac{1}{2}\xi_{r,t}(1+\kappa^2)^{-\frac{1}{2}}\kappa(1-3\cos 2\theta) + \frac{3}{4}\xi_{r,t}(1+\kappa^2)^{-\frac{1}{2}}(1-\kappa^2)\sin 2\theta \right]\right\} \tag{319}$$

$$y'_{r,t} = \left(\frac{3}{2}C_{r,t}\right)^{\frac{2}{3}}(1+\kappa^2)^{\frac{1}{3}}\left\{ A'^{\frac{2}{3}}\sin\theta + \frac{2}{3}A'^{-\frac{1}{3}}(1+\kappa^2)^{-\frac{1}{2}}\left[\sin\frac{3\theta}{2} + \kappa\cos\frac{3\theta}{2} \right.\right.$$
$$\left.\left. - \frac{1}{4}\xi_{r,t}(1-\kappa^2)(1+\kappa^2)^{-\frac{1}{2}}(1+3\cos 2\theta) + \frac{3}{2}\xi_{r,t}\kappa(1+\kappa^2)^{-\frac{1}{2}}\sin 2\theta \right]\right\} \tag{320}$$

with:

$$A' = -\frac{1}{4}\xi_{r,t}(1-\kappa^2)(1+\kappa^2)^{-1}\sin\theta - \frac{1}{2}\xi_{r,t}\kappa(1+\kappa^2)^{-1}\cos\theta + \left\{1 + \frac{1}{4}\xi_{r,t}(1+\kappa^2)^{-2}\right.$$
$$\times\left[7(1+\kappa^2)^{\frac{3}{2}}\left(\sin\frac{\theta}{2} + \kappa\cos\frac{\theta}{2}\right) - (1-3\kappa^2)(1+\kappa^2)^{\frac{1}{2}}\sin\frac{3\theta}{2} \right.$$
$$\left. - \kappa(3-\kappa^2)(1+\kappa^2)^{\frac{1}{2}}\cos\frac{3\theta}{2} \right] + \frac{1}{32}\xi_{r,t}^2(1+\kappa^2)^{-2}$$
$$\left. \times\left[25(1+\kappa^2)^2 + 9(\kappa^4 - 6\kappa^2 + 1)\cos 2\theta - 36\kappa(1-\kappa^2)\sin 2\theta \right]\right\}^{\frac{1}{2}} \tag{321}$$

and the radius of the initial curve becomes:

$$r = r_0 = (\tfrac{3}{2}C_{r,t}A')^{\frac{2}{3}}(1+\kappa^2)^{\frac{1}{3}} \tag{322}$$

Equations (319) and (320) together with Eq. (322) yield the caustic from the cracked specimen made of an optically anisotropic material as it is formed at distance z_0 on the screen Sc. Angle θ varies between $-\pi \leq \theta \leq \pi$. For values of θ between $-\pi$ and π and for ξ positive we obtain one branch of the initial curve and its caustic and for values of θ between $-\pi$ and π and for ξ negative the other branch. For optically inert materials where $\xi = 0$ and $A = 1$ both branches of the initial curve and caustic are coincident. For birefringent materials these two branches are distinct and lie on both sides of the respective curve for $\xi = 0$.

If the ratio κ is zero ($K_{II} = 0$) the axes of symmetry of both curves coincide with the crack axis, whereas for $\kappa \neq 0$ there exists an angular displacement of the initial curve and the caustic relative to the crack axis, so that the axis of symmetry of these curves subtends an angle (-2ω) with the crack axis. Then, it is valid that:

$$\tan\omega = \kappa = K_{II}/K_I \tag{323}$$

with:

$$K_I = \sigma\sqrt{\pi a}\cos^2\omega, \quad K_{II} = \tau\sqrt{\pi a} = \sigma\sqrt{\pi a}\sin\omega\cos\omega \tag{324}$$

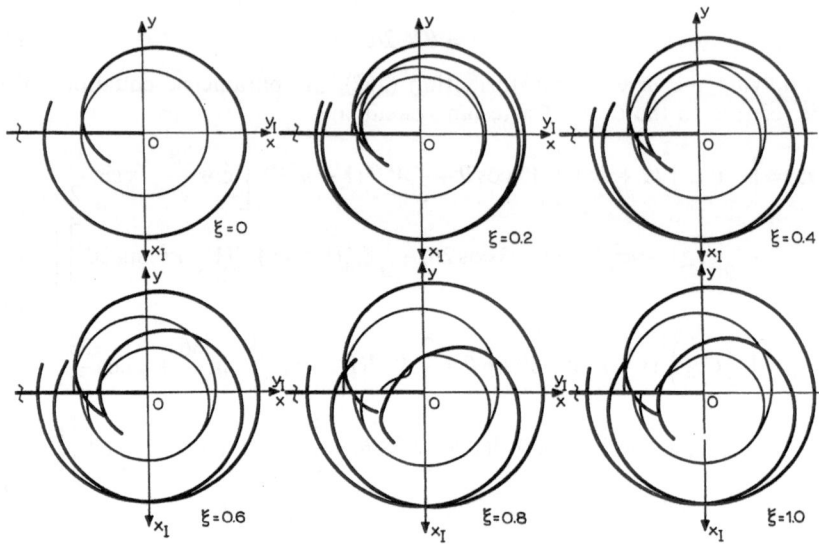

Fig. 12. Initial curves and respective caustics in a cracked plate in tension made of a birefringent material for $\kappa = 1.0$, with different ζ's as plotted by computer for $-\pi \leq \theta \leq \pi$.

Figure 12 presents the forms of the respective caustics as plotted by computer from the above relations for $\kappa = 1.0$, for values of ζ varying between zero and unity, and for angles varying $-\pi \leq \theta \leq \pi$.

In this case, the stress intensity factors K_I and K_{II} are calculated by the Rel. (323) and:

$$K_I = \frac{2(2\pi)^{\frac{1}{2}}}{3\varepsilon z_0 \, d\lambda_m^{\frac{3}{2}} c_{r,t}} \left(\frac{D_{t,l}^{\max}}{\delta_{t,l}^{\max}(\zeta)} \right)^{\frac{5}{2}} \cos \frac{\phi}{2} \tag{325}$$

where ϕ is the angular displacement of the caustic and the correction factors $\delta_{t,l}^{\max}(\zeta)$ are taken from the curves of Fig. 9. Then, the stress intensity factor K_{II} is given by:

$$K_{II} = K_I \tan \frac{\phi}{2} \tag{326}$$

The angle $\phi = 2\omega$ is subtended by the axis of symmetry of the caustic and the axis of crack. The axis of symmetry of the caustic is perpendicular to the tangent of the internal caustic (cusp) (Fig. 6). The calculation of the angle ϕ is difficult for the transmitted caustics, because the internal (cusp) caustic cannot be obtained.

In order to check the potentialities of the method we have applied it to the evaluation of K_I and K_{II} in a thin plate containing an edge crack at an angle $\omega = 45°$ and subjected to pure tension. The material of the plate was polycarbonate of bisphenol A (PCBA) with a $\xi_r = 0.153$ [98]. The specimens were prepared from large sheets of this substance and had the following dimensions: width $w = 50 \times 10^{-3}$ m, thickness $d = 3 \times 10^{-3}$ m, and crack length $a = 10 \times 10^{-3}$ m.

All specimens were subjected to pure tension in an Instron tester at a load $P = 18.639 \times 10^2$ N. The experimental set-up was simple, as in Fig. 10. A coherent light beam from a gas laser impinged normally on the specimen in the vicinity of the crack tip. The reflected and transmitted rays were received at two ground-glass screens placed at distance z_0 from the specimen and parallel to its middle plane. Reflected and transmitted caustics were formed on these screens. The magnification ratio of the light bundles was λ_m given by Rel. (313).

The theoretical values of the stress intensity factors K_I and K_{II} are calculated by Rels (324). For an edge crack, the theoretical values of K_I and K_{II} are given by:

$$K_I = H_I \sigma \sqrt{\pi a}, \quad K_{II} = H_{II} \tau \sqrt{\pi a} \tag{327}$$

with:

$$H_I = 0.809, \quad H_{II} = 0.405 \tag{328}$$

where the correction factors H_I and H_{II} were given by Bowie [102].

Figure 13 presents the experimentally obtained caustics for a cracked polycarbonate plate formed on reference screens placed at distances $z_0 = 1.92$ m for (a) and $z_0 = 2.16$ m for (b) from the specimen. Fig. 13(a) presents both reflected caustics, whereas Fig. 13(b) presents the corresponding transmitted caustic. Both branches of caustics are apparent in both photographs. However, since for caustics reflected from the rear the optical constant c_r is higher than c_t for transmitted light, the caustic in Fig. 13(a) is much larger than that of Fig. 13(b). Moreover, it is worth noting that although both caustics are photographed with the same optical arrangement and conditions, the caustics derived from reflections are much sharper than those derived from transmitted light rays, thus yielding a better and more accurate evaluation of the stress intensity factor.

For the evaluation of K_I and K_{II} the values of the optical constants are necessary. These values c_r and c_t were determined by applying a simple interferometric method developed in Refs [98] and [99] and they were found to be:

$c_r = -2.08 \times 10^{-10}$ m^2/N and $c_t = -1.43 \times 10^{-10}$ m^2/N, with $\xi_r = 0.153$, $\xi_t = 0.223$

In Table 2 the experimental values for K_I and K_{II} are given, as well as the mean values K_I^* and K_{II}^* derived from the respective K_I and K_{II} evaluated either from the transverse diameters D_t^{max} or from the longitudinal diameters D_l^{max} (Fig. 13). By measuring the angle of rotation of the caustic, which was $\phi = -53°$, we find that $\kappa = 0.499$. By measuring $D_{t,l}^{max}$ of these caustics we can evaluate K_I and K_{II} and by taking their average values K_I^* and K_{II}^* we obtain a better approximation. Table 2

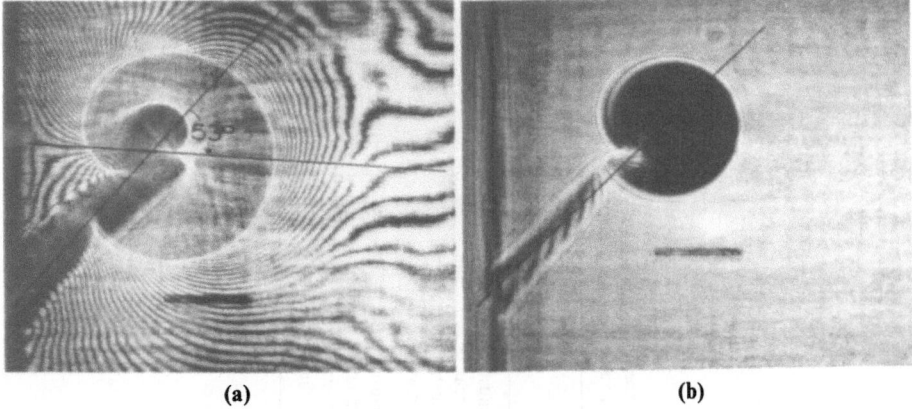

(a)   (b)

Fig. 13. Experimentally obtained caustics for a cracked polycarbonate plate.

Table 2. Experimental determination of stress intensity factors K_I and K_{II} by caustics in polycarbonate. Properties of material: $E = 2.75 \times 10^9$ N/m², $\nu = 0.36$, $\xi_r = 0.153$, $\xi_t = 0.223$. Theoretical values for stress intensity factors: $K_I = 17.82 \times 10^5$ N/m², $K_{II} = 8.92 \times 10^5$ N/m²

	z_0 (m)	z_i (m)	λ_m	c_r (m²/N)	c_i (m²/N)	δ_r^{max}	δ_t^{max}	D_r^{max} (m)	D_t^{max} (m)	K_I (N/m²)	K_{II} (N/m²)	K_I^* (N/m²)	K_{II}^* (N/m²)
External caustic from reflected light from rear face	1.92	0.49	4.9	-2.08×10^{-10}		3.177		5.01×10^{-2}		17.97×10^5	8.95×10^5	15.98×10^5	7.96×10^5
							2.972		4.24×10^{-2}	13.98×10^5	6.97×10^5		
Internal caustic from reflected light from rear face						2.940		4.70×10^{-2}		18.58×10^5	9.26×10^5	17.07×10^5	8.51×10^5
							2.781		4.14×10^{-2}	15.55×10^5	7.75×10^5		
External caustic from transmitted light	2.16	0.49	5.4		-1.43×10^{-10}	3.244		3.57×10^{-2}		16.34×10^5	8.15×10^5	16.13×10^5	8.04×10^5
							3.012		3.28×10^{-2}	15.91×10^5	7.93×10^5		
Internal caustic from transmitted light						2.869		3.20×10^{-2}		16.90×10^5	8.43×10^5	17.13×10^5	8.54×10^5
							2.750		3.10×10^{-2}	17.35×10^5	8.65×10^5		

shows all these values and indicates that although the values of K_I and K_{II} derived either from D_t^{max} or D_l^{max} are in good approximation with the theoretical results, their mean values K_I^* and K_{II}^* are much closer to the theory, their discrepancies being of the order of 1% to 4%.

Moreover, with the method of reflected caustics both types of caustics are obtained, that is those reflected from the front and the rear faces of the specimen. For opaque substances these two types of caustics can be successively obtained by placing the focus of the light beam either in front of or behind the specimen. With the method of transmitted caustics only one caustic is ever obtained.

The two reflected caustics yield much additional information. While the external caustic with its transverse and longitudinal diameters yields the means to evaluate the order of singularity and eventually the stress intensity factor, the internal caustic, which is a cusped curve, yields the additional information of the axis of symmetry of both caustics. Indeed, the normal for the cusp to the common tangent to the two branches of the cusped curve defines the axis of symmetry of the caustics and this yields the ratio $\kappa = K_{II}/K_I$ of the two stress intensity factors in extension and shear modes of deformation. This additional information cannot be obtained with the method of transmitted caustics.

Finally the greater size and sharpness of the reflected caustic yields a much higher accuracy in evaluating stress intensity factors than with transmitted caustics.

The method of reflected caustics has already been applied to cases of isotropic materials [32], as well as to cases with birefringent materials [31, 38, 40, 41].

However, whereas to date isotropic materials were extensively studied in the general case when both stress intensity factors are operative, studies of birefringent materials have been limited to fracture problems where only K_I is operative.

2.5 Application of the Method of Caustics in Mixed-Mode Plane Crack Problems

The optical method of reflected caustics has been applied so far to problems of cracked plates under uniaxial loading. The problem of biaxial tension of the plate has only been considered for the particular case where the crack is transverse to the longitudinal axis of the plate which coincided with the loading axis. In this section the influence of a biaxial loading of the plate on the form and orientation of the caustic is studied in connection with the orientation of the crack. New, modified relations are given for the evaluation of the complex stress intensity factor $K^* = K_I - iK_{II}$ in terms of ϕ, the angular displacement of the caustic axis.

2.5.1 The Stress Field at the Crack Tip for Mixed-Mode Deformation

For a thin elastic and isotropic plate under conditions of generalized plane stress containing a slanting internal crack of length $2a$ and subjected at infinity to a biaxial state of stress defined by the stresses σ and $k\sigma$ along two adjacent sides of the plate

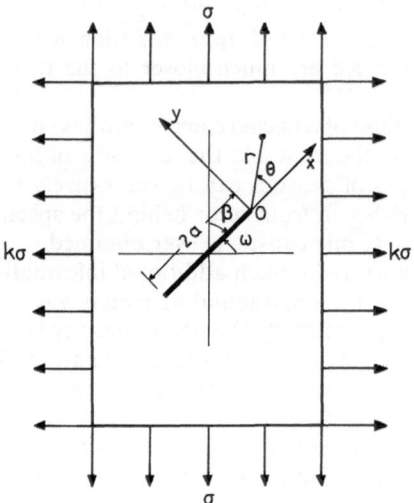

Fig. 14. Geometry of cracked plate.

(Fig. 14) the Muskhelishvili complex stress functions $\Phi(z)$ and $\Omega(z)$ are given by [89]:

$$\Phi(z) = \varphi'(z) = \frac{1}{2}(2\Gamma + \bar{\Gamma}')\frac{z}{(z^2 - a^2)^{\frac{1}{2}}} - \frac{1}{2}\bar{\Gamma}' \qquad (329)$$

$$\Omega(z) = \frac{1}{2}(2\Gamma + \bar{\Gamma}')\frac{z}{(z^2 - a^2)^{\frac{1}{2}}} + \frac{1}{2}\bar{\Gamma}' \qquad (330)$$

where the quantities Γ and $\bar{\Gamma}$ are expressed by:

$$\frac{1}{2}\bar{\Gamma}' = -\frac{\sigma}{4}(1 - k)e^{2i((\pi/2) - \omega)} \qquad (331)$$

$$\frac{1}{2}(2\Gamma + \bar{\Gamma}') = \frac{\sigma}{4}\{[1 - e^{2i((\pi/2) - \omega)}] + k[1 + e^{2i((\pi/2) - \omega)}]\} \qquad (332)$$

In these relations k is the biaxiality factor of the stresses at infinity and ω the angle subtended by the crack-axis and the transverse axis of the specimen ($\omega = 90° - \beta$). The components of stresses at the tip of the crack may be derived by the well-known relations:

$$\sigma_{xx} + \sigma_{yy} = 2[\Phi(z) + \overline{\Phi(z)}] \qquad (333)$$

$$\sigma_{yy} - \sigma_{xx} + 2i\tau_{xy} = 2[(\bar{z} - z)\Phi'(z) + \bar{\Omega}(z) - \Phi(z)] \qquad (334)$$

$$2G(u + iv) = s_{1,2}\varphi(z) - \omega(z) - (z - \bar{z})\overline{\Phi(z)} \qquad (335)$$

where $s_1 = (3 - v)/(1 + v)$ for plane-stress and $s_2 = (3 - 4v)$ for plane-strain conditions, and $\Phi'(z) = d\Phi(z)/dz$. The functions $\varphi(z)$ and $\omega(z)$ are given by:

$$\varphi(z) = \int \Phi(z)dz$$

$$\omega(z) = \int \Omega(z)dz \qquad (336)$$

From Rels (329)–(332) and (336) we obtain [43, 101]:

$$\varphi(z) = \frac{\sigma}{4}\{[(1 - e^{2i((\pi/2) - \omega)}) + k(1 + e^{2i((\pi/2) - \omega)})](z^2 - a^2)^{\frac{1}{2}} + (1 - k)e^{2i((\pi/2) - \omega)}z\}$$

(337)

$$\omega(z) = \frac{\sigma}{4}\{[(1 - e^{2i((\pi/2) - \omega)}) + k(1 + e^{2i((\pi/2) - \omega)})](z^2 - a^2)^{\frac{1}{2}} - (1 - k)e^{2i((\pi/2) - \omega)}z\}$$

(338)

When the angle $\omega = 0°$ (transverse crack) or $\beta = 90°$, Rel. (329) becomes [42]:

$$2\varphi'(z) = \frac{\sigma z}{(z^2 - a^2)^{\frac{1}{2}}} - \frac{1}{2}(1 - k)\sigma$$

(339)

By applying the transformation:

$$z - a = \zeta = re^{i\theta}, \quad z = x + iy$$

(340)

referred to the (x, y) Cartesian coordinate system of Fig. 14, Rels (329), (330), (337) and (338) transform to functions of the complex variable ζ:

$$\Phi(\zeta) = \frac{\sigma}{4}\left\{[(1 - e^{2i((\pi/2) - \omega)}) + k(1 + e^{2i((\pi/2) - \omega)})]\frac{(\zeta + a)}{(\zeta^2 + 2a\zeta)^{\frac{1}{2}}} + (1 - k)e^{2i((\pi/2) - \omega)}\right\}$$

(341)

$$\Omega(\zeta) = \frac{\sigma}{4}\left\{[(1 - e^{2i((\pi/2) - \omega)}) + k(1 + e^{2i((\pi/2) - \omega)})]\frac{(\zeta + a)}{(\zeta^2 + 2a\zeta)^{\frac{1}{2}}} - (1 - k)e^{2i((\pi/2) - \omega)}\right\}$$

(342)

$$\varphi(\zeta) = \frac{\sigma}{4}\left\{[(1 - e^{2i((\pi/2) - \omega)}) + k(1 + e^{2i((\pi/2) - \omega)})](\zeta^2 + 2a\zeta)^{\frac{1}{2}} + (1 - k)e^{2i((\pi/2) - \omega)}(\zeta + a)\right\}$$

(343)

$$\omega(\zeta) = \frac{\sigma}{4}\left\{[(1 - e^{2i((\pi/2) - \omega)}) + k(1 + e^{2i((\pi/2) - \omega)})](\zeta^2 + 2a\zeta)^{\frac{1}{2}} - (1 - k)e^{2i((\pi/2) - \omega)}(\zeta + a)\right\}$$

(344)

By using the well-known relation [42, 103]:

$$K^* = K_I - iK_{II} = \lim_{z \to a}\{\sqrt{2\pi}\sqrt{z - a}\,2\varphi'(z)\}$$

(345)

the stress intensity factors K_I and K_{II} can be calculated. So, from Rels (329), (331) and (332) we obtain:

$$K^* = \lim_{z \to a}\left\{\sqrt{2\pi}\sqrt{z - a}\frac{\sigma}{2}[(1 - e^{2i((\pi/2) - \omega)}) + k(1 + e^{2i((\pi/2) - \omega)})]\frac{z}{(z^2 - a^2)^{\frac{1}{2}}}\right.$$
$$\left. + \sqrt{2\pi}\sqrt{z - a}\frac{\sigma}{2}(1 - k)e^{2i((\pi/2) - \omega)}\right\}$$

(346)

or:

$$K_I - iK_{II} = \frac{\sigma\sqrt{\pi a}}{2}\{[1 + k + (1 - k)\cos 2\omega] - i(1 - k)\sin 2\omega\}$$

(347)

from which we obtain:

$$K_I = \frac{\sigma\sqrt{\pi a}}{2}[(1+k)+(1-k)\cos 2\omega] \qquad (348)$$

$$K_{II} = \frac{\sigma\sqrt{\pi a}}{2}(1-k)\sin 2\omega \qquad (349)$$

When $\omega = 0°$ ($\beta = 90°$), the stress intensity factors are:

$$K_I = \sigma\sqrt{\pi a}, \quad K_{II} = 0 \qquad (350)$$

which are independent of biaxiality factor k.

For uniaxial tension ($k = 0$), Rels (348) and (349) become [103]:

$$K_I = \sigma\sqrt{\pi a}\cos^2\omega, \quad K_{II} = \sigma\sqrt{\pi a}\sin\omega\cos\omega \qquad (351)$$

For an oblique crack subjected to biaxial tension, with biaxiality factor $k = 1$ (tension–tension), the stress intensity factors are given by Rels (350). With biaxiality factor $k = -1$ (tension–compression), the stress intensity factors are given by:

$$K_I = \sigma\sqrt{\pi a}\cos 2\omega, \quad K_{II} = \sigma\sqrt{\pi a}\sin 2\omega \qquad (352)$$

The variation of the K_I and K_{II} stress intensity factors in terms either of the angle ω of inclination of the crack-axis with the transverse direction or of the biaxiality factor k is given in Figs 15, 16 and 17, 18 respectively. It may be concluded from these figures that as the angle ω increases, and therefore angle β decreases, there is also a decrease of K_I. The same phenomenon happens for decreasing values of k. Conversely, the values for K_{II} increase rapidly from zero for $\omega = 0°$, to a maximum at $\omega \simeq 45°$ followed by a decrease to zero again for $\omega = 90°$ ($\beta = 0°$).

The values of the K_I factor for positive values of the biaxiality factor k tend from a positive maximum value which appears at $\omega = 0°$ ($\beta = 90°$) to zero, whereas for

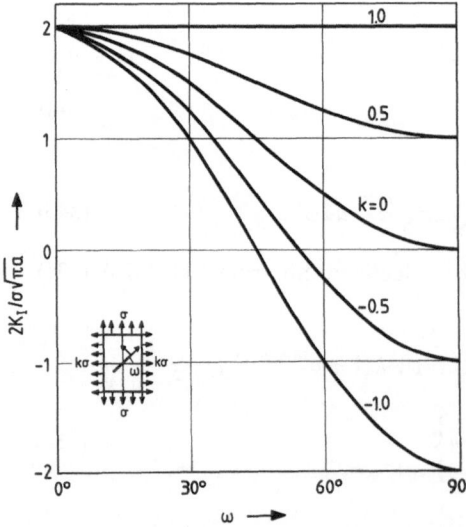

Fig. 15. Variation of K_I versus ω for various parametric values of k.

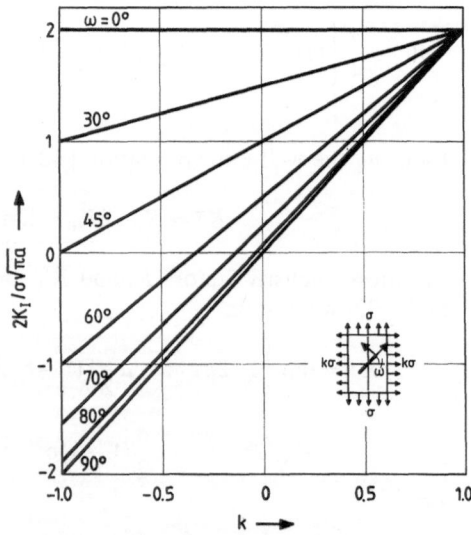

Fig. 16. Variation of K_I versus k for various parametric values of ω.

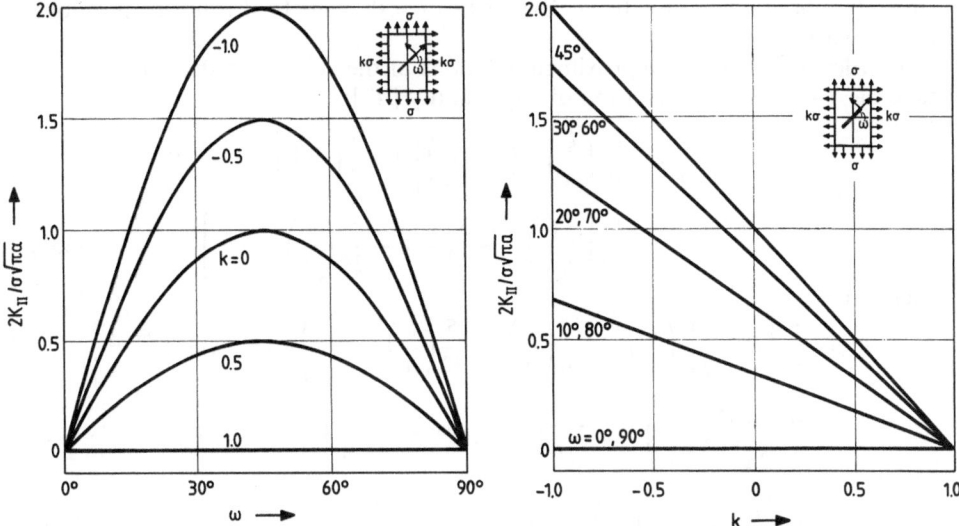

Fig. 17. Variation of K_{II} versus ω for various parametric values of k.

Fig. 18. Variation of K_{II} versus k for various parametric values of ω.

negative values of k this factor takes negative values. In the positions where a change of sign for K_I appears, the state of stress in the vicinity of the crack tip is pure shear ($K_I = 0$, $K_{II} \neq 0$).

In Fig. 19 the positions where the stress distributions at the crack tip are pure shear, with $K_I = 0$ and $K_{II} \neq 0$, i.e. $K_{II}:K_I \to \infty$, are shown. It must be said that this

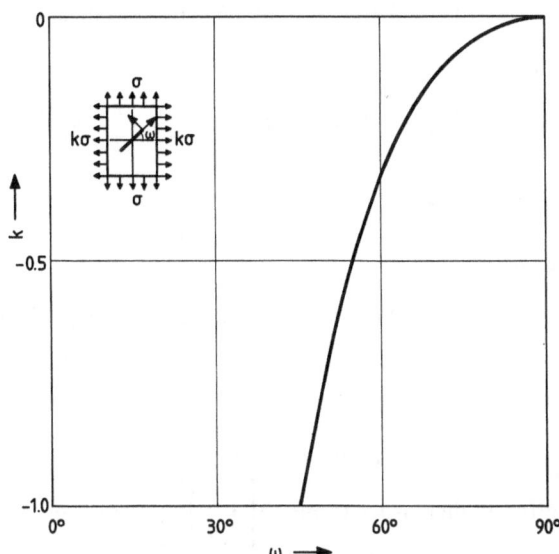

Fig. 19. Positions at which the state of stress in the vicinity of the crack tip is pure shear ($K_I = 0$, $K_{II} \neq 0$) for $k < 0$; for $k > 0$ there are no positions of pure shear.

stress distribution can only be achieved when $k < 0$, while for $k > 0$ such positions do not exist.

In order to derive the approximate relations of the stresses at the crack tip, we take the Taylor series expansion of the quantity [43]:

$$\frac{(\zeta + a)}{(\zeta^2 + 2a\zeta)^{\frac{1}{2}}} = \frac{1}{\sqrt{2}} \left(\frac{\zeta}{a}\right)^{-\frac{1}{2}} \left(1 + \frac{\zeta}{a}\right) \left[1 + \frac{1}{2}\left(\frac{\zeta}{a}\right)\right]^{-\frac{1}{2}}$$

$$= \frac{1}{\sqrt{2}} \left[\left(\frac{\zeta}{a}\right)^{-\frac{1}{2}} + \frac{3}{4}\left(\frac{\zeta}{a}\right)^{\frac{1}{2}} - \frac{5}{32}\left(\frac{\zeta}{a}\right)^{\frac{3}{2}} + \cdots\right] \tag{353}$$

therefore:

$$\Phi(\zeta) = \frac{\sigma}{4} \left\{ D \frac{1}{\sqrt{2}} \left[\left(\frac{\zeta}{a}\right)^{-\frac{1}{2}} + \frac{3}{4}\left(\frac{\zeta}{a}\right)^{\frac{1}{2}} - \frac{5}{32}\left(\frac{\zeta}{a}\right)^{\frac{3}{2}} + \cdots\right] + (1 - k)e^{2i((\pi/2)-\omega)} \right\}$$

$$= \frac{\sigma\sqrt{\pi}}{\sqrt{2\pi}} \frac{D}{2} \left\{ \left(\frac{\zeta}{a}\right)^{-\frac{1}{2}} + \frac{\sqrt{2}}{D}(1 - k)e^{2i((\pi/2)-\omega)}\left(\frac{\zeta}{a}\right)^{0} \right.$$

$$\left. + \frac{3}{4}\left(\frac{\zeta}{a}\right)^{\frac{1}{2}} - \frac{5}{32}\left(\frac{\zeta}{a}\right)^{\frac{3}{2}} + \cdots \right\} \tag{354}$$

where:

$$D = (1 - e^{2i((\pi/2)-\omega)}) + k(1 + e^{2i((\pi/2)-\omega)}) \tag{355}$$

By taking only the constant term of the series expansion, which is independent of the polar distance from the crack tip, Rel. (354) is written:

$$\Phi(\zeta) = \frac{\sigma\sqrt{\pi a}}{\sqrt{2\pi\zeta}} \frac{D}{4} + \frac{\sigma}{4}(1 - k)e^{2i((\pi/2)-\omega)} + O\left[\left(\frac{\zeta}{a}\right)^{\frac{1}{2}}\right] \tag{356}$$

By taking:

$$D = (1 + k) + (1 - k)\cos 2\omega - i(1 - k)\sin 2\omega$$

we have:

$$\frac{\sigma\sqrt{\pi a}}{4} D = \frac{1}{2} \frac{\sigma\sqrt{\pi a}}{2} [(1 + k) + (1 - k)\cos 2\omega] - i\frac{1}{2} \frac{\sigma\sqrt{\pi a}}{2}(1 - k)\sin 2\omega$$

$$= \frac{1}{2}(K_{\mathrm{I}} - iK_{\mathrm{II}})$$

and Rel. (356) is written as:

$$\Phi(\zeta) = \frac{K_{\mathrm{I}} - iK_{\mathrm{II}}}{2\sqrt{2\pi\zeta}} + \frac{\sigma}{4}(1 - k)e^{2i((\pi/2)-\omega)} + O\left[\left(\frac{\zeta}{a}\right)^{\frac{1}{2}}\right] \tag{357}$$

with:

$$\frac{1}{\sqrt{2\pi\zeta}} = \frac{1}{\sqrt{2\pi r}} e^{-i(\theta/2)}$$

and by omitting the higher order terms $O\left[\left(\frac{\zeta}{a}\right)^{\frac{1}{2}}\right]$, we have:

$$2\Phi(\zeta) = 2\varphi'(\zeta) \simeq \frac{K_{\mathrm{I}} - iK_{\mathrm{II}}}{\sqrt{2\pi r}} \left[\cos\frac{\theta}{2} - i\sin\frac{\theta}{2}\right] + \frac{\sigma}{2}(1 - k)e^{2i((\pi/2)-\omega)} \tag{358}$$

Similarly, from Rel. (342) we obtain:

$$2\Omega(\zeta) = 2\omega'(\zeta) \simeq \frac{K_I - iK_{II}}{\sqrt{2\pi r}}\left[\cos\frac{\theta}{2} - i\sin\frac{\theta}{2}\right] - \frac{\sigma}{2}(1-k)e^{2i((\pi/2)-\omega)} \tag{359}$$

Relations (358) and (359) are valid for $0 < \frac{r}{a} \ll 1$.

Similarly we have:

$$(\zeta^2 + 2a\zeta)^{\frac{1}{2}} = a\sqrt{\frac{2\zeta}{a}}\left[1 + \frac{1}{2}\left(\frac{\zeta}{a}\right)\right]^{\frac{1}{2}} = a\sqrt{2}\left[\left(\frac{\zeta}{a}\right)^{\frac{1}{2}} + \frac{1}{2}\left(\frac{\zeta}{a}\right)^{\frac{3}{2}} - \frac{1}{8}\left(\frac{\zeta}{a}\right)^{\frac{5}{2}} + \cdots\right]$$

then:

$$\varphi(\zeta) = \frac{\sigma}{4}Da\sqrt{2}\left[\left(\frac{\zeta}{a}\right)^{\frac{1}{2}} + \frac{1}{2}\left(\frac{\zeta}{a}\right)^{\frac{3}{2}} - \frac{1}{8}\left(\frac{\zeta}{a}\right)^{\frac{5}{2}} + \cdots\right] + \frac{\sigma}{4}(1-k)e^{2i((\pi/2)-\omega)}$$

$$\times a\left(\frac{\zeta}{a}+1\right) = \frac{\sigma a\sqrt{2D}}{4}\left[\left(\frac{\zeta}{a}\right)^{\frac{1}{2}} + \frac{1-k}{\sqrt{2D}}e^{2i((\pi/2)-\omega)}\left(\frac{\zeta}{a}\right) + \frac{1}{2}\left(\frac{\zeta}{a}\right)^{\frac{3}{2}} - \cdots\right]$$

$$+ \frac{\sigma a}{4}(1-k)e^{2i((\pi/2)-\omega)} \tag{360}$$

and by omitting the higher order terms $O\left[\left(\frac{\zeta}{a}\right)^{\frac{3}{2}}\right]$, we have:

$$\varphi(\zeta) \simeq (K_I - iK_{II})\sqrt{\frac{\zeta}{2\pi}} + \frac{\sigma a}{4}(1-k)\left(1 + \frac{\zeta}{a}\right)e^{2i((\pi/2)-\omega)} \tag{361}$$

and from Rel. (344) we obtain:

$$\omega(\zeta) \simeq (K_I - iK_{II})\sqrt{\frac{\zeta}{2\pi}} - \frac{\sigma a}{4}(1-k)\left(1 + \frac{\zeta}{a}\right)e^{2i((\pi/2)-\omega)} \tag{362}$$

By using polar coordinates (r, θ), Rels (361) and (362) are written as:

$$\varphi(\zeta) \simeq (K_I - iK_{II})\sqrt{\frac{r}{2\pi}}e^{i(\theta/2)} + \frac{\sigma a}{4}(1-k)\left[\frac{a + re^{i\theta}}{a}\right]e^{2i((\pi/2)-\omega)} \tag{363}$$

$$\omega(\bar{\zeta}) \simeq (K_I - iK_{II})\sqrt{\frac{r}{2\pi}}e^{-i(\theta/2)} - \frac{\sigma a}{4}(1-k)\left[\frac{a + re^{-i\theta}}{a}\right]e^{2i((\pi/2)-\omega)} \tag{364}$$

with $0 \leq \frac{r}{a} \ll 1$.

By substituting Rels (358) and (359) into Rels (333) and (334), we obtain the stresses at the crack tip. Also, by substituting Rels (361) and (362) into Rel. (335), we obtain the displacements at the crack tip. By setting ζ, $\bar{\zeta}$ and $(\zeta - \bar{\zeta}) = 2ir\sin\theta$ instead of z, $\bar{z}$ and $(z - \bar{z})$, we obtain:

$$\sigma_{xx} \simeq \frac{K_I}{\sqrt{2\pi r}}\cos\frac{\theta}{2}\left[1 - \sin\frac{\theta}{2}\sin\frac{3\theta}{2}\right] - \frac{K_{II}}{\sqrt{2\pi r}}\sin\frac{\theta}{2}\left[2 + \cos\frac{\theta}{2}\cos\frac{3\theta}{2}\right] - \sigma(1-k)\cos 2\omega \tag{365}$$

$$\sigma_{yy} \simeq \frac{K_I}{\sqrt{2\pi r}} \cos\frac{\theta}{2}\left[1 + \sin\frac{\theta}{2}\sin\frac{3\theta}{2}\right] + \frac{K_{II}}{\sqrt{2\pi r}}\sin\frac{\theta}{2}\cos\frac{\theta}{2}\cos\frac{3\theta}{2} \tag{366}$$

$$\tau_{xy} \simeq \frac{K_I}{\sqrt{2\pi r}}\sin\frac{\theta}{2}\cos\frac{\theta}{2}\cos\frac{3\theta}{2} + \frac{K_{II}}{\sqrt{2\pi r}}\cos\frac{\theta}{2}\left[1 - \sin\frac{\theta}{2}\sin\frac{3\theta}{2}\right] \tag{367}$$

$$u \simeq \frac{K_I}{G}\sqrt{\frac{r}{2\pi}}\cos\frac{\theta}{2}\left[\frac{1}{2}(s_{1,2}-1) + \sin^2\frac{\theta}{2}\right] + \frac{K_{II}}{G}\sqrt{\frac{r}{2\pi}}\cos\frac{\theta}{2}\left[\frac{1}{2}(s_{1,2}+1) + \cos^2\frac{\theta}{2}\right]$$

$$-\frac{(1-k)\sigma}{8G}\{r[2\sin\theta\sin 2\omega + \cos(\theta-2\omega) + s_{1,2}\cos(\theta+2\omega)]$$

$$+ (s_{1,2}+1)a\cos 2\omega\} \tag{368}$$

$$v \simeq \frac{K_I}{G}\sqrt{\frac{r}{2\pi}}\sin\frac{\theta}{2}\left[\frac{1}{2}(s_{1,2}+1) - \cos^2\frac{\theta}{2}\right] + \frac{K_{II}}{G}\sqrt{\frac{r}{2\pi}}\cos\frac{\theta}{2}\left[\frac{1}{2}(s_{1,2}-1) + \sin^2\frac{\theta}{2}\right]$$

$$+\frac{(1-k)\sigma}{8G}\{r[2\sin\theta\cos 2\omega + \sin(\theta+2\omega) + s_{1,2}\sin(2\omega-\theta)]$$

$$+ (s_{1,2}+1)a\sin 2\omega\} \tag{369}$$

where the stress intensity factors K_I and K_{II} are given by Rels (348) and (349). For a transverse crack, $\omega = 0°$ ($\beta = 90°$), the relations of the stresses and displacements are written as:

$$\sigma_{xx} \simeq \frac{K_I}{\sqrt{2\pi r}}\cos\frac{\theta}{2}\left[1 - \sin\frac{\theta}{2}\sin\frac{3\theta}{2}\right] - (1-k)\sigma \tag{370}$$

$$\sigma_{yy} \simeq \frac{K_I}{\sqrt{2\pi r}}\cos\frac{\theta}{2}\left[1 + \sin\frac{\theta}{2}\sin\frac{3\theta}{2}\right] \tag{371}$$

$$\tau_{xy} \simeq \frac{K_I}{\sqrt{2\pi r}}\sin\frac{\theta}{2}\cos\frac{\theta}{2}\cos\frac{3\theta}{2} \tag{372}$$

$$u \simeq \frac{K_I}{G}\sqrt{\frac{r}{2\pi}}\cos\frac{\theta}{2}\left[\frac{1}{2}(s_{1,2}-1) + \sin^2\frac{\theta}{2}\right] - \frac{(1-k)(s_{1,2}+1)\sigma}{8G}(r\cos\theta + a) \tag{373}$$

$$v \simeq \frac{K_I}{G}\sqrt{\frac{r}{2\pi}}\sin\frac{\theta}{2}\left[\frac{1}{2}(s_{1,2}+1) - \cos^2\frac{\theta}{2}\right] + \frac{(1-k)(3-s_{1,2})\sigma}{8G}r\sin\theta \tag{374}$$

where:

$$K_I = \sigma\sqrt{\pi a}, \quad K_{II} = 0$$

From Rels (365) and (366) the sum of the principal stresses in the vicinity of the crack tip may be evaluated by:

$$(\sigma_1 + \sigma_2) = (\sigma_{xx} + \sigma_{yy}) \simeq \frac{2}{\sqrt{2\pi r}}\left(K_I\cos\frac{\theta}{2} - K_{II}\sin\frac{\theta}{2}\right) - \sigma(1-k)\cos 2\omega \tag{375}$$

2.5.2 Parametric Equations of Caustics for Biaxial Loading

2.5.2.1 For Optically Isotropic Materials

The parametric equations of the caustic and the initial curve for the case of biaxial loading are of the same form as those in paragraph 2.4.1.2, because it is valid to write [101]:

$$
\text{grad} \left[\frac{2}{\sqrt{2\pi r}} \left(K_I \cos \frac{\theta}{2} - K_{II} \sin \frac{\theta}{2} \right) - \sigma(1-k)\cos 2\omega \right]
$$

$$
= \text{grad} \left[\frac{2}{\sqrt{2\pi r}} \left(K_I \cos \frac{\theta}{2} - K_{II} \sin \frac{\theta}{2} \right) \right] \tag{376}
$$

Then we have:

$$
x'_{r,t,f} = r_0 \cos \theta + C'_{r,t,f} K_I r_0^{-\frac{3}{2}} \cos \frac{3\theta}{2} - C'_{r,t,f} K_{II} r_0^{-\frac{3}{2}} \sin \frac{3\theta}{2} \tag{377}
$$

$$
y'_{r,t,f} = r_0 \sin \theta + C'_{r,t,f} K_I r_0^{-\frac{3}{2}} \sin \frac{3\theta}{2} + C'_{r,t,f} K_{II} r_0^{-\frac{3}{2}} \cos \frac{3\theta}{2} \tag{378}
$$

where r_0 expresses the radius of the generatrix curve (initial curve) given by:

$$
r_0 = (\tfrac{3}{2} C'_{r,t,f})^{\frac{2}{5}} (K_I^2 + K_{II}^2)^{\frac{1}{5}} \tag{379}
$$

with:

$$
C'_{r,t,f} = - \frac{\varepsilon z_0 d c_{r,t,f}}{\lambda_m \sqrt{2\pi}} \tag{380}
$$

Introducing Rels (348) and (349) into Rel. (379), we obtain:

$$
r_0 = \left(\frac{3}{2} C'_{r,t,f} \sigma \sqrt{\pi a} \right)^{\frac{2}{5}} \left(\frac{C}{2} \right)^{\frac{1}{5}} \tag{381}
$$

with:

$$
C = 1 + k^2 + (1-k^2)\cos 2\omega \tag{382}
$$

Relation (381) yields the dependence of the radius r_0 of the initial curve on the angle ω and the biaxiality factor k. The complete variation of r_0 versus ω or k, with the other quantity as a variable in each case is given in Figs 20 and 21 respectively.

The shape and size of the caustics depend on the values of the radius of the initial curve as is clear from Rels (377) and (378). The inclination of the crack axis ω and the biaxiality of the external loading k influence the size of the initial curve as illustrated by Figs 20 and 21. For increasing values of ω and for $k < 0$, a decrease of the magnitude of the radius r_0 of the initial curve appears which, after passing through a minimum value corresponding to $k = 0$, starts to increase continuously for $k > 0$. For $\omega = 0°$ the size of the initial curve remains constant and independent of k. The same phenomenon occurs for $k = \pm 1$ for any angle of inclination ω.

Introducing Rels (348), (349) and (381) into Rels (377) and (378), we obtain the parametric equations of the caustic in terms of ω and k. These equations are:

$$
x'_{r,t,f} = \lambda_m r_0 \left[\cos \theta + \frac{2}{3} A (2C)^{-\frac{1}{2}} \cos \frac{3\theta}{2} - \frac{2}{3} B (2C)^{-\frac{1}{2}} \sin \frac{3\theta}{2} \right] \tag{383}
$$

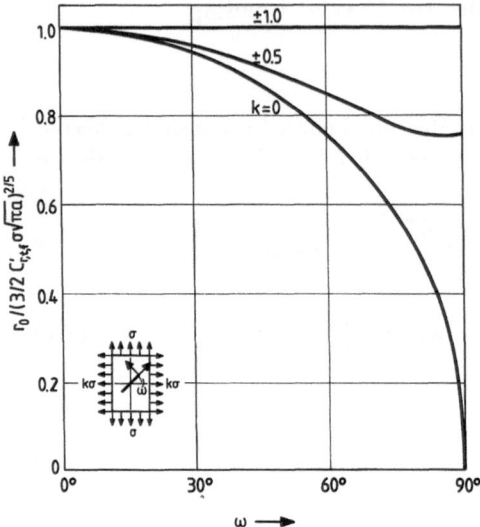

Fig. 20. Variation of the radius r_0 of the initial curve versus ω for various parametric values of k.

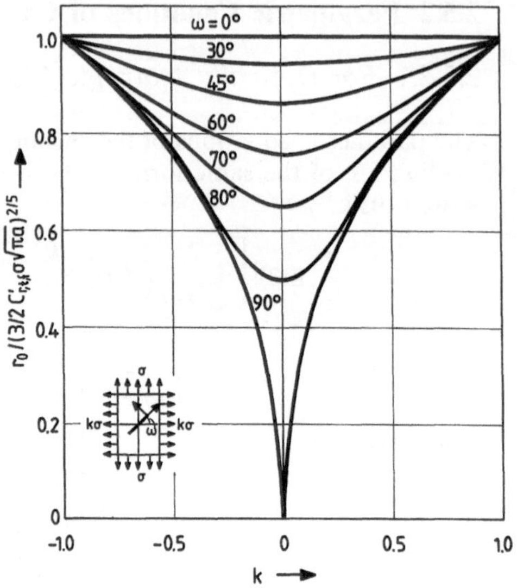

Fig. 21. Variation of the radius r_0 of the initial curve versus k for various parametric values of ω.

$$y'_{r,t,f} = \lambda_m r_0 \left[\sin\theta + \frac{2}{3} A(2C)^{-\frac{1}{2}} \sin\frac{3\theta}{2} + \frac{2}{3} B(2C)^{-\frac{1}{2}} \cos\frac{3\theta}{2} \right] \quad (384)$$

with:

$$A = (1+k) + (1-k)\cos 2\omega \quad (385)$$

$$B = (1-k)\sin 2\omega \quad (386)$$

Relations (383) and (384) are the parametric equations of the caustic referred to the $O'x'y'$ system (Fig. 6). By angularly displacing the $O'x'y'$ system by an angle $(2\pi - \phi)$ we obtain the new $O'x_1,y_1$ system. For this new coordinate system Rels (383) and (384) may be found by the transformation:

$$\begin{bmatrix} x_{1,r,t,f} \\ y_{1,r,t,f} \end{bmatrix} = \begin{bmatrix} \cos\phi & -\sin\phi \\ \sin\phi & \cos\phi \end{bmatrix} \begin{bmatrix} x'_{r,t,f} \\ y'_{r,t,f} \end{bmatrix} \quad (387)$$

These relations in the new coordinate system become:

$$x_{1,r,t,f} = r_0 \left[\cos(\theta + \phi) + \frac{2}{3}\cos\left(\frac{3\theta}{2} + \gamma + \phi\right) \right] \lambda_m \quad (388)$$

$$y_{1,r,t,f} = r_0 \left[\sin(\theta + \phi) + \frac{2}{3}\sin\left(\frac{3\theta}{2} + \gamma + \phi\right) \right] \lambda_m \quad (389)$$

with:

$$\cos\gamma = A(2C)^{-\frac{1}{2}} \quad (390)$$

$$\sin\gamma = B(2C)^{-\frac{1}{2}} \quad (391)$$

and:

$$\sin^2\gamma + \cos^2\gamma = 1 \quad (392)$$

Setting $\gamma = \phi/2$ and $\tau = (\theta + \phi)$ the parametric Eqs (388) and (389) become:

$$x_{l,r,t,f} = r_0 \left[\cos\tau + \frac{2}{3}\cos\frac{3\tau}{2} \right] \lambda_m \tag{393}$$

$$y_{l,r,t,f} = r_0 \left[\sin\tau + \frac{2}{3}\sin\frac{3\tau}{2} \right] \lambda_m \tag{394}$$

It may be deduced from the parametric Eqs (393) and (394) that the caustic in the $O'x_ly_l$-reference system is a symmetric curve having as an axis of symmetry the line subtending an angle $-\phi$ with the $O'x'$-axis of the crack. Relations (390) and (391) for $\gamma = \phi/2$ yield the angle of displacement of the caustic relative to the crack-axis. This angle is given by:

$$\phi = 2\tan^{-1}\frac{(1-k)\sin 2\omega}{(1+k)+(1-k)\cos 2\omega} \tag{395}$$

Comparing Rel. (395) with Rels (348) and (349) we obtain:

$$\phi = 2\tan^{-1}\frac{K_{II}}{K_I} \tag{396}$$

Therefore, Rels (395) and (396) yield the angle of displacement of the caustic in terms of the angle ω of inclination of the crack and the biaxiality factor k. The variations of ϕ with ω and the factor k are given in Figs 22 and 23 respectively, in which it is indicated that for $k \geq 0$ the axis of symmetry of the respective caustic coincides with the crack-axis for $\omega = 0°$, and is displaced by an angle $-\phi$ for other values of ω. For $k \leq 0$, this angle $-\phi$ progressively decreases as ω increases.

The dependence of ϕ on k is as follows: For $k = 1$, ϕ is independent of ω or β. For positive values of k its variation presents a smooth maximum. For $k = 0$

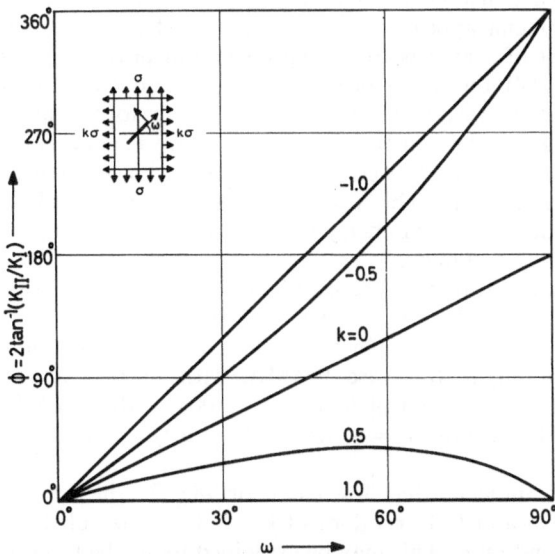

Fig. 22. Variation of the angle ϕ of rotation of caustic versus ω for various parametric values of k.

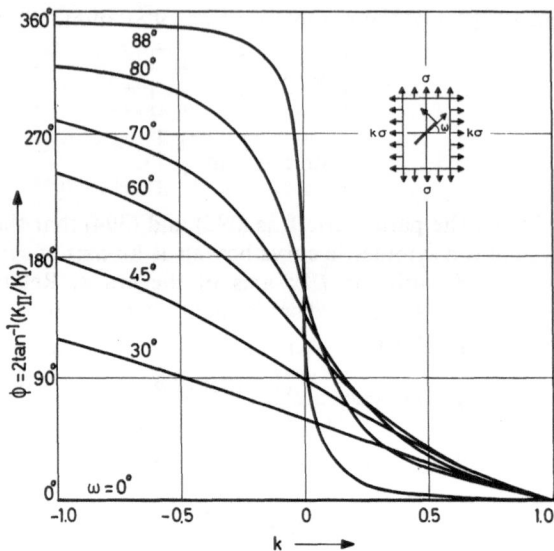

Fig. 23. Variation of the angle ϕ of rotation of caustic versus k for various parametric values of ω.

the dependence of ϕ on ω is linear and for $\omega = 90°$, $\phi = 180°$. For k positive, ϕ always takes positive values.

In the case of pure shear $K_I = 0$ and $K_{II} \neq 0$, an interchange of the branches of the caustics appears, that is the external, almost circular, branch becomes internal and the internal cuspoid branch becomes external. The measurements of diameter for evaluating K_I and K_{II} should be made on the external branch which corresponds to rays reflected from the front face of the plate; therefore for the evaluation of K_I and K_{II} the optical constant c_f should now be taken into consideration instead of c_r.

Similar phenomena appear with the variation of ϕ in terms of k. For $\omega = 0°$, ϕ is independent of k as was to be expected from previous evidence [57]. From Rels (383) and (384), for values of θ between $-\pi$ and π and for various values of the biaxiality factor k we may readily plot the caustics formed around the respective crack tips. For $k = 1$ the caustics plotted are always symmetric to the crack-axis while for $k \neq 1$ and varying between $k = -1$ and $k = 1$ a continuous relative rotation of the caustic is apparent. This rotation is made in such a direction so that the crack-axis and the axis of symmetry of the caustic subtend an angle equal to $-\phi$ given by Rels (395) and (396).

Figures 24 and 25 present a series of caustics plotted by computer for typical values of angle ω ($\omega = 0°$, $30°$, $45°$, $60°$, $70°$, $80°$ and $88°$) and values of k varying between $k = -1$ and $k = 1$. These caustics are typical of an initially isotropic and elastic material, in this case PMMA, with $v = 0.34$ and $E = 3400 \, \text{MN/m}^2$. The thickness of the cracked plate was $d = 0.003 \, \text{m}$, the crack length $2a = 0.02 \, \text{m}$ and the optical constant $c_r = -1.55 \times 10^{-10} \, \text{m}^2/\text{N}$, $z_0 = 2.0 \, \text{m}$, $z_i = 0.5 \, \text{m}$ and $\sigma = 2.0 \, \text{MN/m}^2$.

These figures indicate that for $k = 1$ or $\omega = 0°$ the respective caustics are independent of ω and k. For values of $k = -1$ the size of the caustics and their orientation depend on ω. This may be explained by the fact that while for $k = 1$ the state of stress of the plate at infinity is biaxial tension, which is a symmetric case,

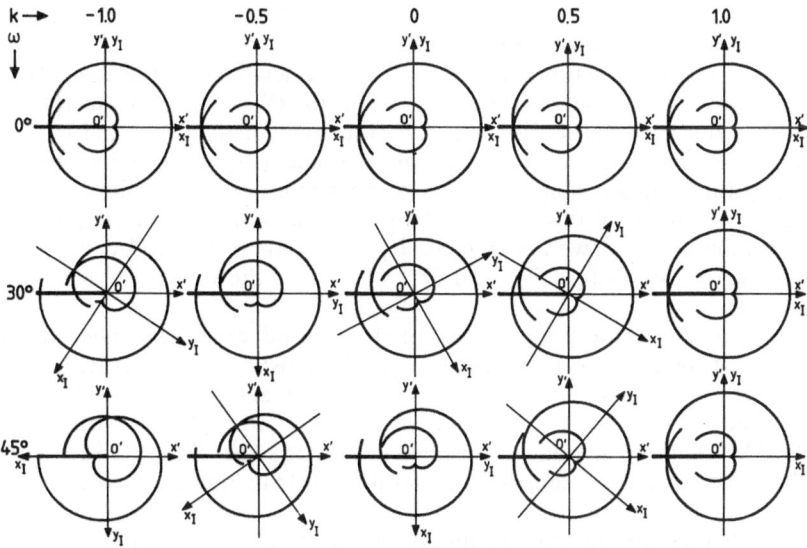

Fig. 24. The caustics from reflected light rays on a cracked plate made of an isotropic elastic material for k between -1 and 1 and for $\omega = 0°$, $30°$ and $45°$ as plotted by computer for $-\pi \leqq \theta \leqq \pi$.

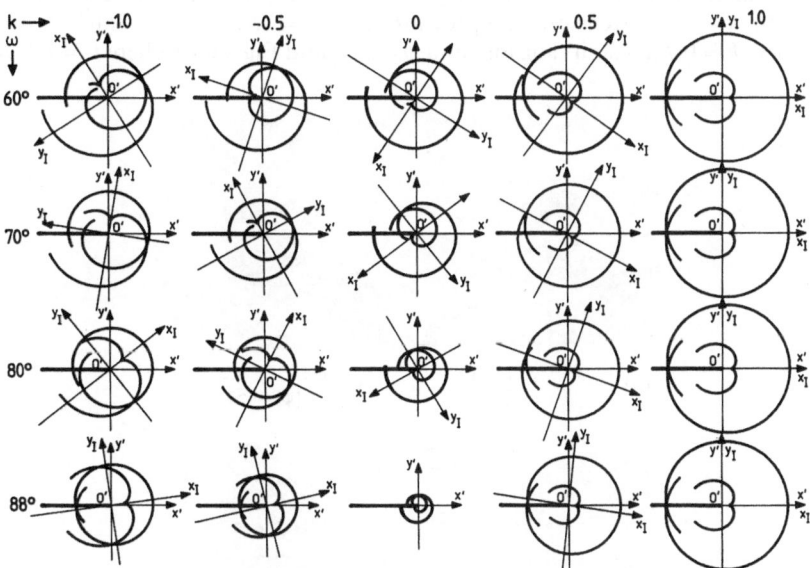

Fig. 25. The caustics from reflected light rays on a cracked plate made of an isotropic elastic material for k between -1 and 1 and for $\omega = 60°$, $70°$, $80°$ and $88°$ as plotted by computer for $-\pi \leqq \theta \leqq \pi$.

for $k = -1$ the state of stress becomes longitudinal tension and transverse compression, which influences weakly the size of the caustics but strongly their orientation. Indeed, as may be seen from Figs 24 and 25, the cusp of the internal caustic has been angularly displaced by $-\pi$ for ω varying between zero and $45°$. For ω varying between zero and $\pi/2$ the cusp is angularly displaced by -2π. On the other hand the external of the caustic is continuously reduced in size while its extremities alternate on both sides of the crack.

The stress intensity factors K_I and K_{II} can be calculated by the Rels (274) and (396), but since the internal caustic always yields the $O' x_I$- axis of symmetry of both caustics, an alternative procedure is to measure the diameters D_t^{max} and D_l^{max} of the caustic. These diameters are along $O' y_I$- and $O' x_I$-axes (Fig. 6) which may be defined by tracing the common tangent to the cuspoid internal caustic and drawing the normal at the middle of this common tangent. This normal is the $O' x_I$-axis of the caustic. In this case the two components K_I and K_{II} are expressed by the Rels (274) and (396) correction factors δ_t^{max} and δ_l^{max} which take the values $\delta_t^{max} = 3.1702$ and $\delta_l^{max} = 3.00$.

In order to check the potentialities of the method we have applied it for the evaluation of the angular displacement ϕ of the caustic in a thin plate containing a central crack subtending an angle $\omega = 45°$ to the loading axis and subjected to a biaxial load at infinity. The material of the plate was plexiglas with $v = 0.34$, $E = 3400\,\text{MN/m}^2$ and an optical constant $c_r = -1.55 \times 10^{-10}\,\text{m}^2/\text{N}$. The specimens had the following average dimensions: width $w = 0.15\,\text{m}$, thickness $d = 0.003\,\text{m}$, and crack length $2a = 0.02\,\text{m}$.

Figure 26 presents the experimentally obtained reflected caustics formed on a reference screen placed at a distance $z_0 = 2.0\,\text{m}$ from the specimen (Fig. 10), for four cases of biaxial load with $\sigma = 1.5\,\text{MN/m}^2$. Figure 26(a) presents the reflected caustic with a biaxiality factor $k = -1.0$ (equal tension–compression) and $\phi = 180°$; in Fig. 26(b), $k = -0.6$ (tension–compression) and $\phi = 150°$. Figure 26(c) shows the case where $k = 0.7$ (tension–tension) and $\phi = 20°$, and Fig. 26(d) where $k = 1.0$ (equal

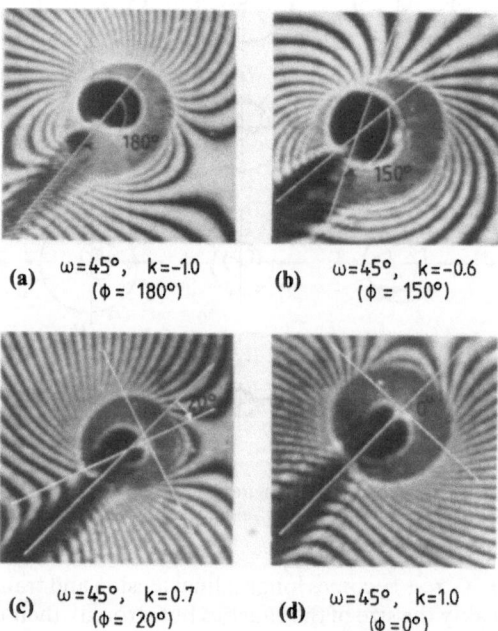

(a) $\omega = 45°, \quad k = -1.0$
$(\phi = 180°)$

(b) $\omega = 45°, \quad k = -0.6$
$(\phi = 150°)$

(c) $\omega = 45°, \quad k = 0.7$
$(\phi = 20°)$

(d) $\omega = 45°, \quad k = 1.0$
$(\phi = 0°)$

Fig. 26. Experimentally obtained reflected caustics for plexiglas plate containing a central crack with angle of inclination $\omega = 45°$ and subjected to biaxial load with $\sigma = 1.5\,\text{MN/m}^2$ and for k **(a)** -1.0, **(b)** -0.6, **(c)** 0.7 and **(d)** 1.0.

Table 3. Comparison of theoretical and experimental values of K_{I} and K_{II}. Properties of plexiglas: $E = 3.4 \times 10^9\,\mathrm{N/m^2}$, $v = 0.34$, $c_r = -1.55 \times 10^{-10}\,\mathrm{m^2/N}$. Loading condition: $\sigma = 1.5 \times 10^6\,\mathrm{N/m^2}$; $\omega = 45°$

k	Theoretical values		ϕ	Experimental values	
	$K_{\mathrm{I}}(\mathrm{N/m^{\frac{3}{2}}})$	$K_{\mathrm{II}}(\mathrm{N/m^{\frac{3}{2}}})$		$K_{\mathrm{I}}(\mathrm{N/m^{\frac{3}{2}}})$	$K_{\mathrm{II}}(\mathrm{N/m^{\frac{3}{2}}})$
-1.0	0	2.66×10^5	$180°$	0	2.9×10^5
-0.6	0.532×10^5	2.13×10^5	$150°$	0.63×10^5	2.35×10^5
0.7	2.26×10^5	0.40×10^5	$20°$	2.21×10^5	0.39×10^5
1.0	2.66×10^5	0	$0°$	2.42×10^5	0

tension–tension) and the angle $\phi = 0°$. All the experimentally measured angles ϕ coincide with the theoretically obtained ones derived from Figs 22 and 23.

The comparison between theoretical and experimental values of the stress intensity factors K_{I} and K_{II} is given in Table 3, for the experiments of Fig. 26. A difference of about ten per cent between theoretical and experimental values of stress intensity factors is observed because the theoretical values have not been corrected for the dimensions of the specimens.

2.5.2.2 For Optically Anisotropic Materials

The influence of biaxiality of loading on the shape of caustics which are obtained at the crack tips in plates of birefringent materials is equally as significant as in the case of optically isotropic materials. The biaxiality factor k and the angle of crack inclination ω influence the size and the rotation of the caustics, whereas the coefficient of optical anisotropy $\pm \xi_{\mathrm{r,t}}$ creates a double caustic with two branches situated on either side of the reference caustic of the corresponding optically isotropic material ($\pm \xi_{\mathrm{r,t}} = 0$).

The parametric equations of caustics for uniaxial tension, with $k = 0$, created from the crack tip of plates made of birefringent materials, are given by Rels (319) and (320). For the case of biaxial load, with $k \neq 0$, where the stress intensity factors K_{I} and K_{II} are given by Rels (348) and (349), the parametric equations of caustics are given by [104]:

$$x'_{\mathrm{r,t}} = \lambda_m r_0 \left\{ \cos\theta + \frac{2}{3} A'^{-1} (2C)^{-\frac{1}{2}} \left[A\cos\frac{3\theta}{2} - B\sin\frac{3\theta}{2} - \frac{1}{2}\xi_{\mathrm{r,t}} \right. \right.$$

$$\left. \left. \times (2C)^{-\frac{1}{2}} AB(1 - 3\cos 2\theta) + \frac{3}{4}\xi_{\mathrm{r,t}}(A^2 - B^2)(2C)^{-\frac{1}{2}}\sin 2\theta \right] \right\} \qquad (397)$$

$$y'_{\mathrm{r,t}} = \lambda_m r_0 \left\{ \sin\theta + \frac{2}{3} A'^{-1} (2C)^{-\frac{1}{2}} \left[A\sin\frac{3\theta}{2} - B\cos\frac{3\theta}{2} - \frac{1}{4}\xi_{\mathrm{r,t}} \right. \right.$$

$$\left. \left. \times (2C)^{-\frac{1}{2}}(A^2 - B^2)(1 + 3\cos 2\theta) + \frac{3}{2}\xi_{\mathrm{r,t}}(2C)^{-\frac{1}{2}} AB\sin 2\theta \right] \right\} \qquad (398)$$

and:

$$r_0 = \left(\frac{3}{2} C'_{\mathrm{r,t}} A' \sigma \sqrt{\pi a} \right)^{\frac{2}{5}} \left(\frac{C}{2} \right)^{\frac{1}{5}} = \left(\frac{3}{2} C'_{\mathrm{r,t}} K_{\mathrm{I}} A' \right)^{\frac{2}{5}} \left(1 + \frac{B^2}{A^2} \right)^{\frac{1}{5}} \qquad (399)$$

where A, B, C are given by Rels (385), (386) and (382) respectively, $C'_{r,t}$ is given by:

$$C'_{r,t} = \frac{\varepsilon z_0 \, dc_{r,t}}{\lambda_m \sqrt{2\pi}} \tag{400}$$

and A' is given by:

$$
\begin{aligned}
A' = & -\frac{1}{4}\xi_{r,t}(A^2 - B^2)(2C)^{-1}\sin\theta - \frac{1}{2}\xi_{r,t}AB(2C)^{-1}\cos\theta \\[6pt]
& + \left\{ 1 + \frac{1}{4}\xi_{r,t}(2C)^{-2}\left[7(2C)^{\frac{3}{2}}\left(A\sin\frac{\theta}{2} + B\cos\frac{\theta}{2} \right) \right.\right. \\[6pt]
& \left. - A(A^2 - 3B^2)(2C)^{\frac{1}{2}}\sin\frac{3\theta}{2} - B(3A^2 - B^2)(2C)^{\frac{1}{2}}\cos\frac{3\theta}{2} \right] \\[6pt]
& + \frac{1}{32}\xi_{r,t}^2(2C)^{-2}[25(2C)^2 + 9(B^4 - 6A^2B^2 + A^4)\cos2\theta \\[6pt]
& \left. - 36\,AB(A^2 - B^2)\sin2\theta] \right\}^{\frac{1}{2}} \tag{401}
\end{aligned}
$$

Equations (397) and (398) are the parametric equations of the caustic referred to the $O'x'y'$-system (Fig. 6) on the reference screen. By angularly displacing the $O'x'y'$-system by an angle $(2\pi - \phi)$ we obtain the new $O'x_1y_1$-system. The caustic in the $O'x'y'$-reference system is a symmetric curve having as an axis of symmetry the line subtending an angle $-\phi$ with the $O'x'$-axis of the crack. The angle ϕ is given by Rels (395) and (396) and nomograms of Figs 22 and 23. The stress intensity factors K_I and K_{II} may be evaluated from the respective Rels (325) and (396), while the correction factors $\delta_{t,l}^{max}(\pm\xi_{r,t})$ may be derived from the nomograms of Fig. 9. The correction factors $\delta_{t,l}^{max}(\pm\xi_{r,t})$ correspond to the external branch of the double caustic, while the correction factors $\delta_{t,l}^{max}(-\xi_{r,t})$ correspond to the internal branch.

By setting $\xi_{r,t} = 0$ in Rels (397) and (398), we get the parametric equations of the biaxial caustics for optically isotropic materials. By setting $k = 0$ and $\xi_{r,t} = 0$ in Rels (397) and (398), we get the parametric equations of the uniaxial caustics for optically isotropic materials. Finally, by setting $k = 0$ and $\xi_{r,t} \neq 0$ in Rels (397) and (398), we get the parametric equations of uniaxial caustics for optically anisotropic (birefringent) materials.

Figures 27 and 28 present a series of caustics plotted by computer for values of $k = -1$ to $+1$ and values of angle ω ($\omega = 0°, 30°, 45°, 60°, 70°, 80°, 88°$) and for $\xi_r = 0.153$. This value of ξ_r corresponds to the value of optical anisotropy of PCBA (Lexan), for which Poisson's ratio $v = 0.36$, the elastic modulus $E = 2750 \, \text{MN/m}^2$ and the optical constants are: $c_r = -2.08 \times 10^{-10} \, \text{m}^2/\text{N}$ and $c_f = v/E = 1.31 \times 10^{-10} \, \text{m}^2/\text{N}$ and $\xi_t = 0.223$ [98].

From Figs 27 and 28 we observe that the caustics obtained by reflected rays from the rear face of the plate are double curves. For $\xi_r > 0$ we obtain one branch of the caustic and for $\xi_r < 0$ the other one; while for caustics which are obtained by reflected rays from the front face of the plate the caustics are simple, because the birefringent constant ξ does not influence this caustic. The influence of k and ω on the size and the shape of caustics is the same as for optically isotropic materials [101].

The influence of ξ is evident from the formation of a double caustic, which is independent of k and ω. Indeed, for $k = -1$ and for ω from 0° to 88° a rotation

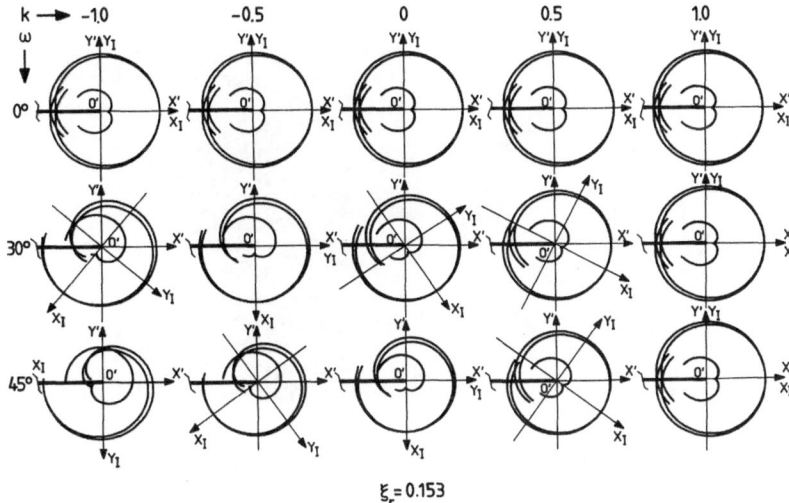

Fig. 27. The caustics from reflected light rays on a cracked plate made of an anisotropic elastic material with $\xi_r = 0.153$ (Lexan) for k between -1 and 1 and $\omega = 0°$, $30°$ and $45°$ as plotted by computer for $-\pi \leq \theta \leq \pi$.

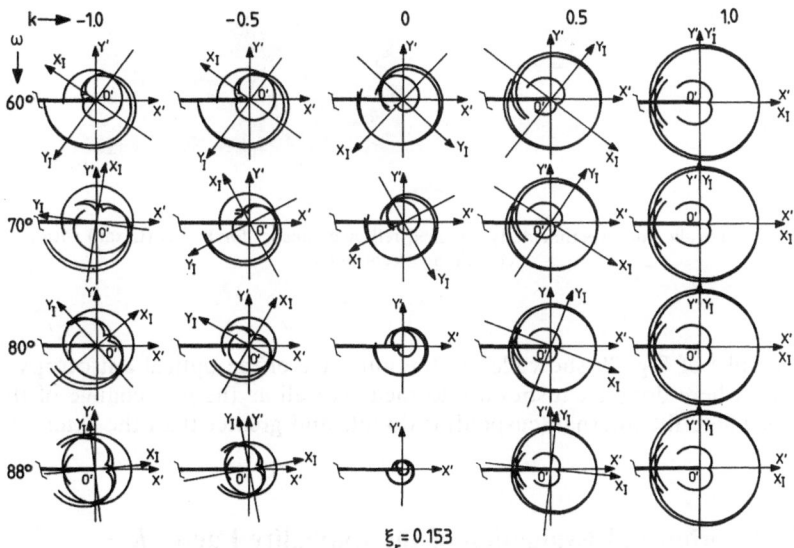

Fig. 28. The caustics from reflected light rays on a cracked plate made of an anisotropic elastic material with $\xi_r = 0.153$ (Lexan) for k between -1 and 1 and $\omega = 60°$, $70°$, $80°$ and $88°$ as plotted by computer for $-\pi \leq \theta \leq \pi$.

and an interchange of internal (cuspoid) and external (smooth) caustic is observed. For $\omega = 88°$ the external double caustic becomes an internal double (cuspoid) and the internal simple (cuspoid) becomes an external simple and smooth curve. This phenomenon does not appear for $k > 0$.

Figure 29 presents the theoretical and the corresponding experimental caustics, which were obtained from a cracked plate made of PCBA with $\xi_r = 0.153$. The

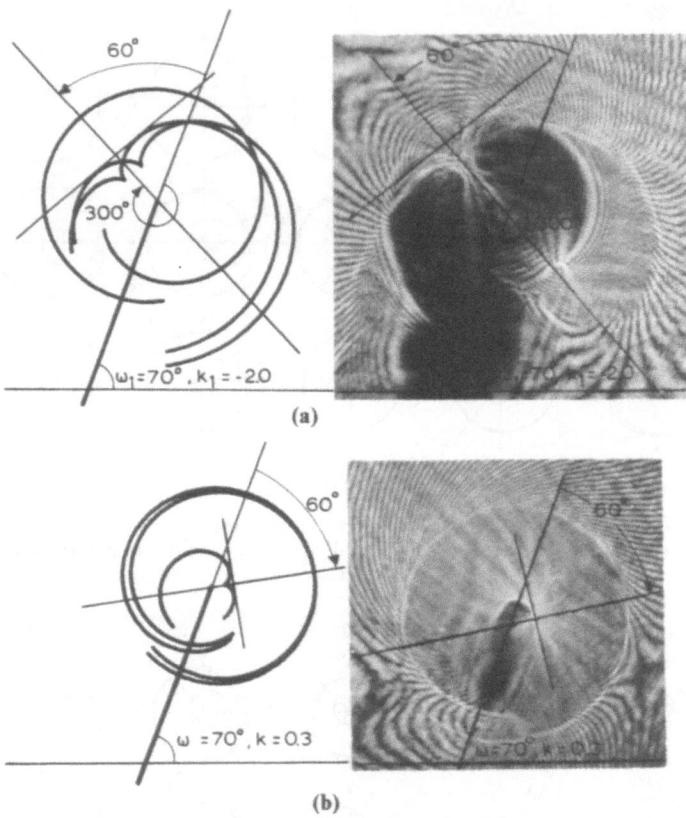

Fig. 29. Theoretically and experimentally obtained reflected caustics for PCBA (Lexan) with $\xi_r = 0.153$, containing a central crack.

photographs of Fig. 29 show clearly the influence of the optical anisotropy of the material, where double caustics are formed, as well as the interchange of the two caustics where the internal (cuspoid) is double and greater than the external.

2.5.3 Experimental Evaluation of the Biaxiality Factor k

The above study of the influence of the biaxiality factor k of loading on the form of caustics was considered only for ω varying between $0°$ and $90°$ and for k varying between -1 and 1 for optically isotropic or anisotropic materials. For values of k lying outside this region $[-1, 1]$ we can obtain an equivalent situation of biaxial loading as is clear from Fig. 30; from the states with ω_1 varying between $0°$ and $90°$ and $|k_1| \geqq 1$ we can go to the corresponding states with $\omega = 90° - \omega_1$ and $k = 1/k_1$ where $|k| \leqq 1$.

From this equivalence the relation between the angle of rotation of caustics is obtained as $\phi_1 = (360° - \phi)$. This relation may be derived from Eq. (395). Indeed,

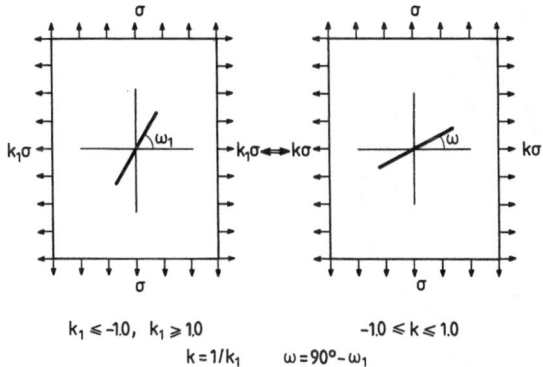

$$k_1 \leqslant -1.0, \quad k_1 \geqslant 1.0 \qquad\qquad -1.0 \leqslant k \leqslant 1.0$$
$$k = 1/k_1 \qquad \omega = 90° - \omega_1$$

Fig. 30. The equivalence of two states of biaxial loading.

it can be readily shown that [105]:

$$
\tan\frac{\phi_1}{2} = \frac{(1-k_1)\sin 2\omega_1}{(1+k_1)+(1-k_1)\cos 2\omega_1} = \frac{\left(1-\dfrac{1}{k}\right)\sin 2(90° - \omega)}{\left(1+\dfrac{1}{k}\right)+\left(1-\dfrac{1}{k}\right)\cos 2(90° - \omega)}
$$

$$
= -\frac{(1-k)\sin 2\omega}{(1+k)+(1-k)\cos 2\omega} = -\tan\frac{\phi}{2} \tag{402}
$$

Relation (402) in the region of the tangent included in the interval $[0°, 90°]$ yields:

$$\frac{\phi_1}{2} + \frac{\phi}{2} = 180° \tag{403}$$

and therefore:

$$\phi_1 = 360° - \phi \tag{404}$$

Therefore, from the equivalence of the quantities:

$$(\omega_1, k_1, \phi_1) \leftrightarrow \left(\omega = 90° - \omega_1, \quad k = \frac{1}{k_1}, \quad \phi = 360° - \phi_1 \right) \tag{405}$$

it is easy to calculate experimentally the coefficient k_1 from measurements of the angle ϕ of the caustic and the biaxiality factor k, using the respective nomograms of Figs 22 and 23 and the above-indicated equivalence.

In order to check the potentialities of the method we have applied it to evaluate the angular displacement ϕ of the caustics in plates under conditions of plane stress containing central inclined cracks at angles (a) $\omega_1 = 60°$, (b) $\omega_1 = 80°$ and (c) $\omega = 80°$ (Fig. 31). The plates were subjected to a biaxial load at infinity. The material of the plates was an optically isotropic plexiglas (PMMA) with $v = 0.34$, and $E = 3400\,MN/m^2$. Its optical constant is $c_r = -1.55 \times 10^{-10}\,m^2/N$. Specimen dimensions were: width, $w = 0.15\,m$, thickness, $d = 0.003\,m$ and crack length, $2a = 0.02\,m$. We have also applied the method to a thin plate made of an optically anisotropic material (polycarbonate, PCBA), containing a central crack with angles of inclination (a) $\omega_1 = 70°$ and (b) $\omega = 70°$ (Fig. 29) and subjected to a biaxial load at infinity. Specimen dimensions were as for PMMA.

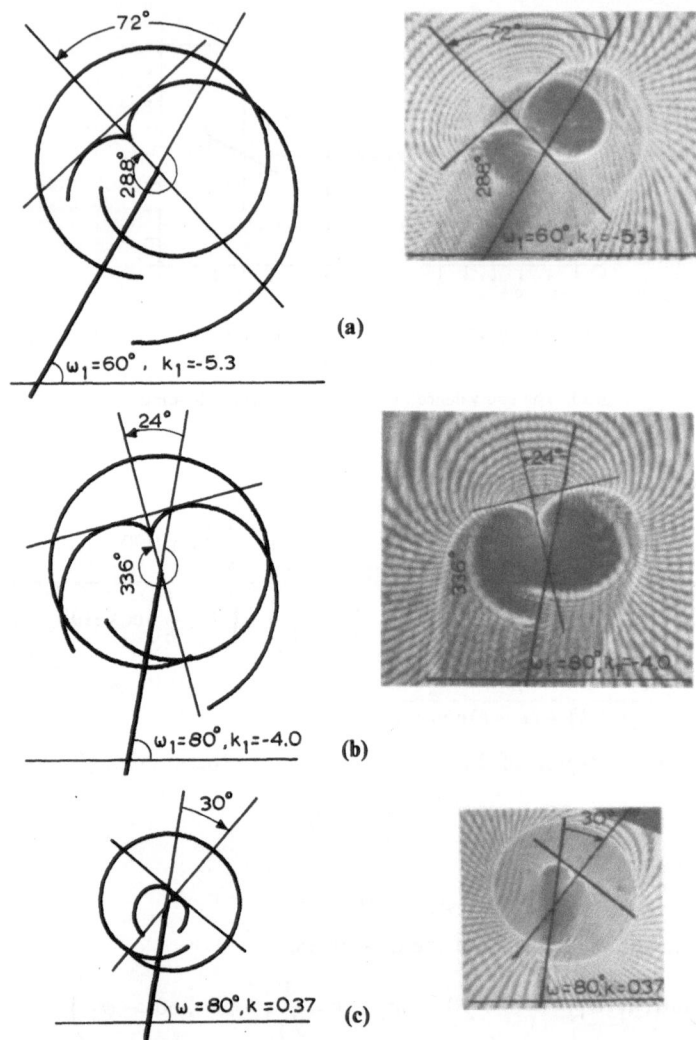

Fig. 31. Theoretically and experimentally obtained reflected caustics for PMMA (plexiglas) plates containing a central crack with **(a)** $\omega_1 = 60°$ and $k_1 = -5.3$, **(b)** $\omega_1 = 80°$ and $k_1 = -4.0$ and **(c)** $\omega = 80°$ and $k = 0.37$.

Figures 29 and 31 present both theoretically and experimentally obtained caustics: Figure 29, for PCBA with $\xi_r = 0.153$, clearly shows the influence of the optical anisotropy of the material, where double caustics are formed, as well as the interchange of the two caustics, where the internal (cuspoid) is double and greater than the external (Fig. 29a). On the other hand, the influence of ω and k for isotropic materials (PMMA) is apparent in Fig. 31. The validity of the equivalence (405) is exemplified in Figs 29 and 31, in conjunction with the nomograms of Figs 22 and 23. Indeed, the following equivalence may be derived from Fig. 31(a):

$$(\omega_1 = 60°, \quad k_1 = -5.3, \quad \phi_1 = 288°) \leftrightarrow (\omega = 30°, \quad k = -0.19, \quad \phi = 72°) \quad (406)$$

Similarly, from Fig. 31(b) we have the equivalence:

$$(\omega_1 = 80°, \quad k_1 = -4, \quad \phi_1 = 336°) \leftrightarrow (\omega = 10°, \quad k = -0.25, \quad \phi = 24°) \quad (407)$$

and likewise from Fig. 31(c) we have the equivalence:

$$(\omega = 80°, \quad k = 0.37, \quad \phi = 30°) \leftrightarrow (\omega_1 = 10°, \quad k_1 = 2.7, \quad \phi_1 = 330°) \quad (408)$$

From the above examples of application of the equivalence of states, we may readily deduce the values of the biaxiality factor k from the values of ω and ϕ, by using the nomograms of Figs 22 and 23.

2.6 Experimental Determination of the Stress-Optical Constants

The method of reflected caustics can be used for the experimental determination of the stress-optical constants of an isotropic or anisotropic elastic material using a simple tension test with a specimen having a circular hole [106, 107] or crack [108].

The stress-optical constants c_r and c_t are related to the variation of the optical paths of light rays impinging almost normally to a thin-plane specimen of thickness d made of the material under consideration and reflected from the rear face of the specimen or traversing the specimen, by the well-known formulae (149) and (150). From these relations we easily obtain the Rels (161):

$$c_r = c_t - \frac{v}{2E}, \quad \frac{\xi_r}{\xi_t} = \frac{c_r}{c_t} \quad (409)$$

for $\eta_0 = 1$.

Kartalopoulos and Raftopoulos [108] made a simple tension test with a cracked specimen and obtained two caustics on two screens (Fig. 10) at equal distances from the specimen. The ratio of the transverse diameters $D_{t(r)}/D_{t(t)}$ of these caustics was found to be directly related to the ratio c_r/c_t:

$$\frac{D_{t(r)}}{D_{t(t)}} = \left(\frac{c_r}{c_t}\right)^{\frac{2}{5}} \quad (410)$$

where $D_{t(r)}$ refers to the caustic formed on the screen in front of the specimen and $D_{t(t)}$ to that behind the specimen. Equations (409) and (410) permit the direct experimental determination of both c_r and c_t if the ratio v/E is known. For optically anisotropic, mechanically isotropic elastic media the stress-optical constants c_r and c_t determined using the method of Kartalopoulos and Raftopoulos do not remain unchanged. In this case Eq. (410) is no longer valid because the relation between the stress intensity factor K_I and diameters of the caustic depends on the coefficient of optical anisotropy of the material of the specimen [38, 40, 41].

Therefore, experimental determination of the stress-optical constants c_r and c_t using the method of caustics in cracked specimens presents disadvantages: (a) The use of two screens may introduce errors in the numerical results; (b) The caustics formed at the crack tip are greatly dependent on the geometry of the crack near its tip and are very sensitive to small imperfections; (c) It does not take into account the exact elasticity formulae of the stress field near a crack tip to establish the relation between the stress intensity factor K_1 at the crack tip and the transverse diameter of the caustic D_t [109]; and (d) The results are influenced by the plastic zone present in the near vicinity of the crack tip even if the initial curve of the caustic on the specimen lies outside this zone.

For these reasons another technique for the estimation of the stress-optical constants c_r and c_t, also based on caustics was proposed by Theocaris [106]. According to this technique, a thin-plane specimen with a circular hole of radius R much smaller than the dimensions of the specimen is used in the tension test. In this technique, only the reflected caustics on one screen lying in front of the specimen were taken into account.

Consider an optically and mechanically isotropic elastic thin infinite plate containing a central circular hole of radius R and subjected to a tensile stress σ_∞ at infinity [106]. A Cartesian coordinate system Oxy is related to the plate with its origin at the centre of the hole (Fig. 32). In this case, the function $\Phi(z)$ of the complex variable $z = x + iy$ is given by [89]:

$$\Phi(z) = \frac{\sigma_\infty}{2}\left(\frac{1}{2} + \frac{R^2}{z^2}\right) \tag{411}$$

The biaxial state of stress in the vicinity of the hole is expressed in complex form by:

$$4\,\mathrm{Re}\,\Phi(z) = (\sigma_1 + \sigma_2) = (\sigma_r + \sigma_\theta) \tag{412}$$

where σ_r and σ_θ are the polar components of stress, given by [89, 110]:

$$\sigma_r = \frac{\sigma_\infty}{2}\left[\left(1 - \frac{R^2}{r^2}\right)\left[1 + \left(\frac{3R^2}{r^2} - 1\right)\cos 2\theta\right]\right] \tag{413}$$

$$\sigma_\theta = \frac{\sigma_\infty}{2}\left[\left(1 + \frac{R^2}{r^2}\right) + \left(1 + \frac{3R^4}{r^4}\right)\cos 2\theta\right] \tag{414}$$

$$\tau_{r\theta} = \frac{\sigma_\infty}{2}\left[\left(1 + \frac{3R^2}{r^2}\right)\left(1 - \frac{R^2}{r^2}\right)\sin 2\theta\right] \tag{415}$$

The equation of the intial curve of the caustic is given by Rel. (208) while the equation of the caustic is given by Rel. (210). Differentiating the function $\Phi(z)$ with respect to z, we obtain:

$$\Phi'(z) = -\frac{\sigma_\infty R^2}{z^3} \tag{416}$$

$$\Phi''(z) = \frac{3\sigma_\infty R^2}{z^4} \tag{417}$$

$$\overline{\Phi'(z)} = \frac{\sigma_\infty R^2}{\bar{z}^3} \tag{418}$$

By substituting Rel. (417) into Rel. (208), we obtain:

$$r = r_0 = (12R^2 C^*_{r,t,f}\sigma_\infty)^{\frac{1}{4}} \tag{419}$$

where $C^*_{r,t,f}$ is given by Rel. (182). Relation (419) indicates that the initial curve is a circle, whose radius r_0 depends on the radius R of the hole, the applied stress σ_∞ and the global constant $C^*_{r,t,f}$.

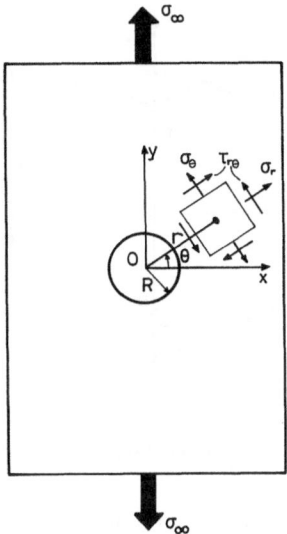

Fig. 32. Geometry of the specimen with central hole.

By substituting Rel. (418) into Rel. (210), we obtain:

$$\frac{W}{\lambda_m} = z + 4C^*_{r,t,f}\frac{\sigma_\infty R^2}{\bar{z}^3} \tag{420}$$

Relation (420) is the equation of the caustic.
   By setting:

$$z = r\,e^{i\theta} = r(\cos\theta + i\sin\theta) \tag{421}$$

and:

$$\bar{z}^3 = r^3\,e^{-i3\theta} = r^3(\cos 3\theta - i\sin 3\theta) \tag{422}$$

we can derive the parametric equations of the caustic. These equations are given by:

$$x'/\lambda_m = r\cos\theta + 4C^*_{r,t,f}\sigma_\infty R^2 r^{-3}\cos 3\theta \tag{423}$$

$$y'/\lambda_m = r\sin\theta + 4C^*_{r,t,f}\sigma_\infty R^2 r^{-3}\sin 3\theta \tag{424}$$

By introducing Rel. (419) into Rels (423) and (424), we obtain:

$$x' = \lambda_m r_0(\cos\theta + \tfrac{1}{3}\cos 3\theta) \tag{425}$$

$$y' = \lambda_m r_0(\sin\theta + \tfrac{1}{3}\sin 3\theta) \tag{426}$$

The angle θ varies between 0 and 2π, and the caustic image has the shape shown in Fig. 33.
   By using Eqs (425) and (426) for $\theta = 0°$ and $\theta = 90°$, we find:

$$D_l^{\max} = \tfrac{8}{3}\lambda_m r_0 = \tfrac{8}{3}(12\lambda_m^4 C^*_{r,t,f}R^2\sigma_\infty)^{\frac{1}{4}} \tag{427}$$

and:

$$D_t^{\min} = \tfrac{4}{3}\lambda_m r_0 = \tfrac{4}{3}(12\lambda_m^4 C^*_{r,t,f}R^2\sigma_\infty)^{\frac{1}{4}} \tag{428}$$

where $D_l^{\max}$ and $D_t^{\min}$ denote the major and minor diameters of each caustic on the screen respectively (Fig. 33).

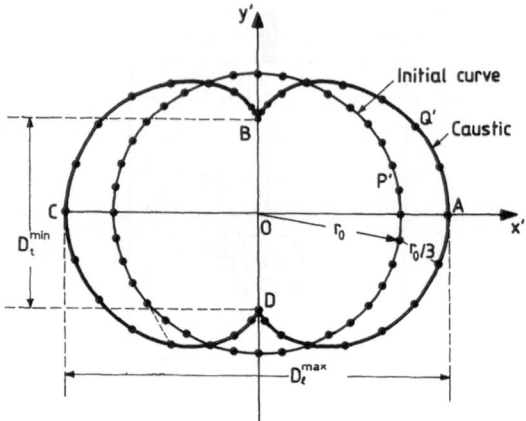

Fig. 33. Theoretical form of the caustic around a hole.

From Rels (182), (427) and (428), we obtain:

$$c_{r,t,f} = -(\tfrac{3}{4})^4(D_l^{max})^4/12\varepsilon z_0 d\lambda_m^3 R^2\sigma_\infty \tag{429}$$

$$c_{r,t,f} = -(\tfrac{3}{4})^4(D_t^{min})^4/12\varepsilon z_0 d\lambda_m^3 R^2\sigma_\infty \tag{430}$$

For the reflected caustics from the rear face (r) of the specimen, Rels (429) and (430) become:

$$c_r = -(\tfrac{3}{8})^4(D_l^{max})^4/24z_0 d\lambda_m^3 R^2\sigma_\infty \tag{431}$$

and:

$$c_r = -(\tfrac{3}{4})^4(D_t^{min})^4/24z_0 d\lambda_m^3 R^2\sigma_\infty \tag{432}$$

For the transmitted caustics (t), Rels (429) and (430) become:

$$c_t = -(\tfrac{3}{8})^4(D_l^{max})^4/12z_0 d\lambda_m^3 R^2\sigma_\infty \tag{433}$$

and:

$$c_t = -(\tfrac{3}{4})^4(D_t^{min})^4/12z_0 d\lambda_m^3 R^2\sigma_\infty \tag{434}$$

For the reflected caustics from the front face (f) of the specimen, we have:

$$c_f = v/E \tag{435}$$

and the Rels (429) and (430) become:

$$c_f = -(\tfrac{3}{8})^4(D_l^{max})^4/12z_0 d\lambda_m^3 R^2\sigma_\infty = \frac{v}{E} \tag{436}$$

and:

$$c_f = -(\tfrac{3}{4})^4(D_t^{min})^4/12z_0 d\lambda_m^3 R^2\sigma_\infty = \frac{v}{E} \tag{437}$$

Thus, both stress-optical constants can be determined. From the above equations, we obtain:

$$c_r = \frac{v}{2E}\left(\frac{D_{l(r)}^{max}}{D_{l(f)}^{max}}\right)^4 \tag{438}$$

and:

$$c_r = \frac{v}{2E}\left(\frac{D_{t(r)}^{min}}{D_{t(f)}^{min}}\right)^4 \tag{439}$$

From Rels (438) and (439), we obtain:

$$c_r = \frac{v}{2E}\left(\frac{D_{l(r)}^{max} D_{t(r)}^{min}}{D_{l(f)}^{max} D_{t(f)}^{min}}\right)^2 \tag{440}$$

Likewise, we have:

$$c_t = \frac{v}{E}\left(\frac{D_{l(t)}^{max}}{D_{l(f)}^{max}}\right)^4, \quad c_t = \frac{v}{E}\left(\frac{D_{t(t)}^{min}}{D_{t(f)}^{min}}\right)^4, \quad c_t = \frac{v}{E}\left(\frac{D_{l(t)}^{max} D_{t(t)}^{min}}{D_{l(f)}^{max} D_{t(f)}^{min}}\right)^2$$

Figure 34 presents (a) two caustics formed by the reflected rays at the front and rear faces of the plexiglas plate in the vicinity of the hole and (b) the caustic formed by traversing light rays. It is clear from these photographs that the caustic formed by reflections at the front face in Fig. 34(a) is angularly displaced by $\pi/2$ to the caustic formed by reflections at the rear face.

Consider a material which is elastically and mechanically isotropic but optically anisotropic, containing a central circular hole of radius R and subjected to a tensile stress σ_∞ at infinity (Fig. 32); the influence of the coefficients of optical anisotropy $\xi_{r,t}$ should be taken into account if a higher accuracy of results is to be obtained. In this case, the deviation vector $\mathbf{W}$ is given by Rels (290) and (279):

$$\mathbf{W} = \mathbf{r} + C_{r,t}^* \mathrm{grad}(\sigma_1 + \sigma_2) \pm C_{r,t}^* \xi_{r,t}\mathrm{grad}(\sigma_1 - \sigma_2) \tag{441}$$

The sum and difference of principal stresses, by using the polar coordinates (413)–(415), are:

$$\sigma_1 + \sigma_2 = \sigma_r + \sigma_\theta = \sigma_\infty\left(1 + \frac{2R^2}{r^2}\cos 2\theta\right) \tag{442}$$

$$\sigma_1 - \sigma_2 = \sqrt{(\sigma_r - \sigma_\theta)^2 + 4\tau_{r\theta}^2} = \sigma_\infty\left[1 + \frac{R^4}{r^4} + \left(\frac{2R^2}{r^2} - \frac{3R^4}{r^4}\right)^2\right.$$

$$+ \left(\frac{2R^2}{r^2} - \frac{4R^4}{r^4} + \frac{6R^6}{r^6}\right)\cos 2\theta + 2\left(\frac{2R^2}{r^2} - \frac{3R^4}{r^4}\right)$$

$$\left. \times (\sin^2 2\theta - \cos^2 2\theta)\right]^{\frac{1}{2}} \tag{443}$$

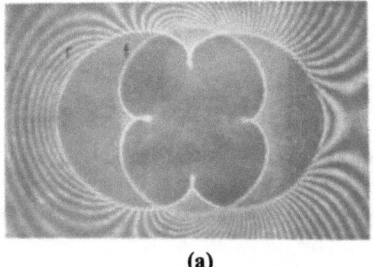
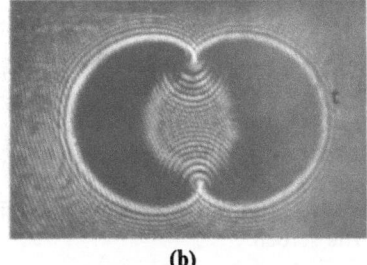

(a)   (b)

Fig. 34. Caustics formed **(a)** by reflected rays at the front and rear faces of the plexiglas plate and **(b)** by rays traversing the plate in the vicinity of the hole.

Along the symmetry directions, $\theta = 0°, 90°, 180°$ and $270°$, $\tau_{r\theta} = 0$. Along the direction $\theta = \pm 90°$, the sum and the difference of the principal stresses σ_1 and σ_2, are:

$$\sigma_1 + \sigma_2 = \sigma_r + \sigma_\theta = \sigma_\infty\left(1 - \frac{2R^2}{r^2}\right) \tag{444}$$

$$\sigma_1 - \sigma_2 = \sigma_r - \sigma_\theta = \sigma_\infty\left(1 - \frac{3R^2}{r^2} + \frac{3R^4}{r^4}\right) \tag{445}$$

The gradients in the Cartesian coordinates (u, v) shown in Fig. 5, are:

$$\text{grad}_{u,v}(\sigma_1 + \sigma_2) = \frac{\partial}{\partial r}\left(1 - \frac{2R^2}{r^2}\right)\mathbf{j} = \frac{4R^2}{r^3}\mathbf{j} \tag{446}$$

and:

$$\text{grad}_{u,v}(\sigma_1 - \sigma_2) = \frac{\partial}{\partial r}\left(1 - \frac{3R^2}{r^2} + \frac{3R^4}{r^4}\right)\mathbf{j} = \left(\frac{6R^2}{r^3} - \frac{12R^4}{r^5}\right)\mathbf{j} \tag{447}$$

and Rel. (441), gives:

$$\mathbf{W} = \mathbf{r} + C_{r,t}^*\left[\frac{4R^2}{r^3} \pm \xi_{r,t}\left(\frac{6R^2}{r^3} - \frac{12R^4}{r^5}\right)\right]\mathbf{j} \tag{448}$$

or:

$$y'_{r,t}/\lambda_m = r + C_{r,t}^*\left[\frac{4R^2}{r^3} \pm \xi_{r,t}\left(\frac{6R^2}{r^3} - \frac{12R^4}{r^5}\right)\right] \tag{449}$$

where r is the initial curve of the caustic.

Along the direction $\theta = 0°$, the sum and the difference of the principal stresses σ_1 and σ_2, are:

$$\sigma_1 + \sigma_2 = \sigma_r + \sigma_\theta = \sigma_\infty\left(1 + \frac{2R^2}{r^2}\right) \tag{450}$$

$$\sigma_1 - \sigma_2 = \sigma_\infty\left(1 - \frac{R^2}{r^2} + \frac{3R^4}{r^4}\right) \tag{451}$$

and the gradients are:

$$\text{grad}_{u,v}(\sigma_1 + \sigma_2) = \frac{\partial}{\partial r}\left(1 + \frac{2R^2}{r^2}\right)\mathbf{i} = -\frac{4R^2}{r^3}\mathbf{i} \tag{452}$$

and:

$$\text{grad}_{u,v}(\sigma_1 - \sigma_2) = \frac{\partial}{\partial r}\left(1 - \frac{R^2}{r^2} + \frac{3R^4}{r^4}\right)\mathbf{i} = \left(\frac{2R^2}{r^3} - \frac{12R^4}{r^5}\right)\mathbf{i} \tag{453}$$

and Rel. (441) gives:

$$x'_{r,t}/\lambda_m = r + C_{r,t}^*\left[-\frac{4R^2}{r^3} \pm \xi_{r,t}\left(\frac{2R^2}{r^3} - \frac{12R^4}{r^5}\right)\right] \tag{454}$$

From Eqs (449) and (454) we can find the diameters of the caustic (double caustic) on the screen:

$$D_t^{\min} = 2y'_{r,t} = 2\lambda_m\left(r + C_{r,t}^*\left[\frac{4R^2}{r^3} \pm \xi_{r,t}\left(\frac{6R^2}{r^3} - \frac{12R^4}{r^5}\right)\right]\right) \tag{455}$$

Table 4. Stress-optical constants [163]. Static stress of fracture for plexiglas: $\sigma_u = 880$ kp/cm^2. Birefringent constants for Lexan (static and dynamic): $\xi_r = 0.153$, $\xi_t = 0.223$

	Poisson's ratio v	Elastic modulus E(kp/cm^2)	Stress-optical constants		
			c_r(cm^2/kp)	c_t(cm^2/kp)	c_f(cm^2/kp)
Static values Plexiglas	0.34	34000	-1.70×10^{-5}	-1.21×10^{-5}	1×10^{-5}
Static values Lexan	0.36	28000	-2.04×10^{-5}	-1.4×10^{-5}	1.286×10^{-5}
Dynamic values Plexiglas	0.34	43000	-1.13×10^{-5}	-0.74×10^{-5}	0.79×10^{-5}
Dynamic values Lexan	0.36	28000	-2.2×10^{-5}	-1.55×10^{-5}	1.286×10^{-5}

and:

$$D_l^{\max} = 2x'_{r,t} = 2\lambda_m \left(r + C^*_{r,t} \left[-\frac{4R^2}{r^3} \pm \xi_{r,t} \left(\frac{2R^2}{r^3} - \frac{12R^4}{r^5} \right) \right] \right) \qquad (456)$$

By differentiating Eqs (449) and (454) with respect to r, the values of the initial curve of the caustic along the $O'y'$- and $O'x'$-axes respectively, are determined. So, we obtain:

$$\frac{\mathrm{d}y'_{r,t}}{\mathrm{d}r} = 0$$

or:

$$1 - \frac{6C^*_{r,t}}{R^2} \left[(2 \pm 3\xi_{r,t}) \left(\frac{R}{r} \right)^4 \mp 10\xi_{r,t} \left(\frac{R}{r} \right)^6 \right] = 0 \qquad (457)$$

and:

$$\frac{\mathrm{d}x'_{r,t}}{\mathrm{d}r} = 0$$

or:

$$1 - \frac{6C^*_{r,t}}{R^2} \left[(-2 \pm \xi_{r,t}) \left(\frac{R}{r} \right)^4 \mp 10\xi_{r,t} \left(\frac{R}{r} \right)^6 \right] = 0 \qquad (458)$$

respectively.

From the above equations, the constants c_r, c_t and $\xi_{r,t}$ can be determined by using Eqs (409). For the materials Plexiglas (PMMA) and Lexan (PCBA), the stress-optical constants, for static and dynamic loading, are given in Table 4.

2.7 Influence of Geometry of Edge-Cracked Plates on Stress Intensity Factors

The stress intensity factor is the governing quantity in the proximity of the crack-tip region [103], and it is of fundamental importance in the prediction of brittle failure using linear elastic fracture mechanics principles. It is a function of both the plate geometry and the associated loading. It is common practice to present K solutions in dimensionless form, normalized with respect to an appropriate infinite-plate solution.

There are analytical solutions for the determination of stress intensity factors, such as those using Westergaard stress functions and the equivalents of Muskhelishvili's complex stress function (see Chap. 1.7). For mixed-mode problems, the general form of equation for a crack orientated along the Ox-axis with its tip located at z_1, [9], is:

$$K = K_I - iK_{II} = 2(2\pi)^{\frac{1}{2}} \lim_{z \to z_1} (z - z_1)^{\frac{1}{2}} Z(z) \qquad (459)$$

where $Z(z)$ is the Westergaard stress function. By using conformal mapping we obtain the stress intensity factors K_I and K_{II}.

Another technique uses Green's functions [111]. According to this solution, the stress intensity factor K_I may be written in terms of the Green's function $G(x)$:

$$K_I = \frac{1}{(\pi a)^{\frac{1}{2}}} \int_a p(x) G(x) dx \qquad (460)$$

where $p(x)$ is the pressure acting normal to the crack surface.

Other techniques for determination of stress intensity factors are, the weight function technique [112, 113], the boundary collocation technique [2, 12], finite element methods [114], the integral equations technique [115, 116], boundary methods and the compounding method. Finally, the Wiener–Hopf technique was used for the determination of stress intensity factor K_I in a cracked-plane orthotropic [117, 118, 119] or viscoelastic [120] strip.

The experimental method of reflected caustics has been used to study the influence of the geometry of an edge-cracked plate on stress intensity factors K_I and K_{II} [121]. This influence was determined for various values of the ratio a/w, of the crack length to the width of the specimen. The influence of the geometry of the specimens on the form and orientation of the caustic was studied in relation to the orientation of the edge crack. The complex stress intensity factor $K = K_I - iK_{II}$ was evaluated in terms of the angular displacement ϕ of the caustic axis. The experimental values of the stress intensity factors K_I and K_{II} were compared with corresponding theoretical values. From this comparison, correction factors are derived and nomograms given in terms of the angle ω of inclination of the crack.

An analysis of the determination of stress intensity factors K_I and K_{II} for the case of finite edge-cracked plates, loaded in tension, has been presented by Bowie and Neal [122], Andersson [123] and Bowie [102]; use was made in these theoretical analyses of the conformal mapping technique. Bowie [102] gave nomograms of the stress intensity factor K_I and K_{II} for various values of the ratio, a/w, of crack length to width of the specimen.

In our experimental studies the optical method of reflected caustics was used for the determination of stress intensity factors K_I and K_{II} and of the influence of the geometry of the specimen [32, 36].

The relationship between caustics and the law of reflection and catastrophe theory is given by Theocaris and Michopoulos [124], in a paper which contains an extensive bibliography, including the most important contributions of this general theory to problems of instabilities and bifurcations. The geometry of the specimen influences the form, the size and the orientation of the caustics. The stress intensity factors K_I and K_{II} were determined from the size and the angular displacement of the caustics and they were then compared with their respective theoretical values. The experimental arrangement of the method is simple and well known (Fig. 10); the caustic obtained on the reference plane is shown in Fig. 6.

It has been shown that for uniaxial tension the parametric equations of the caustic are given by Eqs (265) and (266), or:

$$x'_{r,f}/\lambda_m = r_0 \cos \theta + C'_{r,f} K_I r_0^{-\frac{3}{2}} \cos \frac{3\theta}{2} - C'_{r,f} K_{II} r_0^{-\frac{3}{2}} \sin \frac{3\theta}{2} \qquad (461)$$

$$y'_{r,f}/\lambda_m = r_0 \sin \theta + C'_{r,f} K_I r_0^{-\frac{3}{2}} \sin \frac{3\theta}{2} + C'_{r,f} K_{II} r_0^{-\frac{3}{2}} \cos \frac{3\theta}{2} \qquad (462)$$

with:

$$r_0 = (\tfrac{3}{2} C'_{r,f})^{\frac{2}{5}} (K_I^2 + K_{II}^2)^{\frac{1}{5}} \qquad (463)$$

$$C'_{r,f} = \frac{\varepsilon z_0 \, dc_{r,f}}{\lambda_m (2\pi)^{\frac{1}{2}}} \qquad (464)$$

The angular displacement of the caustics for oblique central cracks in infinite plates is given by:

$$\phi = 2 \tan^{-1}(\tan \omega) \quad \text{or} \quad \phi = 2\omega \qquad (465)$$

where ω is the crack inclination angle (Fig. 10). For oblique edge cracks the angular displacement ϕ is given by:

$$\phi = 2 \tan^{-1}(K_{II}/K_I) \qquad (466)$$

The stress intensity factors K_I and K_{II} are given by the Rels (274) and (396).

In the elastic region of deformation the ratio of diameters of the caustic remains constant and is:

$$D_t^{max}/D_l^{max} = \delta_t^{max}/\delta_l^{max} = 1.057 \qquad (467)$$

So, as the applied stress increases, the caustic increases and remains self-similar for the whole elastic region of deformation. Then the stress intensity factors K_I and K_{II} are determined by this variation of the caustic. When the value of the factor K_I reaches the critical value, K_{Ic}, of crack propagation, the shape of the caustic changes because elasticity no longer dominates the stress field around the crack tip, and the ratio of its diameters becomes other than 1.057. At this critical value the deformation passes from a purely elastic to an elastic–plastic mode. In this region K_{Ic} can be approximately calculated from the caustic and the size of the plastic zone in front of the crack tip [125] by using the simple Dugdale–Barenblatt model or its modified version [30].

For a thin elastic and isotropic plate, containing a slant central crack of length $2a$ and subjected to a uniaxial tension at infinity, under plane-stress conditions, the stress intensity factors K_I and K_{II} are given by the relations:

$$K_I = \sigma \sqrt{\pi a} \cos^2 \omega \qquad (468)$$

$$K_{II} = \sigma \sqrt{\pi a} \sin \omega \cos \omega \qquad (469)$$

while, for a slant edge crack, Rels (468) and (469) become:

$$K_I = f_I \sigma \sqrt{\pi a} \cos^2 \omega \qquad (470)$$

$$K_{II} = f_{II} \sigma \sqrt{\pi a} \sin \omega \cos \omega \qquad (471)$$

where f_1 and f_{II} are nondimensional quantities which depend on the geometry of the specimen and the crack length, and ω denotes the crack inclination angle.

If the stress intensity factors K_I and K_{II} are experimentally determined by using Rels (238) and (396) and data from the caustics, then Rels (470) and (471) yield:

$$f_1 = K_1/\sigma\sqrt{\pi a}\cos^2\omega \tag{472}$$

$$f_{II} = K_{II}/\sigma\sqrt{\pi a}\sin\omega\cos\omega \tag{473}$$

or:

$$g_1 = f_1\cos^2\omega = K_1/\sigma\sqrt{\pi a} \tag{474}$$

$$g_{II} = f_{II}\sin\omega\cos\omega = K_{II}/\sigma\sqrt{\pi a} \tag{475}$$

Relations (474) and (475) give the nondimensional quantities depending on the geometry of the specimen, the crack length and the crack inclination angle, ω.

Relations (473) and (475), by introducing Rel. (396), yield:

$$f_{II} = f_1\cot\omega\tan\frac{\phi}{2} \tag{476}$$

$$g_{II} = g_1\tan\frac{\phi}{2} \tag{477}$$

The nondimensional quantities give the ratio of experimental values of stress intensity factors K_I and K_{II} to their theoretical values. The experimental values are calculated from the caustic, while the theoretical values are calculated from the mathematical theory of elasticity.

The method of caustics has been applied to a thin plate containing an edge crack with angle of inclination ω and subjected to a uniaxial load at infinity. The material of the plate was PMMA with $v = 0.34$ and $E = 2610$ MN/m²; its optical constant was $c_r = -2.02 \times 10^{-10}$ m²/N [106]. The specimens had the following dimensions: width $w = 0.09$ m, thickness $d = 0.003$ m and free length $h = 0.18$ m. Five different crack lengths were selected, $a = 0.009$ m, 0.018 m, 0.027 m, 0.036 m and 0.045 m; and six different inclination angles for the crack, $\omega = 0°, 15°, 30°, 45°, 60°$ and $75°$. The cracks were produced by very fine saw cuts of thickness not exceeding 0.3×10^{-3} m.

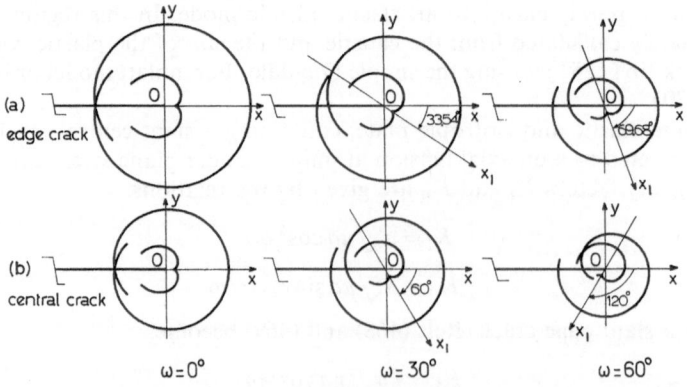

Fig. 35. Theoretical caustics for inclined cracks with angles $\omega = 0°$, 30° and 60° and for $h/w = 2.0$ and $a/w = 0.4$. **(a)** The case of edge cracks and **(b)** the equivalent central cracks.

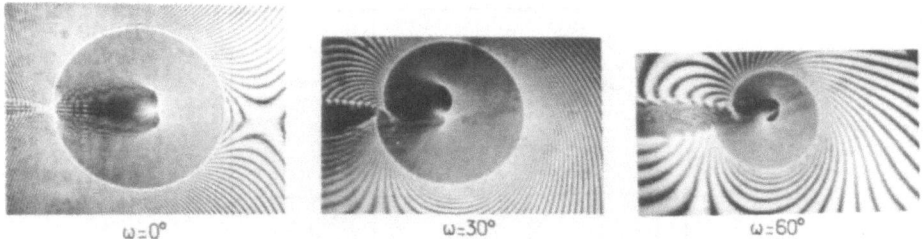

$\omega=0°$ $\omega=30°$ $\omega=60°$

Fig. 36. Experimental forms of the caustics obtained on a screen for inclined cracks with angles $\omega = 0°$, 30°, and 60° and for $h/w = 2.0$ and $a/w = 0.4$.

Figures 35 and 36 present the theoretically and the experimentally obtained caustics for angles of inclination $\omega = 0°$, 30°, 60°, $h/w = 2.0$ and $a/w = 0.4$. If we assign the correction factors H_I and H_{II} for the two stress intensity factors K_I and K_{II} respectively [102], we can derive from these tests that for $\omega = 0°$, $H_I = 2.155$ and $H_{II} = 0$; for $\omega = 30°$, $H_I = 1.556$, $H_{II} = 0.469$; and for $\omega = 60°$, $H_I = 0.596$ and $H_{II} = 0.415$. Figure 35(a) shows the shape and size of caustics for edge cracks of different slant angles, whereas Fig. 35(b) presents the same caustics under the same conditions but for central cracks. Finally, Fig. 36 shows experimental caustics for edge cracks.

The influence of the geometry of the specimen and the crack inclination angle, ω, on the size and the angular displacement ϕ of the caustics is strong, as may be seen in Fig. 35. For an edge crack, the angle ϕ of rotation of the caustic is smaller than the corresponding angle for a central crack. The variation of the angle $(-\phi)$ versus the angle ω, for various ratios a/w of the dimensions of the cracked plates

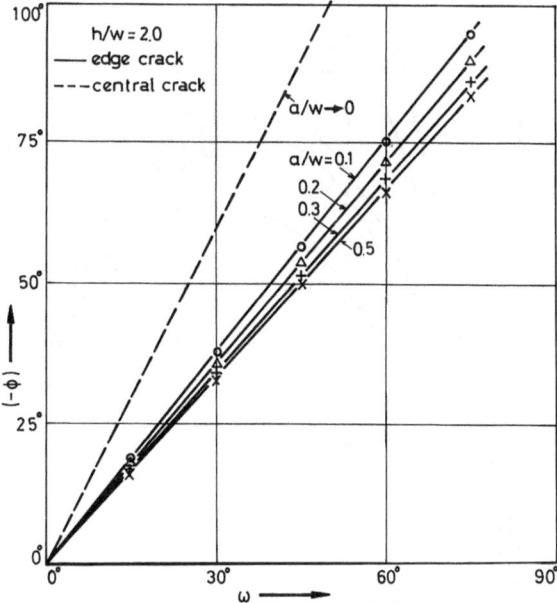

Fig. 37. The variation of the angular displacement $(-\phi)$ of the caustics, versus the crack inclination angle ω for various values of the ratio a/w and for $h/w = 2.0$. Full curves, edge crack; broken curve, central crack.

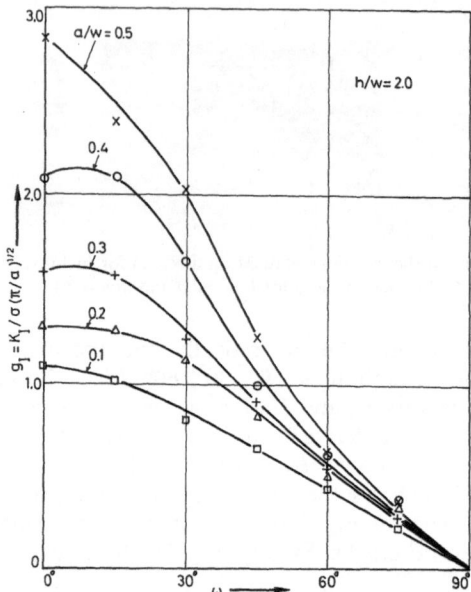

Fig. 38. The variation of the nondimensional quantities $g_1 = K_1/\sigma(\pi a)^{\frac{1}{2}}$, versus the crack inclination angle ω for $h/w = 2.0$ and $a/w = 0.1, 0.2, 0.3, 0.4, 0.5$.

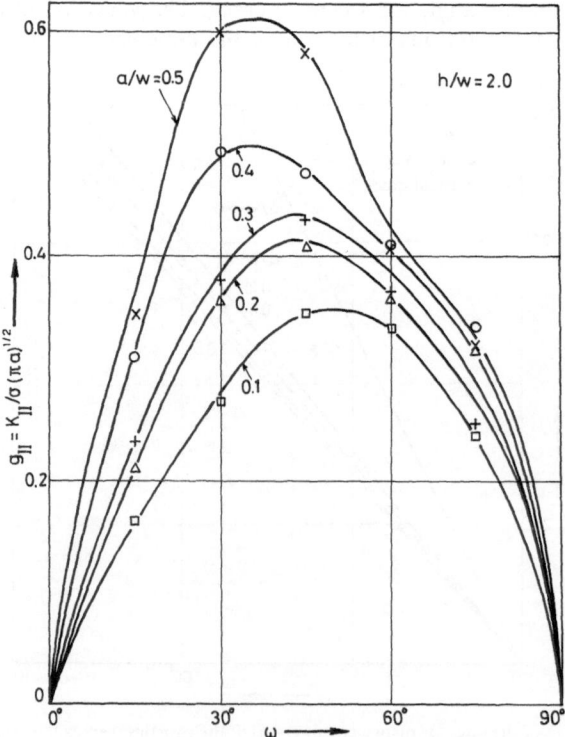

Fig. 39. The variation of the nondimensional quantities $g_{II} = K_{II}/\sigma(\pi a)^{\frac{1}{2}}$, versus the crack inclination angle ω for $h/w = 2.0$ and $a/w = 0.1, 0.2, 0.3, 0.4, 0.5$.

and for the ratio, h/w, of free length of specimen to its width is presented in Fig. 37. As may be seen in this figure, an increase of the ratio a/w results in a corresponding decrease of the angular displacement $(-\phi)$ of the caustics. This angle $(-\phi)$ always remains smaller than the corresponding angular displacement $(\phi = 2\omega)$ for the case of central cracks in infinite plates $(a/w \rightarrow 0)$. This variation of the angle $(-\phi)$ versus ω is linear; as the ratio a/w increases, the quantity $\lambda = \delta(-\phi)/\delta\omega$ initially decreases rapidly; but beyond $a/w = 0.25$ this decrease becomes weaker. This relationship is convenient to determine the angular displacement $(-\phi)$ of the caustic for any cracked plate in which the ratio a/w and the angle ω are known.

The influence of the geometry of the specimen, the size of the crack and the crack inclination angle ω, on the values of K_I and K_{II} are presented in Figs 38 and 39. It may be seen from Fig. 38 that, as the angle ω increases, K_I decreases and tends to zero for $\omega = 90°$. Also, as the ratio a/w increases K_I-values increase. On the other hand, it can be seen from Fig. 39 that K_{II} values increase rapidly from zero, for $\omega = 0°$, to a maximum in the region $30° < \omega < 55°$, before decreasing again to zero for $\omega = 90°$. Also, that as the ratio a/w increases K_{II} increases.

In Figs 40 and 41 the influence of the geometry of the specimen and the crack length on the values of K_I and K_{II} stress intensity factors is presented. In these figures the f_I and f_{II} nondimensional quantities are given as the values of the stress intensity factors for special geometries of the specimens, normalized to their theoretical values for infinite plates. It may be seen from Fig. 40 that, as the angle

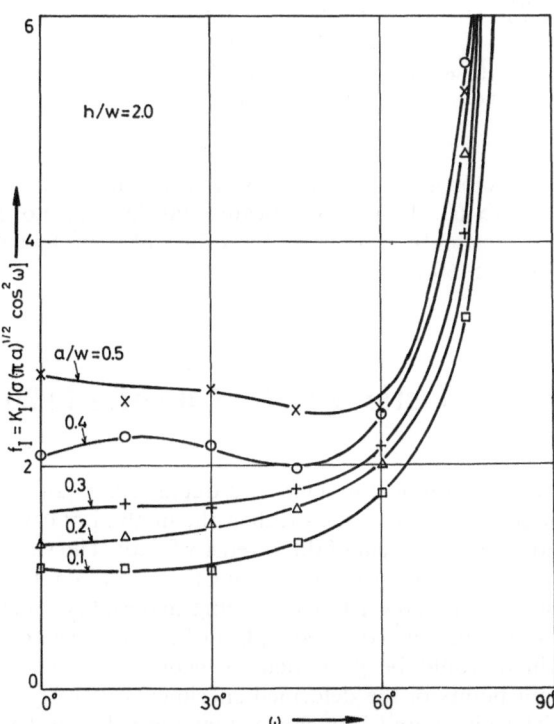

Fig. 40. The variation of the nondimensional quantities $f_I = K_I/\sigma(\pi a)^{\frac{1}{2}}\cos^2\omega$ versus the crack inclination angle ω for $h/w = 2.0$ and $a/w = 0.1, 0.2, 0.3, 0.4, 0.5$.

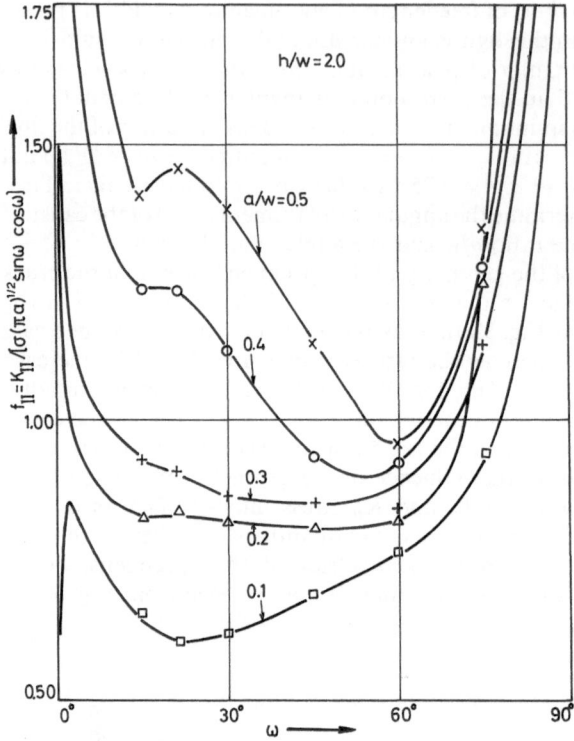

Fig. 41. The variation of the nondimensional quantities $f_{II} = K_{II}/\sigma(\pi a)^{\frac{1}{2}} \sin \omega \cos \omega$ versus the crack inclination angle ω for $h/w = 2.0$ and $a/w = 0.1, 0.2, 0.3, 0.4, 0.5$.

ω increases, the K_I factor increases weakly in the region $0° < \omega < 60°$, followed by a rapid increase for $\omega > 60°$. Figure 41 indicates that the K_{II} factor increases rapidly for ω close to $0°$, followed by a decrease for values of ω up to $60°$, and then with a rapid increase for $\omega > 60°$.

2.8 Analysis of Elastic–Plastic Caustics [126]

If the crack tip deformation field in a cracked ductile solid can be characterized by means of a single plastic intensity factor, then the method of caustics has potential as a means for direct measurement of this intensity factor. The value of the J-integral is adopted as a plasticity intensity factor, and the lateral contraction of a planar specimen of elastic–plastic power-law hardening material is calculated in terms of J from the HRR asymptotic field of elastic–plastic fracture mechanics. The theoretical caustic curve which would be generated by geometrical reflection of normally incident light from points of the deformed specimen surface lying well within the plastic zone is determined, and it is shown that the value of J is proportional to the maximum transverse diameter of the caustic to the power $(3n + 2)/n$, where n is the power hardening exponent. Synthetic caustics obtained by numerical simulation of the optical reflection process are also shown.

The use of the optical method of caustics for measuring the stress intensity factor is now quite common in experimental work underlying the field of linear elastic fracture mechanics. It was suggested in a recent paper by Rosakis and Freund [127] that the method may have potential for application in elastic–plastic fracture testing as well. This suggestion was based on the following observation. When a large plate that contains a long through-crack and which responds in a nominally elastic manner is loaded so that crack opening occurs in the tensile mode, the stress and deformation fields very near to the crack tip assume the familiar universal spatial distributions. Only the magnitude of the near-tip field varies with load and geometry, and this magnitude is customarily the mode-I stress intensity factor. Within the framework of plane stress analysis, the deformed shape of the specimen surface near the crack tip is thus also known up to a scalar amplitude that is equivalent to the stress intensity factor. The success of the method of caustics is based on the fact that, with a suitable optical arrangement, the light pattern obtained by reflecting parallel incident light from the specimen surface near the crack tip provides a direct measure of the stress intensity factor.

Once the idea of the method is described in this way, it becomes clear that applicability of the method does not hinge on the material in the crack-tip region responding in an elastic manner. Instead, the key feature is that the deformed shape of the specimen surface (that is, the reflecting surface) in the crack-tip region is known up to a scalar amplitude. Although the mechanics of elastic–plastic fracture is not as fully developed as elastic fracture mechanics, the available asymptotic analyses of near-tip fields in power-law hardening materials suggest that this situation may prevail for these cases as well. Within the framework of plane stress analysis, with small strains and proportional stress histories for stationary cracks, the value of Rice's J-integral has been proposed as a plastic intensity factor. The viewpoint is adopted here that J provides a suitable scalar amplitude for the deformed shape of the surface of an elastic–plastic fracture specimen, and a means of directly measuring this amplitude is discussed. While there are methods available for measuring J values for ductile materials, they apply only for rate-independent materials subjected to quasi-static loading. While the direct optical method has drawbacks of its own, its use is not subject to the same limitations on rate of loading or material response.

Most structural materials that contain cracks undergo substantial plastic deformation under rising load prior to onset of crack growth. In many cases, including standard specimen configurations, a large plastic zone develops around the crack tip and there is no region of the body over which a stress intensity factor controlled elastic field prevails. The linear elastic fracture toughness approach to fracture resistance characterization is then not applicable and other criteria, such as the critical J criterion must be adopted. Methods for measuring values of J for ductile fracture specimens are available but the methods are indirect in general, in the sense that values of J are inferred from values of other measured quantities, typically load and deflection data. In this section, attention is focused on points deep within the crack tip plastic zone, and a means of inferring values of J from the local deformation field is suggested.

Consider a large plate of elastic–plastic material that exhibits power-law hardening behaviour. Suppose the plate is initially of uniform thickness d and that it contains a long through-crack. The plate is subjected to edge loading which results in a plane stress opening mode of deformation. If the midplane of the plate undergoes no transverse displacement, then the normal displacement of the plate surface $w(x, y)$ is:

$$w(x, y) = f(x, y) = \frac{d}{2}\varepsilon_{zz} \tag{478}$$

Hutchinson [128] and Rice and Rosengren [129] showed that the strain components in the crack tip region scale with the value of J for a power-law hardening material. They considered a monotonically loaded stationary crack in an incompressible material described by a J_2-deformation theory of plasticity and a relationship between post-yield strain ε_{ij} and stress σ_{ij} of the form:

$$\frac{\varepsilon_{ij}}{\varepsilon_0} = \frac{3}{2}\left(\frac{\sigma_e}{\sigma_0}\right)^{n-1}\frac{S_{ij}}{\sigma_0} \tag{479}$$

where:

$$S_{ij} = \sigma_{ij} - \tfrac{1}{3}\delta_{ij}\sigma_{kk}, \quad \sigma_e^2 = \tfrac{3}{2}S_{ij}S_{ij} \tag{480}$$

and σ_0 is the tensile yield stress, ε_0 is the equivalent tensile yield strain, and n is the hardening exponent. By introducing the preceding assumptions, they observed that, within a small strain formulation, a possible asymptotic strain distribution in the crack tip region is:

$$\varepsilon_{ij} \to \varepsilon_0\left(\frac{J}{\sigma_0\varepsilon_0 l_n r}\right)^{n/(n+1)} E_{ij}(n, \theta) \tag{481}$$

as $r \to 0$ in a polar coordinate system with origin at the crack tip. The angular factors E_{ij} in (481) depend on the mode of loading and on the hardening exponent. The dimensionless quantity l_n, which is defined in Ref. [128], decreases from 5 for $n = 1$ to 2.57 for infinitely large values of n for cases of plane stress. The intensity factor J in Rel. (481) is the value of Rice's J-integral. The singular field (481) is customarily referred to as the HRR singularity. The asymptotic result (481) was derived under the further assumption that the dependence of the local field on the polar coordinates r and θ is indeed separable.

For plane deformation, the J-integral is defined for any path of intergration C by [130]:

$$J = \int_C [Wn_1 - n_i\sigma_{ij}u_{j,1}]\,dC \tag{482}$$

where W is the local stress work density, n_i is a unit vector normal to C, and u_i is the particle displacement vector. The integral has the well-known property of path independence, that is, $J = 0$ for any simple closed path in the body in the absence of body forces. This implies that J has the same value for all paths that begin on one traction-free face of a crack in a plane with normal in the y-direction and that end on the opposite traction-free face of the crack. Because the path of integration can be chosen to be arbitrarily close to the crack tip, J has been interpreted as a measure of the strength of the crack-tip singular field, a role that is obvious from the form of Rel. (481). Based on the observation that J is a characterizing parameter for the crack-tip field, it has been suggested that a condition for onset of crack growth is the attainment of a critical value of J. This seems reasonable, provided that the one parameter characterization remains valid and that the one parameter field prevails over a region large in size compared to the fracture process zone. The interest in measuring values of J for ductile materials stems from the potential usefulness of this suggested criterion.

In view of the assumed incompressibility of material response, the strain in the thickness direction which appears in Rel. (478) is related to the in-plane strains by:

$$\varepsilon_{zz} = -(\varepsilon_{xx} + \varepsilon_{yy}) = -(\varepsilon_{rr} + \varepsilon_{\theta\theta}) \tag{483}$$

For specified material parameters and a given value of J, the in-plane strains are known from Rel. (481), the out-of-plane strain is computed from Rel. (483), and the shape of the reflecting surface is then given by Rel. (478). Thus, for points near the crack tip, the shape of the reflecting surface is known up to a scalar amplitude, namely the plastic strain intensity J.

It should be noted that Rel. (483) will provide an overestimate of lateral contraction for given in-plane strains because, for most materials, the elastic part of the local deformation is not incompressible, although the plastic part may be so. For points close enough to the crack tip for the plastic strain to dominate the elastic strain, however, the total strain is expected to satisfy Rel. (483) to an acceptable degree. An estimate of the ratio of total strain ε_{yy} to tensile yield strain ε_0 as a function of distance ahead of a crack tip in a low-hardening (large n) material may be obtained in the following way. For very large n, $E_{yy}(n, 0) \simeq 0.75$ and $l_n \simeq 2.6$. For contained plastic deformation in plane stress, the value of J is commonly related to the maximum extent of the plastic zone r_p by $J = \pi\sigma_0\varepsilon_0 r_p$. Thus, from Rel. (481) for large values of n:

$$\varepsilon_{yy}/\varepsilon_0 \simeq 0.9 r_p/r \tag{484}$$

so that Rel. (483) is expected to provide a reasonably accurate relationship if the distance from the crack tip to the initial curve along $\theta = 0°$ is less than one-half or one-third of the extent of the plastic zone along $\theta = 0°$.

Consider a specimen of uniform thickness d in the undeformed state occupying a region of the (x, y) plane. When boundary loads that tend to open the crack are applied, the resulting change in thickness is non-uniform and the equation of the deformed specimen surface is assumed to be given by:

$$z + f(x, y) = 0 \tag{485}$$

The shape of the reflecting surface $f(x, y)$ is given by:

$$f(x, y) = \frac{1}{2}\varepsilon_0 d \left(\frac{J}{\sigma_0\varepsilon_0 l_n r}\right)^{n/(n+1)} (E_{rr} + E_{\theta\theta}) \tag{486}$$

Numerical values of the singular variations E_{ij} are available for many values of hardening exponent n from the work of Shih [131]. Again, Eq. (486) represents the normal displacement of points on the initially plane specimen surface due to deformation. For an initial curve that is well within the crack tip plastic zone where the HRR field can be expected to dominate, the mapping takes the form (Rel. 2.6 in Ref. [126]):

$$X_i = x_i + G \frac{\partial}{\partial x_i}\left[\frac{\psi(\theta, n)}{r^{n/(n+1)}}\right] \tag{487}$$

where:

$$G = \varepsilon_0 z_0 d \left(\frac{J}{\varepsilon_0\sigma_0 l_n}\right)^{n/(n+1)}, \quad \psi = E_{rr} + E_{\theta\theta} \tag{488}$$

Relation (487) is written in polar form:

$$X = r\cos\theta + Gr^{-(2n+1)/(n+1)}\left[\frac{n}{n+1}\psi\cos\theta + \psi'\sin\theta\right]$$

$$Y = r\sin\theta + Gr^{-(2n+1)/(n+1)}\left[\frac{n}{n+1}\psi\sin\theta - \psi'\cos\theta\right] \qquad (489)$$

where the prime denotes the derivative with respect to θ.

If the determinant of the Jacobian matrix of the transformation (489) is set equal to zero, then the result is a quadratic equation for $r^{(3n+2)/(n+1)}$ in which the coefficients depend on ψ, ψ' and ψ'', as well as on the hardening exponent n. The root of the quadratic equation is:

$$r(\theta, n)^{(3n+2)/(n+1)} = GR(\theta, n) \qquad (490)$$

where:

$$2R(\theta, n) = \left[\left(\frac{n}{n+1}\right)^2\psi + \psi''\right] + \left(\left[\left(\frac{n}{n+1}\right)^2\psi + \psi''\right]^2\right.$$
$$\left. + 4\frac{2n+1}{n+1}\left[\left(\frac{n}{n+1}\right)^2\psi^2 - \frac{n}{n+1}\psi\psi'' + \frac{2n+1}{n+1}(\psi')^2\right]\right)^{\frac{1}{2}} \qquad (491)$$

Equation (490) gives the initial curve on the specimen surface for a given intensity of local deformation field. Then, substitution of Rel. (490) into Rel. (487) yields the equation of the corresponding caustic curve in the (X, Y) plane, parametric in the angle θ:

$$X = G^{(n+1)/(3n+2)}R^{(n+1)/(3n+2)}\left[\cos\theta + R^{-1}\left(\frac{n}{n+1}\psi\cos\theta + \psi'\sin\theta\right)\right]$$

$$Y = G^{(n+1)/(3n+2)}R^{(n+1)/(3n+2)}\left[\sin\theta + R^{-1}\left(\frac{n}{n+1}\psi\sin\theta - \psi'\cos\theta\right)\right] \qquad (492)$$

for values of θ in the range $-\pi < \theta < \pi$.

The shape of the caustic curve depends only on the distribution of plastic strain in the crack-tip region. The absolute size of the caustic curve, on the other hand, depends on the strength of the plastic strain singularity, the bulk material properties, the geometrical parameters, and the optical parameters. In fact, Eq. (492) is a relationship among all of these parameters. Thus, if the values of the material, geometrical and optical parameters are known, then Rel. (492) provides a relationship between the size of the caustic curve D and the strength of the plastic singularity J. Adopting the maximum transverse diameter $D_t^{max} = 2Y^{max}$ as the characteristic dimension, the intensity of the strain singularity can be expressed as:

$$J = S_n\varepsilon_0\sigma_0\left(\frac{1}{\varepsilon_0 z_0 d}\right)^{(n+1)/n}(D_t^{max})^{(3n+2)/n} \qquad (493)$$

Relation (493) is the main result of this section. It provides a simple relationship between the size of an observed caustic and the strength of the plastic singularity near the tip of the crack in an elastic–plastic power-law hardening material under conditions of plane stress.

All of the parameters that appear in Rel. (493) are determined from prior knowledge of the material behaviour or from the experimental set-up. The coefficient S_n is a function of n only and is given by:

$$S_n = \frac{l_n}{\max_\theta \left(R^{(n+1)/n} \left[2\sin\theta + \frac{2}{R}\left(\frac{n}{n+1}\psi\sin\theta - \psi'\cos\theta \right) \right]^{(3n+2)/n} \right)} \qquad (494)$$

The expression in Rel. (494) has been evaluated numerically for a range of values of n from the data in Ref. [131], and the results are shown in Table 5. It is noted that for values of n greater than about 4 or 5, the value is almost indistinguishable from the asymptotic value 0.072 for large n. Further, it is noted that the limiting case of Rel. (493) for $n \to \infty$ is consistent with the corresponding result for nonhardening materials which was reported in Ref. [127]. Likewise, for an elastic material ($n = 1$), it is well known that the stress intensity factor K is proportional to the caustic diameter raised to the power $5/2$. That Rel. (493) is also consistent with this result is evident once the relationship $J = K^2/E$, which applies for elastic cracks under plane stress conditions, is recalled.

In order to more fully understand the phenomenon of caustic formation for a power-law hardening material a companion study was carried out in which the full reflected optical field was simulated numerically. In this study, a square array of light rays was assumed to be normally incident on the surface of the specimen. The pattern in which these light rays pierce the plane reflecting surface of the undeformed specimen is shown in Fig. 42(a). These light rays were then "reflected" from a surface deformed according to Rel. (486), and the pattern in which the lines of the reflected rays pierce a screen positioned behind the reflecting surface was constructed. The patterns obtained for $n = 1, 3, 9, 13$ and 25 are shown in Fig. 42. The anticipated features, such as the caustic curve, are evident in these constructions of the reflected optical fields. Further insight into the physical process of caustic formation can be gained by comparing a straight row or column of rays in Fig. 42(a) to the positions into which these points map in Figs 42(b)–(f). These rows and columns map into curves that are tangent to the caustic curve at their common point, as is evident from the synthetic optical patterns.

Figure 43 presents the experimentally obtained caustic. This photograph was produced by using light reflected from deep within the crack-tip plastic zone of a

Table 5. Values of the numerical coefficient S_n, which appears in Rel. (493), from evaluation (494)

n	S_n
1	0.0277
2	0.0513
3	0.0611
4	0.0660
5	0.0687
6	0.0701
7	0.0710
8	0.0715
9	0.0718
10	0.0719
15	0.0719
20	0.0717
25	0.0714

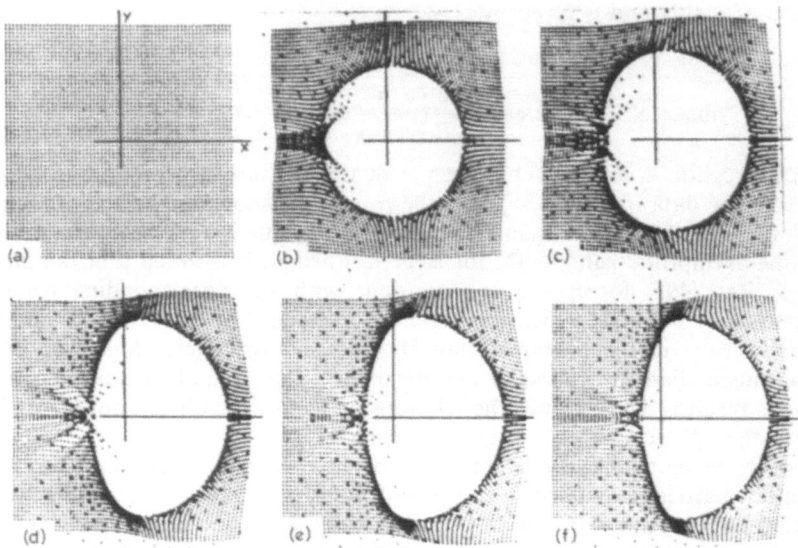

Fig. 42. (a) Regular array of incident light rays assumed in generating the synthetic caustic in (b–f), based on the HRR asymptotic field. The numerical simulation of the reflection process is shown for the following hardening exponents: (b) $n = 1$, (c) $n = 3$, (d) $n = 9$, (e) $n = 13$, and (f) $n = 25$.

tool steel with a low rate of strain hardening in the double cantilever beam configuration. It is clear from the photograph that surface evidence of plastic deformation extends far beyond the caustic. Some limited quantitative data are reported in Ref. [127].

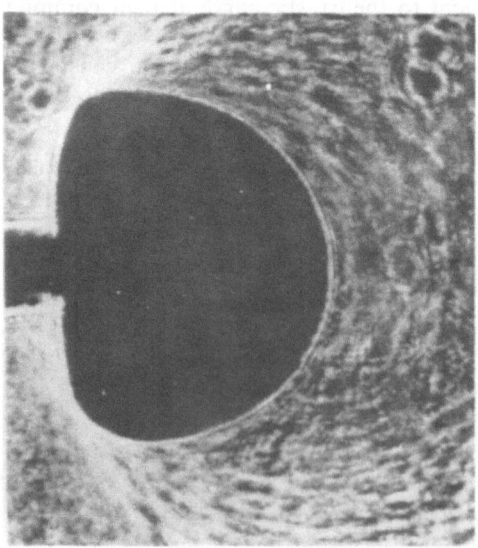

Fig. 43. Photograph of a caustic, surrounded by surface evidence of plastic deformation. The material is a tool steel with a low hardening rate [127].

2.9 Equations of Caustics at V-Notch Tip [132]

The evaluation of the complex stress intensity factors K at the apices of sharp V-notches in homogeneous, isotropic and elastic plates, symmetrically loaded in their planes, is treated by applying the experimental method of reflected caustics [132, 133]. The symmetry in loading and in position of the notch allows the expression of the complex Muskhelishvili stress function $\Phi(z)$ in the close vicinity of the notch as a single power term with exponent of the order of singularity.

The analytical solution of the problem is based on the method of complex potentials [89] together with the powerful method of conformal mapping. Other analytical solutions have been based on collocation and on the method of finite elements [134] and on integral equations [135].

The complex stress function $\Phi(z)$ for problems of cracks and notches is expressed by:

$$\Phi(z) = K_1 z^{\lambda_1} + K_2 z^{\lambda_2} + \sum_{i=3}^{n} K_i z^{(i-3)} \tag{495}$$

where the exponents $\lambda_{1,2}$ lie in the range $-0.5 \leqq \lambda_{1,2} < 0$. This expression in the close vicinity of the apex of the wedge simplifies to:

$$\Phi(z) = K_1 z^{\lambda_1} + K_2 z^{\lambda_2} \tag{496}$$

where $z = x + iy = re^{i\theta}$, $K_{1,2}$ are generalized complex stress intensity factors and $\lambda_{1,2}$ are the respective orders of singularities at the apex of the wedge, which are given by:

$$\lambda_{1,2} = \eta_{1,2} - 1 \tag{497}$$

where the quantities $\eta_{1,2}$ depend on angle ω of the notch and are given by [136]:

$$\sin \eta_{1,2} \omega = \pm \eta_{1,2} \sin \omega \tag{498}$$

The stress intensity factor K_I for problems of notches loaded symmetrically by the load σ_∞ at infinity is given by [137]:

$$K_I = 2\sigma_\infty a^{-\lambda} \left(\sin^2 \frac{\pi a}{2w} + \sec^2 \frac{\pi a}{2w} \right) \sec^{\frac{4}{3}} \left(\frac{\omega}{2} \right) \tag{499}$$

while the relation yielding the stress intensity factor K_I through the generalized complex stress intensity factor K_1 of Eq. (496), is given by [135]:

$$K_I = \pm (\lambda + 2)(2\pi)^{\frac{1}{2}} \left(1 - \frac{\cos(\lambda \tau)}{\cos[(\lambda + 2)\tau]} \right) K_1 \tag{500}$$

with:

$$\tau = \pi - \frac{\omega}{2} \tag{501}$$

For experimental evaluation of the stress intensity factor at the apex of the wedge, the method of reflected caustics will be applied.

It has been shown in Sect. 2.4 that the equations of the initial curve and of the caustic are given by Eqs (208) and (210) respectively. For problems where $\lambda_2 = 0$, as is the case for notch angles $\frac{\pi}{2} \leqq \omega < \pi$, Eq. (496) simplifies to:

$$\Phi(z) = K_1 z^{\lambda_1} = Kz^\lambda = (K_I - iK_{II})z^\lambda \tag{502}$$

By substituting Eqs (502) into Eqs (208) and (210) we obtain the parametric equations of the caustic [132]:

$$X = \lambda_m r_0 \left(\cos\theta \mp \frac{1}{|\lambda - 1|} \cos[\theta(1 - \lambda) - \phi] \right) \tag{503}$$

$$Y = \lambda_m r_0 \left(\sin\theta \mp \frac{1}{|\lambda - 1|} \sin[\theta(1 - \lambda) - \phi] \right) \tag{504}$$

and the radius r_0 of the initial curve of the caustic, is given by [138]:

$$r_0 = |C' K \lambda(\lambda - 1)|^{1/(2-\lambda)} \tag{505}$$

where $C' = C'_{r,t,f}$ is given by Rel. (380) and the polar coordinate θ varying as:

$$-\left(\pi - \frac{\omega}{2} \right) \leqq \theta \leqq \left(\pi + \frac{\omega}{2} \right) \tag{506}$$

Angle ϕ is given by:

$$\tan(\lambda\phi) = \kappa = \frac{K_{II}}{K_I} \tag{507}$$

For $K_{II} = 0$, or $\phi = 0°$, the caustic is symmetric to the axis of symmetry of the notch and the parametric Eqs (503) and (504) of the caustic become:

$$X = \lambda_m r_0 \left(\cos\theta \mp \frac{1}{|\lambda - 1|} \cos(1 - \lambda)\theta \right) \tag{508}$$

$$Y = \lambda_m r_0 \left(\sin\theta \mp \frac{1}{|\lambda - 1|} \sin(1 - \lambda)\theta \right) \tag{509}$$

Figure 44 presents the experimental caustics (Fig. 44(a)) obtained in a plexiglas notched specimen ($\omega = 90°$), under tension with a load $\sigma_\infty = 4.91 \times 10^6 \, N/m^2$ and the corresponding theoretical caustics (Fig. 44(b)).

The complex stress intensity factor K for the case of a sharp V-notch is given by [132, 138, 139]:

$$|K| = \delta_\lambda(\omega) \left(\frac{D_t^{max}}{\lambda_m} \right)^{2-\lambda} \frac{1}{|C'|} \tag{510}$$

where $\delta_\lambda(\omega)$ is a correction factor, λ is the order of singularity which depends on the angle of the notch, ω, D_t^{max} is the maximum transverse diameter of the caustic, which is formed around the apex of the notch, λ_m is the magnification ratio of the optical set-up and C' is an overall constant, which is given by Rel. (380).

The correction factor $\delta_\lambda(\omega)$ is given by [132]:

$$\delta_\lambda(\omega) = \frac{2^{\lambda-2}}{\lambda(\lambda - 1)} \left(\sin\frac{\pi}{2 - \lambda} + \frac{1}{|\lambda - 1|} \sin\frac{\pi(1 - \lambda)}{2 - \lambda} \right)^{\lambda - 2} \tag{511}$$

The stress intensity factors K_I and K_{II} are given by:

$$K_I = \frac{|K|}{\sqrt{1 + \kappa^2}}, \quad K_{II} = \kappa K_I = \frac{\kappa|K|}{\sqrt{1 + \kappa^2}} \tag{512}$$

with:

$$\kappa = \tan(\lambda\phi) \tag{513}$$

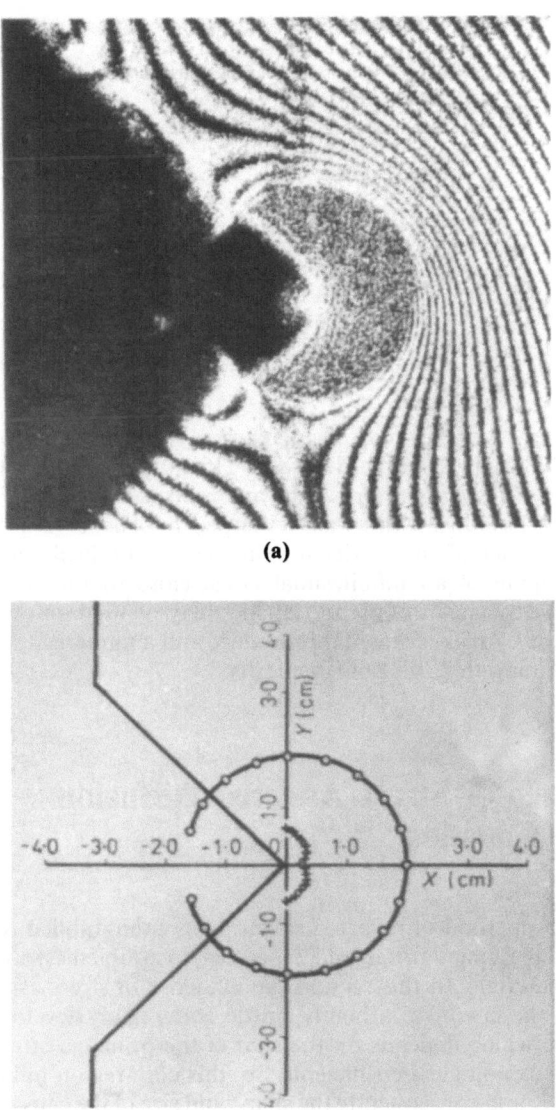

(a)

(b)

Fig. 44. Experimental **(a)** and theoretical **(b)** forms of the caustics obtained at the tip of a V-notched tension specimen under mode-I conditions ($\omega = 90°$, $\sigma_\infty = 4.91 \times 10^6$ N/m^2) [132].

where ϕ is the angular displacement of the axis of the caustic relative to the axis of symmetry of the crack or notch. The quantities λ and $\delta_\lambda(\omega)$ are given in Table 6 [132, 140, 141].

When a crack nucleates at the apex of the notch the order of singularity λ, which is different from $\lambda = -0.5$, changes abruptly to -0.5 which corresponds to the singularities of crack tips. Then, the stress intensity factors K_I and K_{II} are calculated according to the theory of caustics at crack tips.

Table 6. Typical values for the order of singularity, λ, and the correction factor, $\delta_\lambda(\omega)$, for sharp V-notches the lips of which subtend an angle ω [132]

ω(deg.)	λ	$\delta_\lambda(\omega)$
0	-0.50000	0.0745
30	-0.49855	0.0748
60	-0.48778	0.0771
90	-0.45552	0.0846
120	-0.38427	0.1064
150	-0.24802	0.1874
180	-0.00000	∞

An extensive analysis of the study of singularities by the optical method of caustics is given in Ref. [138], Chap. XI by Theocaris. This study is based on the parametric equations of caustics at V-notch tips. Also, an experimental technique based on the optical method of pseudocaustics was developed by Theocaris [142, 143] for the evaluation of the order of singularity for singular elastic fields. According to this method, the mapping of an infinitesimal circle close to the singularity into an infinitesimal ellipse by means of pseudocaustics may be used for evaluating directly the stress singularity. An experimental procedure and a numerical scheme were used for accurately defining the order of singularity.

2.10 Influence of Stress-Assisted Diffusion on the Caustics

So far, the optical method of reflected caustics has been applied to various elastic problems containing singularities and especially to problems with cracked plates from isotropic materials. In this section the influence of stress-assisted diffusion is studied. Around the crack tip a highly brittle core-region develops due to stress-assisted diffusion, which depends on the sum of the principal stresses at the crack tip and on phenomenological coefficients. So, this core-region influences the initial curve of the caustic and consequently the shape and size of the caustic [144, 145, 146].

Stress-corrosion phenomena and crack initiation are influenced by many interdependent variables; one significant factor is stress-assisted diffusion. A stress-assisted diffusion theory has been put forward recently by Aifantis and co-workers [147–151], who applied it to various problems of stress-corrosion cracking and hydrogen embrittlement. According to the stress-assisted diffusion theory, the distribution of the concentration of the diffusion species ρ around the crack tip is given by:

$$\rho = \rho_0 [1 + B(\sigma_{xx} + \sigma_{yy})]^A \tag{514}$$

where ρ_0 denotes the concentration on the boundary, and A, B are phenomenological coefficients. In the case of stress-assisted diffusion, it turns out that $A = M/N$ and $B = N/D$ with the coefficients D, M, N having a precise physical interpretation. Relation (514) is valid for points which lie outside a small region (process zone) surrounding the crack tip and defined by a "critical distance" determined by the particular constitutive structure of the material. For perfectly elastic materials the "critical distance" approaches zero [152]. According to Oriani's theory [153], stress-assisted diffusion creates a brittle region around the crack tip. This brittle region is defined by Rel. (514). So, we suggest that a new core-region is developed around the crack tip. In this new core-region there are many microcracks and voids which are created by movement of the crack tip.

The parametric equations of the caustic, when the crack is normal to the tensile load, $\beta = 90°$ (Fig. 5), for reflected light rays from the front face of the specimen, are:

$$x_f = \lambda_m r_0 \left(\cos\theta + \frac{2}{3}\cos\frac{3\theta}{2} \right)$$ (515)

$$y_f = \lambda_m r_0 \left(\sin\theta + \frac{2}{3}\sin\frac{3\theta}{2} \right)$$ (516)

with:

$$r_0 = (\tfrac{3}{2} C_f)^{\frac{2}{3}}$$ (517)

and:

$$C_f = \frac{z_0 dv K_I}{E\lambda_m (2\pi)^{\frac{1}{2}}}$$ (518)

Figure 45(a) presents the caustic and its initial curve according to the parametric Eqs (515) and (516).

The stress intensity factor K_I is determined by the relation:

$$K_I = \frac{2(2\pi)^{\frac{1}{2}} E}{3z_0 d\lambda_m^{\frac{3}{2}} v} \left(\frac{D_t^{max}}{\delta_t^{max}} \right)^{\frac{5}{2}}$$ (519)

where D_t^{max} is the maximum transverse diameter of the caustic, $\delta_t^{max} = 3.1702$ is the correction factor of the diameter D_t^{max}.

For $\beta = 90°$ and for singular solution (Eqs (82)–(85) of stress field), Eq. (514) becomes:

$$r_d = C_d \cos^2 \frac{\theta}{2}$$ (520)

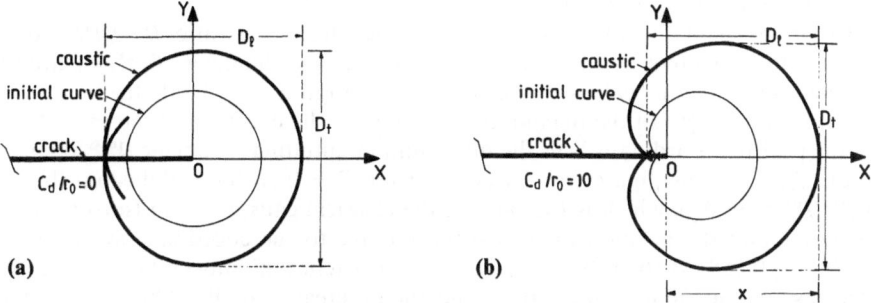

Fig. 45. Caustics and their initial curves at crack tip for singular solution and for ratio (a) $C_d/r_0 = 0$ (without diffusion) and (b) $C_d/r_0 = 10$ (under diffusion).

with:

$$C_d = \left\{ \frac{2BK_1}{\sqrt{2\pi}\left[\left(\frac{\rho}{\rho_0}\right)^{1/A} - 1\right]} \right\}^2 \tag{521}$$

With the assumption that the initial curve of the caustic r_0 is influenced by stress-assisted diffusion, a new initial curve of the caustic is developed. This new initial curve is given by:

$$r = r_0 + r_d = r_0 + C_d \cos^2 \frac{\theta}{2} \tag{522}$$

Then the parametric equations of the caustic are given by:

$$x_f = \lambda_m \left(r_0 + C_d \cos^2 \frac{\theta}{2} \right) \left(\cos \theta + \frac{2}{3} \cos \frac{3\theta}{2} \right) \tag{523}$$

$$y_f = \lambda_m \left(r_0 + C_d \cos^2 \frac{\theta}{2} \right) \left(\sin \theta + \frac{2}{3} \sin \frac{3\theta}{2} \right) \tag{524}$$

Figure 45(b) presents the caustic and its initial curve according to Eqs (523) and (524) for ratio $C_d/r_0 = 10$. In this figure we can observe that the two caustics are different in form and position related to the coordinate system Oxy with origin at the crack tip. In this case the stress intensity factor K_1 is given by:

$$K_1 = \frac{2(2\pi)^{\frac{1}{2}} E}{3z_0 d\lambda_m^{\frac{3}{2}} v} \left(\frac{D_t^{\max}}{\delta_t^{\max}(C_d/r_0)} \right)^{\frac{3}{2}} \tag{525}$$

where the correction factor $\delta_t^{\max}(C_d/r_0)$ depends on the ratio C_d/r_0. This correction factor is:

$$\delta_t^{\max}(C_d/r_0) = \left(1 + \frac{C_d}{r_0} \cos^2 \frac{\theta^{\max}}{2} \right) \left(\sin \theta^{\max} + \frac{2}{3} \sin \frac{3\theta^{\max}}{2} \right) \tag{526}$$

where $\theta^{\max}$ is the angle where the transverse diameter D_t of the caustic becomes maximum. This angle is calculated by the condition:

$$\frac{\partial y_f}{\partial \theta} = 0 \tag{527}$$

for various values of the ratio C_d/r_0.

Comparison of the two caustics is obtained from the ratio $D_t^{\max}/D_l^{\max}$ of the maximum diameters of the caustic ($D_t^{\max}$ transverse diameter, $D_l^{\max}$ longitudinal diameter) for various values of the ratio C_d/r_0. Figure 46 presents the variation of the ratio $D_t^{\max}/D_l^{\max}$ of the maximum diameters of the caustic and the ratio $X/D_l^{\max}$ of the positive coordinate X to the maximum longitudinal diameter $D_l^{\max}$ versus the ratio C_d/r_0. In this figure we observe that for $C_d = 0$ (without diffusion) the ratio $D_t^{\max}/D_l^{\max} = 1.056$, which is the ratio of the classical caustic for an isotropic elastic material, and the position of the caustic relative to the coordinate system Oxy is $X/D_l^{\max} = 0.558$. As the ratio C_d/r_0 increases (existence of diffusion) the above ratios decrease to the value $C_d/r_0 \simeq 10^{-2}$ and then increase rapidly. This means that in the first region ($.... \leq 10^{-2}$) the caustic becomes elongated in the x-direction, while in the second region ($10^{-2} \leq ...$) the caustic becomes elongated in the y-direction.

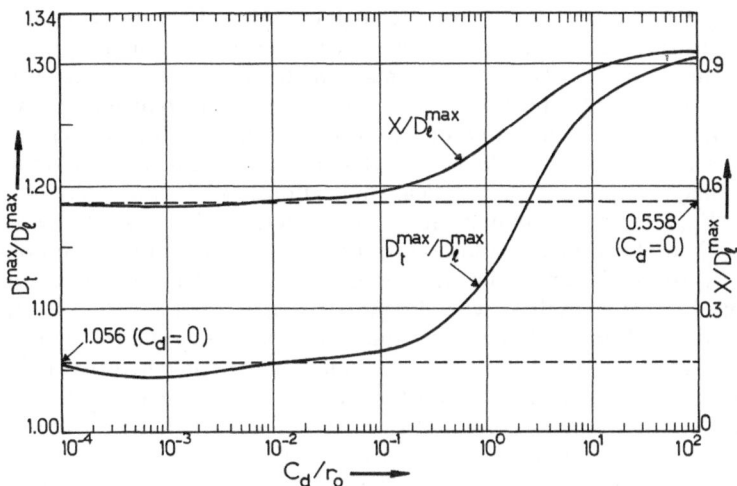

Fig. 46. Variation of the ratio D_t^{max}/D_l^{max} of the maximum diameters of the caustic and the ratio X/D_l^{max} of the positive coordinate X to the maximum longitudinal diameter versus the ratio C_d/r_0.

According to Ref. [151] the critical distance r_c of diffusion is of order about 10^{-7} m, while the radius of the initial curve r_0 is about 2×10^{-3} m, then the ratio C_d/r_0 reaches a maximum value of about 0.5×10^{-4}. This means that oval caustics are formed by the influence of stress-assisted diffusion because the ratio C_d/r_0 takes values in the first region of the curve of Fig. 46. Such caustics were experimentally observed in Ref. [39].

From this analysis, it is concluded that, from the caustics and according to Fig. 46, C_d and hence the diffusion can be calculated if the phenomenological coefficients are known. Also, from the variation of the diameter D_t^{max} of the caustics the stress intensity factor K_I is determined by Rel. (525).

2.11 Study of Stress-Corrosion Crack Growth in Aluminium Alloys by Caustics

A macroscopic study of the growth of a stress-corrosion crack in an aluminium alloy was undertaken by using the optical method of caustics. The dependence of the crack tip stress intensity factor K_I on the applied stress, as well as on the molarity of a NaCl aqueous solution was studied in detail. The calculation of this quantity was made by taking into account the thickness variation of the specimen due to the corrosive environment. A criterion of crack growth in terms of the caustics was formulated and the corresponding values of the threshold stress intensity factor K_{ISCC} were accurately determined. Finally, the creation and evolution of the pits developed in the vicinity of the crack tip was studied and their significance to the mechanism of fracture of the specimens was examined [39].

A comprehensive discussion of the various factors affecting the kinetics of a corrosion crack in an aqueous solution under sustained loads has been given by

Speidel [154]. As stated in this study, referring to the threshold stress intensity K_{ISCC}, below which crack growth does not occur, the existence of this limit is due to rather arbitrary cut-off times of the experiments and there is no experimental evidence supporting such a limit value for K_I. It is, however, possible to define a conventional value for K_{ISCC}, that is the value of K_I corresponding to a definite crack-growth rate (for example 10^{-10} m/s). As is self-evident, the experimental determination of K_{ISCC} is of major importance in practical applications. In this study the conventional values of K_{ISCC} for an aluminium alloy were determined accurately by using the optical method of caustics. In addition to the determination of the threshold values K_{ISCC} of K_I, the variation of the stress intensity K_I for the same aluminium alloy with applied stress, as well as with the molarity of a NaCl aqueous solution, was also studied. In all these studies the values of K_I were determined by taking into account the thickness variation of the specimen due to the corrosive environment.

Finally, the study was directed towards the mechanism of crack growth by studying the creation, evolution and growth of pits developed near the crack tip due to the corrosive environment and the applied stress. The significance of the role of pits on stress-corrosion cracking has been emphasized by several authors [155, 156], who pointed out the dominant role of pitting in the field of stress corrosion. In a way, stress corrosion may be regarded as an extreme case of pitting, leading to a slow mechanism of growth of the primary crack by the progressive coalescence with it of the deeply eroded pits in the vicinity of the crack tip.

The stress intensity factor, in this case, is given by Rel. (519). As is well known, the value of K_I for a tension specimen with a transverse crack of length a is equal to [103]:

$$K_I = f_1 \frac{P}{wd} \sqrt{a} \tag{528}$$

where P is the applied load, w and d are the width and thickness of the specimen, respectively and f_1 is a constant expressing the influence of specimen geometry on K_I.

When the load P applied to the specimen is kept constant, then from Rel. (528) it is concluded that, if the crack does not propagate, the quantity $(K_I d)$ is constant, which by taking into account Rel. (519) means that the diameter D_t of the caustic is constant. Thus, any increment of the diameter D_t means that the crack starts to propagate. This provides another means of visualizing crack growth, as an alternative to conventional measurements made on the crack tip by a travelling microscope. By comparing these two methods of crack growth verification and localization of the propagating crack tip, it can be observed that the method of caustics gives a twofold magnification of the crack propagation, while the magnification of crack growth by the conventional method is unity. Indeed, the crack growth, which influences the crack-tip stress field, is depicted on the caustic generated by illuminating the specimen by a light beam, thus providing a first magnification since the caustic can be highly magnified according to the particular arrangement of the experimental apparatus used. Furthermore the increase of the diameter of the caustic can be measured afterwards by a travelling microscope, thus providing a second magnification and higher precision.

Besides this twofold magnification, the method of caustics presents a further advantage over conventional methods. In the latter methods, where the crack-tip propagation is observed through a microscope, it is frequently very difficult to determine the exact position of the crack tip, due to crack blunting, caused by the corrosive environment, and to the presence of particles of the aqueous solution

Table 7. Composition and mechanical properties of 57S aluminium alloy

Composition (%)	2.2–2.8 Mg	0.15–0.35 Cr	0.45 Fe + Si	0.10 Cu	0.10 Mn	0.10 Zn
Mechanical properties	Elastic modulus $E(\mathrm{kp/mm^2})$ 7200	Poisson's ratio ν 0.33		Yield stress $\sigma_0(\mathrm{kp/mm^2})$ 20.8	Ultimate stress $\sigma_u(\mathrm{kp/mm^2})$ 22.5	

which may fill the crack tip region. In the method of caustics these factors do not influence the measurement of the crack length. Indeed, in this method the crack growth influences the stress field near to the crack tip, which is transformed into an optical curve.

Very accurate determination of the crack tip growth as described above enables the precise calculation of the limiting value K_{ISCC}, below which the crack does not propagate. This can be achieved by applying to the specimen small increments of load and determining the particular load for which we have a prescribed crack growth velocity. The corresponding value of K_I represents the threshold value K_{ISCC}. Such experiments were undertaken in a 57 S aluminium alloy whose composition and mechanical properties are given in Table 7. Tension specimens of width $w = 0.069$ m and length $l = 0.390$ m were prepared from a large plate of uniform thickness $d = 0.001$ m. Two symmetrical artificial V-notches of angle and length $\varphi = 30°$ and $a = 0.010$ m respectively were accurately sawn. One lateral surface of each specimen was mechanically polished to become reflective by using 0-type emery paper. The cracked tension specimens thus prepared were subjected to a constant load by using a creep universal testing machine. The non-reflective surface of the specimens was exposed in a NaCl aqueous solution while the reflective surface was covered by greasing substance to protect it from the corrosive environment. The molarity M of the NaCl aqueous solution was varied for different batches of specimens and had the following values: $M = 0.6, 1.0$ and 2.0 m. The specimens were also loaded with an anodic current of density $j = 2 \,\mathrm{mA/cm^2}$ in order to accelerate the corrosion process. The temperature T of the environment was kept constant at 23°C during the experiments. The optical set-up was as shown in Fig. 10, with a magnification factor $\lambda_m = 6.5$.

Three different initial stress levels were applied to the specimens, so that the stress intensity factor K_I, defined by Rel. (528), takes three initial values $K_I^i = 15.62$, 21.68 and 27.86 kp/mm$^{\frac{3}{2}}$. The load applied to the specimen was kept constant and the caustic formed on a reference screen by illuminating the specimen was photographed at defined intervals of time.

Figure 47 shows the patterns obtained on the reference screen by illuminating a cracked tension specimen with a coherent He–Ne laser light beam. The values of the initial stress intensity factor K_I^i of the specimens and the molarity M of the NaCl aqueous solution were: $K_I^i = 21.68$ kp/mm$^{\frac{3}{2}}$ and $M = 2$ m. The photographs correspond to times t from application of the load (a) $t = 0$ and (b) $t = 16$ hr. It can be observed from these two photographs that at the tip of the V-notch a strongly illuminated curve, the so-called caustic is formed. By measuring the transverse diameter of this curve the values of the stress intensity factor K_I can be calculated through Rel. (523), as described above. It can be remarked from these two photographs, that the diameter of the caustic is large in the second photograph corresponding to a time interval $t = 16$ hr.

We can also note on the surface of the corroded specimen some other strongly illuminated lines, tending to form closed curves as the time interval increases. These

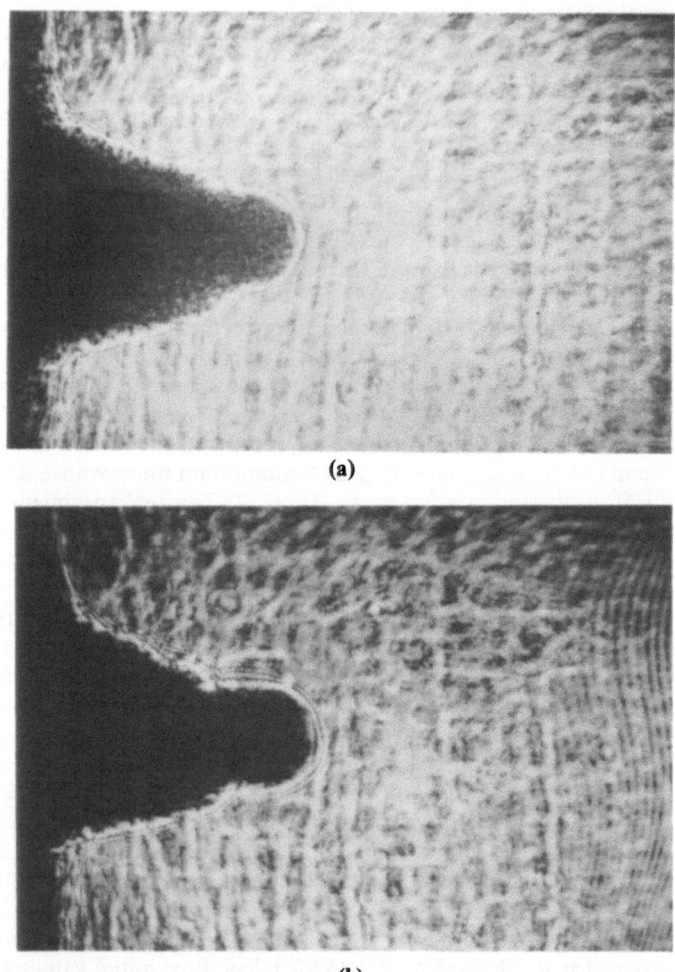

(a)

(b)

Fig. 47. Caustic patterns obtained on a reference screen by illuminating a double notched aluminium specimen with He–Ne laser light. Specimen is immersed in a 2 m NaCl aqueous solution and is under constant load $P = 473$ kp, so that the initial value of the stress intensity factor $K_{\mathrm{I}}^{i} = 21.68$ kp/mm$^{\frac{3}{2}}$. Cases (a) and (b) correspond to times $t = 0$ and 16 hours from the beginning of the test.

marks are more distinct in the following two photographs (Fig. 48) corresponding to times from application of the load $t = 38.25$ and 49.00 hr respectively. As time increases, these closed, strongly illuminated curves, become more and more distinguishable, while their size increases. These secondary caustics are created by local irregularities, called pits, formed on the illuminated surface of the specimen, during exposure to the corrosive environment. In addition to the creation of these secondary caustics we can see that the main caustic, formed at the crack tip, becomes less distinct and merges with the secondary caustics.

Finally, Fig. 49 shows the primary and the secondary caustics in an aluminium specimen with an initial stress intensity factor $K_{\mathrm{I}}^{i} = 15.62$ kp/mm$^{\frac{3}{2}}$ immersed in a NaCl aqueous solution with molarity $M = 2$ m. Photograph 49(a) corresponds to

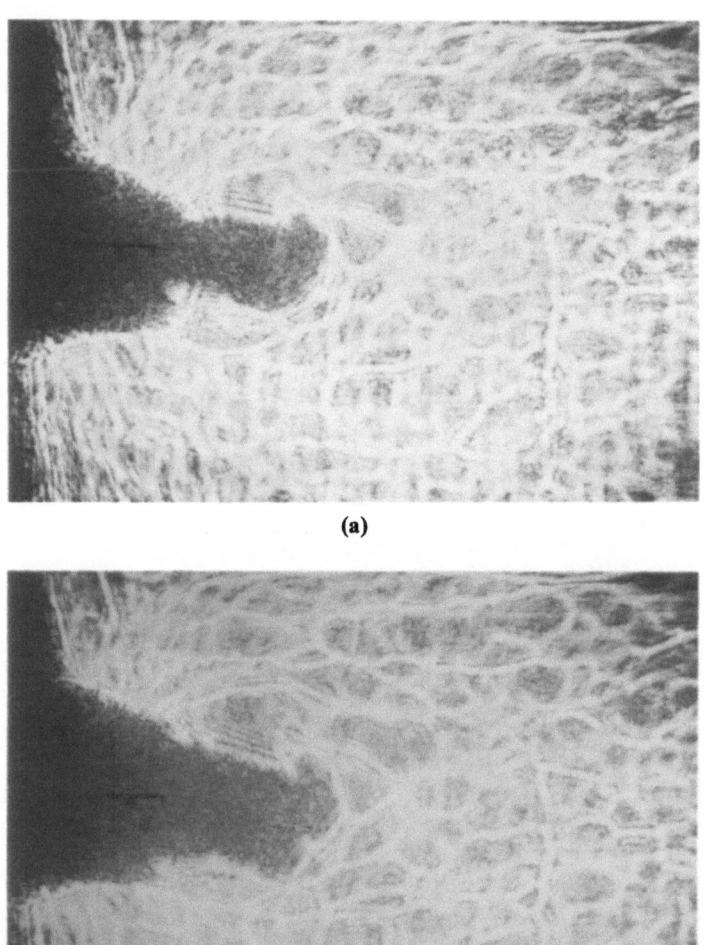

(a)

(b)

Fig. 48. As Fig. 47, for times (a) $t = 38.25$ hours and (b) 49.00 hours. The secondary caustics formed by pits become more distinguishable and increase in size as time t increases. The shape of the main caustic formed at crack tip is different to the typical epicycloid of brittle materials, which indicates that some plasticity is introduced at the crack tip.

$t = 49.76$ hr, and photograph 49(b) to $t = 76.24$ hr. Both these photographs correspond to a near-to-fracture stage of the specimen. We can see that the secondary caustics formed from the pits have been significantly enlarged, while their shapes become elongated in the direction of crack propagation. In addition, the secondary caustics are confused with the primary caustic formed at the tip of the V-notch.

From the experimentally obtained caustics and by using Rel. (523) the crack tip stress intensity factors K_I were determined. Since the thickness d of the specimens varies considerably during the corrosion process, stress intensity factors were

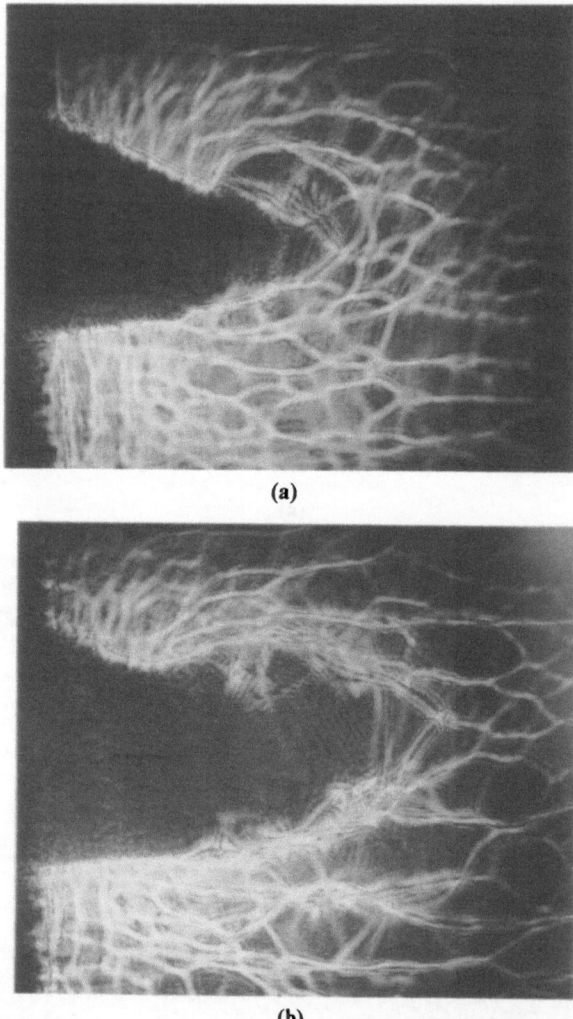

(a)

(b)

Fig. 49. Caustic patterns on a reference screen for a doubly notched aluminium specimen with: $K_I^i = 15.62\,\text{kp/mm}^{\frac{3}{2}}$, $M = 2\,\text{m}$ at times $t = 49.76$ hours (a) and 76.24 hours (b) respectively. Secondary caustics become elongated in the direction of primary crack propagation, while the primary caustic at the crack tip is coalescing with secondary caustics.

calculated by determining actual specimen thickness during the experiments. This determination was made by using a microscope, measuring the specimen thickness at a number of points in the vicinity of the crack tip and taking the mean value of all measurements.

The variation of the gross specimen thickness d near the crack tip during the corrosion process for a precracked specimen in a $M = 2\,\text{m}$ NaCl aqueous solution for three different values of the initial stress intensity factor, $K_I^i = 15.62$, 21.68, and $27.86\,\text{kp/mm}^{\frac{3}{2}}$ respectively is shown in Fig. 50. It will be observed that thickness varies rapidly with time t. We can also remark that at any one time the gross

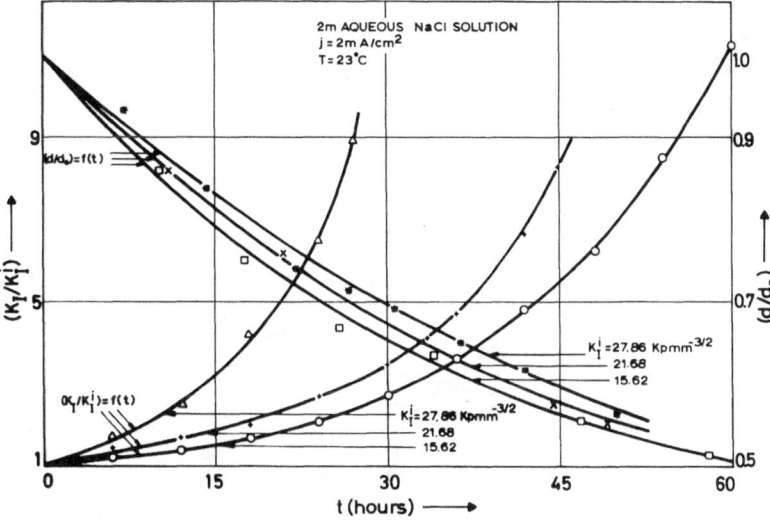

Fig. 50. Variation of the gross thickness d of the specimen near the crack tip, normalized to the initial thickness d_0, as well as of the stress intensity factor K_1, normalized to its initial value K_1^i, versus time t from the beginning of the test for a doubly notched aluminium tension specimen in a 2 m NaCl aqueous solution for three different values of the initial stress intensity factor $K_1^i = 15.62$, 21.68 and 27.86 kp/mm$^{\frac{3}{2}}$.

thickness variation is lower for higher values of the initial stress intensity. This phenomenon is contrary to the conventional thickness variation when loading the specimen by in-plane forces due to Poisson's effect. The thickness variation of the specimen is the present case, however, is not due to Poisson's effect only but also to the existence of pits created on the face of the specimen exposed to the corrosive environment. Thus, it can be concluded that the higher stress level increases the area of pits rather than their depths. In Fig. 50, the variation of the stress intensity factor K_1 versus time for the above three initial values of the stress intensity factor K_1^i is also given. The corresponding curves terminate at the points where the specimen breaks. We can see that the critical time of fracture for each specimen is reduced as the initial stress intensity factor increases. It is worth noting here the rapid rate of increase of K_1 with time t.

The influence of the molarity of the NaCl aqueous solution on the gross-thickness variation of the specimen and on the crack-tip stress intensity factor K_1 is shown in Fig. 51. Again, note the strong thickness variation of the specimen, as well as the progressive increase of the stress intensity factor during the evolution of the experiment. Increasing the molarity of the corrosive environment increases the rate of change of both the thickness of the specimen and the stress intensity factor.

Finally, Fig. 52 shows the variation of the diameter of the caustic versus time t for four different values of the initial stress intensity factor $K_1^i = 11.94$, 15.62, 21.68 and 27.86 kp/mm$^{\frac{3}{2}}$. As already indicated above, the increment of the diameter of the caustic ensures that the threshold value K_{ISCC} of K_1 has been reached for the particular cut-off time of the experiment considered. We can see that for the smallest value of the initial stress intensity factor $K_1^i = 11.94$ kp/mm$^{\frac{3}{2}}$ there is a large interval of time t for which the diameter of the caustic is kept constant. This interval extends up to $t = 45$ hr. Thus, if the cut-off time of the experiment is smaller than $t = 45$ hr the value of K_{ISCC} is equal to $K_{\text{ISCC}} = 11.94$ kp/mm$^{\frac{3}{2}}$. By increasing the initial stress level,

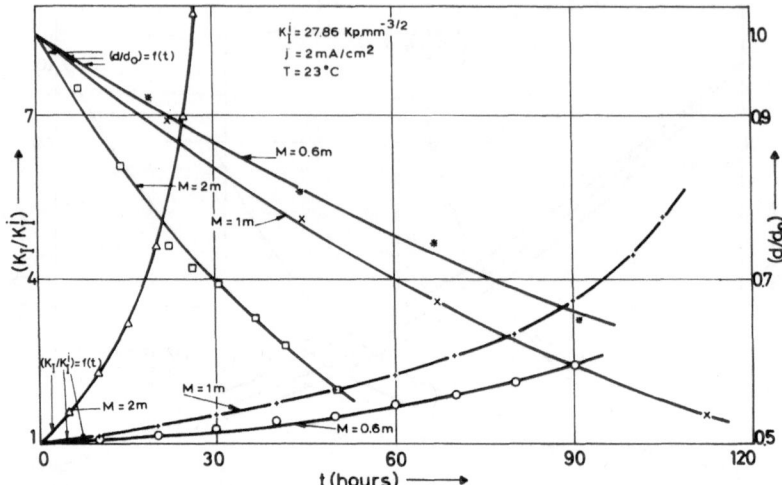

Fig. 51. Variation of (d/d_0) and (K_1/K_1^i) versus time t for a doubly notched aluminium tension specimen with $K_1^i = 27.86$ kp/mm$^{\frac{3}{2}}$ for three values of the molarity of the NaCl aqueous solution $M = 0.6$ m, 1.0 m and 2.0 m.

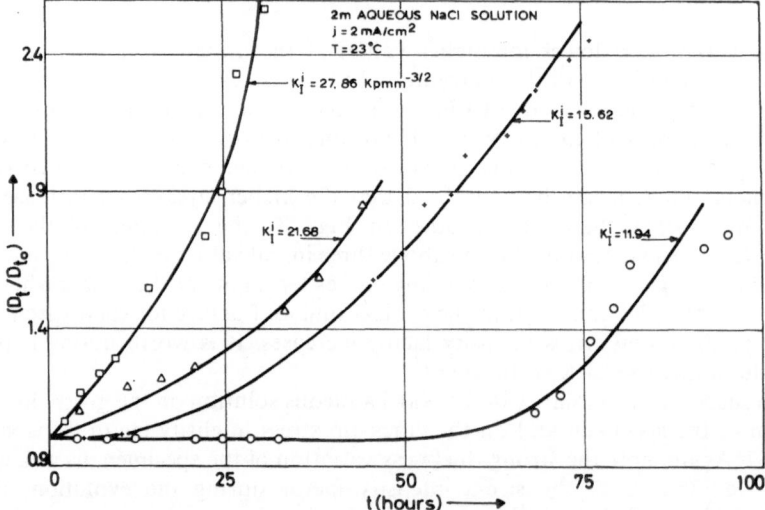

Fig. 52. Variation of the diameter D_t of the primary caustic formed at the crack tip, normalized to its initial value for time $t = 0$, versus time t for a doubly notched aluminium tension specimen, with $M = 2$ m for four different values of the initial stress intensity factor $K_1^i = 11.94$, 15.62, 21.68 and 27.86 kp/mm$^{\frac{3}{2}}$.

the time interval over which the diameter of the caustic is constant is reduced. Thus, for $K_1^i = 21.68$ kp/mm$^{\frac{3}{2}}$ there is no such interval.

As has already been proved above, high accuracy localization of crack growth is achieved by studying the variation of the diameter of the caustic formed by illuminating the specimen with a light beam. Thus, in each of the $D_t/D_{t_0} = f(t)$ curves in Fig. 52 there is a limit after which the diameter of the caustic starts to increase. This limit gives the appropriate time limit for the initiating of propagation of the crack.

As has previously been indicated (Figs 47–49), besides the primary caustic formed at the tip of the notch and after a definite time interval from the beginning of the experiment, some other highly illuminated curves are created on the image of the surface of the specimen. These secondary caustics are generated by the light rays reflected from the surface of the specimen in the vicinity of the pits which are created on the specimen. As can be seen from Figs 47–49 the secondary caustics are blurred in the early stages of the pit initiation, but as the time increases these caustics become progressively more distinct and coalesce to form larger caustics.

The successive evolution of the secondary caustics for a cracked aluminium specimen with an initial stress intensity factor $K_i^i = 15.62 \, \text{kp/mm}^{\frac{3}{2}}$ in a $2 \, \text{m}$ NaCl aqueous solution is shown in Fig. 53. Five successive stages of the secondary caustics, corresponding to times $t = 43.83, 67.83, 89.58, 100.58$ and $114.58 \, \text{hr}$ from the beginning of the test are drawn with identical scales. It can be remarked that in the early stages of the corrosion process the shape and orientation of the secondary caustics is rather random (Figs 53(a) and (b)), while as time t increases these caustics coalesce to form larger and more intensive caustics which are progressively more elongated with their longer axes oriented toward the direction of propagation of the primary crack.

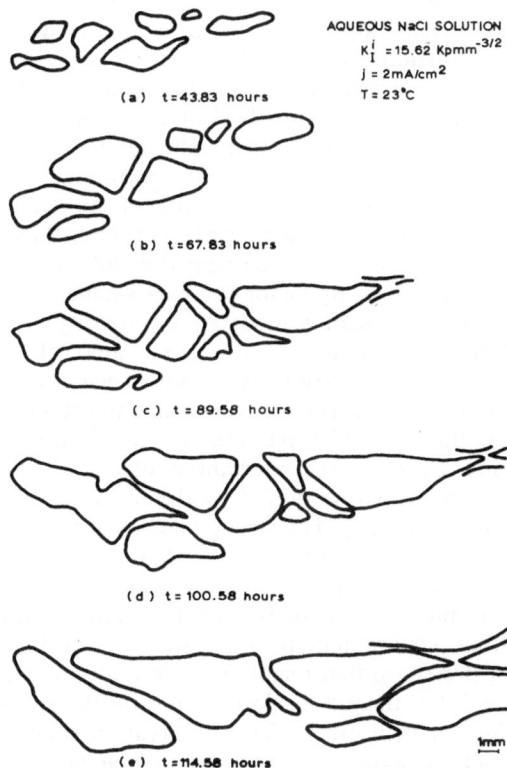

Fig. 53. Evolution of secondary caustics formed by pits for a doubly notched aluminium specimen with $K_i^i = 15.62 \, \text{kp/mm}^{\frac{3}{2}}$, and $M = 2 \, \text{m}$ at indicated times t from the beginning of the test.

From Figs 50 and 51 it can be concluded that the rate of variation of K_I versus time t is not constant, but increases with time. This increment becomes larger as the initial stress intensity values or the molarities of the corrosive environment increase. All these results indicate the significant role of mechanical (expressed by the stress intensity factor) as well as environmental (expressed by the molarity of the solution) factors on the crack tip stress field. As is well known from the theory of elastic fracture mechanics, the stress components at the crack tip region are directly related to stress intensity factor K_I, so that the value of the critical stress for which the material of the specimen fails can be determined.

Study is then directed towards determination of the particular time at which the crack starts to propagate. This is done with high accuracy and simplicity by measuring the diameter of the caustic formed on a reference screen. As has been already proved, any increment of the caustic means that a crack has started to propagate. The crack does not propagate with high velocity as occurs with brittle or semi-brittle fracture cases. This is due to the fact that crack propagation is largely induced by the corrosive environment and not by the applied load. The corrosive environment contributes mainly in breaking the bonds in the molecular structure of the material, while any increase of stress is of secondary importance.

It is worthwhile observing from Figs 48 and 49 that the main caustic, formed at the crack tip, is not symmetric to the crack axis, as in Fig. 47; there is distortion of the axis of symmetry of the caustic. This ensures, as has been previously pointed out, that shear stresses are introduced at the crack tip. Thus, it is concluded that the crack does not propagate along its initial axis, but follows a zig-zag path.

From the angle of distortion of the axis of symmetry of the caustic the contribution of the sliding-mode stress intensity factor K_{II} can be readily determined. This characterizes the near to crack-tip stress field of a cracked plate subjected to shear forces. The variation of K_{II}/K_I^i with time indicates that shear stresses become appreciable after a definite time interval from the beginning of the test and that the ratio K_{II}/K_I^i takes larger values as the initial value of the stress intensity factor K_I^i decreases. This peculiar behaviour can be explained by the fact that as the applied stresses increase microcracks and pits are oriented mainly parallel to the crack axis, so that the shearing stress intensity factor becomes smaller.

Moreover, the value of the sliding-mode stress intensity factor K_{II} increases as the molarity M of the solution increases; for large values of M the time at which appreciable shear stresses are introduced is shortened. This indicates that the molarity of the solution plays a role of particular importance for the creation of microcracks and pits through which the primary crack propagates.

Another interesting aspect in the mechanism of crack growth with corrosive environment is the significant role played by the pits developed on the specimen surface after a certain test time t_a. The strong dependence of t_a on the values of the stress intensity factor K_I^i and on the molarity M of the NaCl solution proves the significant role of K_I^i and M in the mechanism of crack growth, which is initiated with the assistance of pits. The major role of stress intensity in creation of pits is shown by the fact that pits are first created in the region near to the crack tip due to the high value of stresses in that region. However, although pits are first formed very near to the crack tip, there are selective sites on the specimen at which they develop more rapidly than in others. This observation shows the increasing role of the crystallographic structure of the material of the specimen on its resistance to stress corrosion cracking. It can also be observed (Figs 47–49) that, while early in their creation, pits are arbitrarily oriented, as time increases they become elongated in the direction of crack propagation and progressively coalesce with the primary caustic formed at the crack tip. From photographs of reflected images of the specimen on the screen we can see that, while at the beginning of the experiment the main caustic at the crack tip has the usual epicycloid form of brittle or semi-brittle materials [33], after a while the caustic becomes extended in the direction of the crack. This shows that some plasticity is introduced near the crack tip. As has already been shown [30, 35] the primary caustic, formed at the crack tip, can be successfully used to derive the form of the stress distribution in the cracked specimen, as well as for the determination of the crack opening displacement and the elastic–plastic boundary. This is in accordance with Swann's [157] predictions that in a corrosive specimen some initial fracture occurs, which is accompanied by small plastic deformation. The variation of the ratio of the longitudinal (D_l) to the transverse (D_t) diameter of the primary caustic is shown in Fig. 54 for three different values of the initial stress intensity factor K_1^i (Fig. 54(a)) and two values of molarity M (Fig. 54(b)) versus time t. It is clear from these two figures that there is an important role played by the applied stress and the molarity of the corrosion environment on the plasticity introduced at the crack tip.

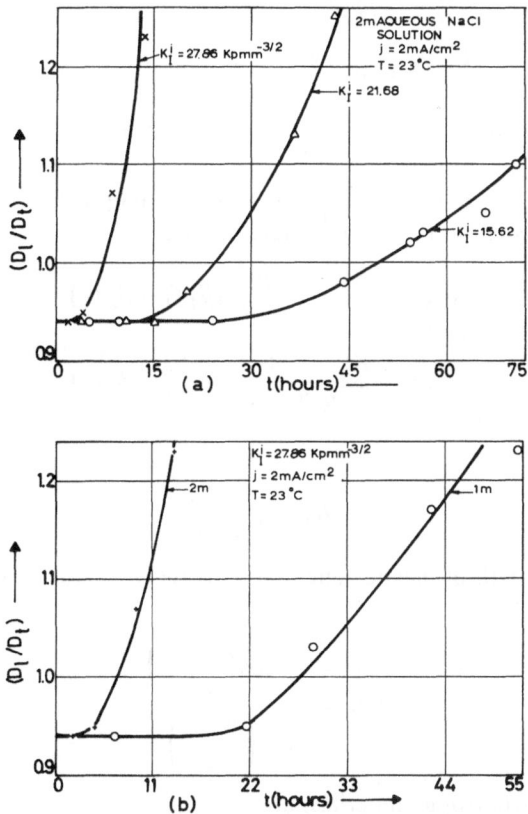

Fig. 54. Variation of the ratio (D_l/D_t) of the longitudinal (D_l) to the transverse (D_t) diameter of the primary caustic versus time t for (a) $K_1^i = 15.62$, 21.68 and 27.86 kp/mm$^{\frac{3}{2}}$ and $M = 2$ m, and (b) $M = 1$ m and 2 m for $K_1^i = 27.86$ kp/mm$^{\frac{3}{2}}$.

Another important role played by the applied stress is stress-assisted diffusion (Chap. 2.10) because the experimental caustics are elongated, as are those predicted by the influence of stress-assisted diffusion.

2.12 Influence of Orthotropy of Ductile Materials on the Caustics

The optical method of caustics was extended to determine experimentally the stress intensity factor around the crack tip in orthotropic plates in the presence of a Dugdale–Barenblatt (D–B) plastic zone [146].

According to the D–B model [17, 158, 159] (i) yielding occurs in a narrow wedge-shaped zone of a length R, (ii) the material in the plastic zone is assumed to be under uniform tensile stress equal to the yield stress σ_0 of the material under simple tension; (iii) the material outside the line plastic-zone is considered as elastic and (iv) a Tresca yield criterion is obeyed.

For orthotropic material with two mutually orthogonal axes of elastic symmetry in the planes, we have [160, 161]:

$$\varepsilon_{zz} = -S_{31}\sigma_{xx} - S_{32}\sigma_{yy} = -S_{13}\sigma_{xx} - S_{23}\sigma_{yy} \tag{529}$$

with:

$$S_{13} = -\frac{v_{13}}{E_1} = -\frac{v_{31}}{E_3}, \quad S_{23} = -\frac{v_{23}}{E_2} = -\frac{v_{32}}{E_3} \tag{530}$$

The stresses σ_{xx}, σ_{yy} for the orthotropic material and for the D–B model, relative to the physical system in polar coordinates (r, θ), are:

$$\sigma_{xx} = \frac{K_I^p}{\sqrt{8\pi R}} \frac{\beta_1 \beta_2}{\beta_1 - \beta_2} [\beta_1 f_1 - \beta_2 f_2] \tag{531}$$

$$\sigma_{yy} = \frac{K_I^p}{\sqrt{8\pi R}} \frac{1}{\beta_1 - \beta_2} [\beta_1 f_2 - \beta_2 f_1] \tag{531}$$

with:

$$K_I^p = \sigma_0 \sqrt{\frac{8R}{\pi}} \tag{532}$$

where R is the length of plastic zone and σ_0 is the yield stress. The plastic zone R is given by:

$$R = \frac{1 - \cos\left(\dfrac{\pi}{2}\dfrac{\sigma_\infty}{\sigma_0}\right)}{\cos\left(\dfrac{\pi}{2}\dfrac{\sigma_\infty}{\sigma_0}\right)} \tag{533}$$

and β_1, β_2 for orthotropic materials, are:

$$\beta_{1,2} = \frac{1}{2}\left[\left(\frac{E_1}{E_2}\right) + 2\left(\frac{E_1}{E_2}\right)^{\frac{1}{4}} + 1\right]^{\frac{1}{2}} \pm \left[\left(\frac{E_1}{E_2}\right) - 2\left(\frac{E_1}{E_2}\right)^{\frac{1}{4}} + 1\right]^{\frac{1}{2}} \tag{534}$$

The functions f_1, f_2 are:

$$f_j = \tan^{-1}\left(\frac{2\sqrt{A}\cos(B)}{1-A}\right), \quad j = 1,2 \tag{535}$$

with:

$$A = \frac{R}{[(r\cos\theta - c)^2 + \beta_j^2 r^2 \sin^2\theta]^{\frac{1}{2}}} \tag{536}$$

$$B = \frac{1}{2}\tan^{-1}\left(\frac{\beta_j r \sin\theta}{r\cos\theta - c}\right) \tag{537}$$

where:

$$c = \begin{cases} 0, & \text{at plastic zone-tip} \\ R, & \text{at crack-tip} \end{cases}$$

The parametric equations of the front-face caustic are:

$$x_f = \lambda_m x - z_0 d \frac{\partial \varepsilon_{zz}}{\partial x}$$

$$y_f = \lambda_m y - z_0 d \frac{\partial \varepsilon_{zz}}{\partial y} \tag{538}$$

with:

$$x = r\cos\theta, \quad y = r\sin\theta$$

By substituting Eq. (529) into Eq. (538), we have:

$$x_f = \lambda_m r \cos\theta + z_0 d \frac{\partial}{\partial x}\left(\frac{v_{13}}{E_1}\sigma_{xx} + \frac{v_{23}}{E_2}\sigma_{yy}\right)$$

$$y_f = \lambda_m r \sin\theta + z_0 d \frac{\partial}{\partial y}\left(\frac{v_{13}}{E_1}\sigma_{xx} + \frac{v_{23}}{E_2}\sigma_{yy}\right) \tag{539}$$

By substituting Eqs (531) and (535) into Eqs (539), we have:

$$x_f/\lambda_m = r\cos\theta + C_f\left[R\cos\theta - \frac{1}{r}\Theta\sin\theta\right]$$

$$y_f/\lambda_m = r\sin\theta + C_f\left[R\sin\theta + \frac{1}{r}\Theta\cos\theta\right] \tag{540}$$

where:

$$C_f = \frac{z_0 d K_I^p}{\sqrt{8\pi R}\,\lambda_m(\beta_1 - \beta_2)} \tag{541}$$

$$R = \frac{v_{13}}{E_1}\beta_1\beta_2(\beta_1 \dot{f}_1^{(r)} - \beta_2 \dot{f}_2^{(r)}) + \frac{v_{23}}{E_2}(\beta_1 \dot{f}_2^{(r)} - \beta_2 \dot{f}_1^{(r)}) \tag{542}$$

$$\Theta = \frac{v_{13}}{E_1}\beta_1\beta_2(\beta_1 \dot{f}_1^{(\theta)} - \beta_2 \dot{f}_2^{(\theta)}) + \frac{v_{23}}{E_2}(\beta_1 \dot{f}_2^{(\theta)} - \beta_2 \dot{f}_1^{(\theta)}) \tag{543}$$

where $\dot{f}_{1,2}^{(r)}$ are the first derivative of $f_{1,2}$ with respect to r and $\dot{f}_{1,2}^{(\theta)}$ are the first derivative of $f_{1,2}$ with respect to θ.

The initial curve, r, of the caustic is derived by zeroing the Jacobian determinant:

$$J = \frac{\partial(x_f, y_f)}{\partial(r, \theta)} = 0 \tag{544}$$

Therefore, the equation of the initial curve of the caustic is:

$$r + C_f r \dot{R}^{(r)} + \frac{1}{r} C_f^2 \left(\frac{1}{r} \Theta - \dot{\Theta}^{(r)} \right) \left(\dot{R}^{(\theta)} - \frac{1}{r} \Theta \right) + C_f \left(R + \frac{1}{r} \dot{\Theta}^{(\theta)} \right)$$

$$+ C_f^2 \dot{R}^{(r)} \left(R + \frac{1}{r} \dot{\Theta}^{(\theta)} \right) = 0 \tag{545}$$

Relation (545) is the equation of the initial curve of the caustic.

From the parametric Eqs (540) we get the caustic at the plastic zone-tip for $c = 0$ (non-singular stress field) and the caustic at the crack tip for $c = R$ (singular stress field).

Chapter 3
The Optical Method of Dynamic Caustics

3.1 General Aspects

Many static crack problems were solved by the experimental method of reflected caustics. The application of static caustics to dynamic crack problems yielded satisfactory results at low crack propagation velocities, but not when the cracks propagated with high velocities. In all experimental studies of crack propagation based on caustics, existence of a static solution for the stress field in the vicinity of the crack tip is assumed to be valid also for the dynamic case, as are assumptions of the approximate invariability of the mechanical and optical properties of the material used in the tests.

Since it was established that the mechanical and optical properties of the materials used in the tests changed considerably with the rate of application of external loading or propagation of the crack, a basic improvement of the method was the introduction not only of some standard dynamic values for the moduli and optical constants, but, more accurately, their exact values found experimentally for the specific dynamic conditions of each particular test [60, 61, 162, 163].

The dynamic stress intensity factor K_I^d and the evaluation of the velocity c of propagation of a transverse crack under mode-I in specimens made of optically inert material (PMMA) and the influence of the propagation velocity c on the shape and size of the caustics formed by reflection were studied by Theocaris and Papadopoulos [164–168].

3.2 Dynamic Stress Field Around the Crack Tip

We consider an elastic and isotropic medium subjected to a uniaxial dynamic load at infinity and containing a propagating crack of initial length $2a$ with a constant velocity c ($c < c_2$, where c_2 is the shear wave velocity) along the Ox-axis (Fig. 55). After a period of transition and when a steady motion of the crack has been established, a steady-state stress field develops around the crack tip and transient effects may be omitted.

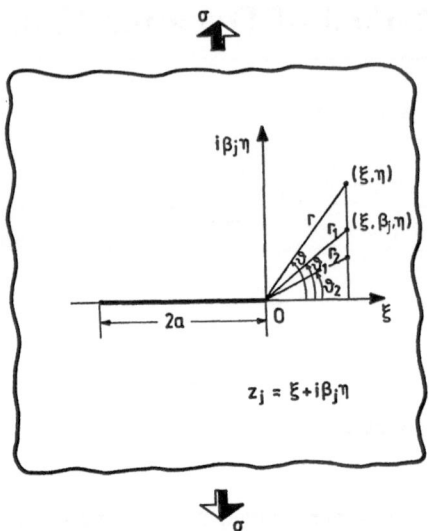

Fig. 55. Geometry of the specimens; the $O\xi\eta$-coordinate system moves with the crack.

A moving coordinate system, $O\xi\eta$, which is attached to the moving crack, is introduced with $\xi = x - ct$ and $\eta = y$, where the Oxy-system is a fixed coordinate system. The stress field at the crack tip is given by [169, 170]:

$$\sigma_\xi = 2\mu[(2\beta_1^2 - \beta_2^2 + 1)\operatorname{Re}\Omega_1'(z_1) - 2\beta_2\operatorname{Re}\Omega_2'(z_2)] \tag{546}$$

$$\sigma_\eta = -2\mu[(1 + \beta_2^2)\operatorname{Re}\Omega_1'(z_1) - 2\beta_2\operatorname{Re}\Omega_2'(z_2)] \tag{547}$$

$$\tau_{\xi\eta} = -2\mu[2\beta_1\operatorname{Im}\Omega_1'(z_1) - (1 + \beta_2^2)\operatorname{Im}\Omega_2'(z_2)] \tag{548}$$

where μ is the shear modulus and:

$$\beta_j = (1 - c^2/c_j^2)^{\frac{1}{2}}, \quad j = 1, 2 \tag{549}$$

In these relations c_j are the velocities of the longitudinal, c_1, and the shear, c_2, waves which are given by:

$$c_1 = \left(\frac{E}{\rho(1 - v^2)}\right)^{\frac{1}{2}} = \left(\frac{2}{1 - v}\right)^{\frac{1}{2}} c_2, \qquad \text{for plane stress}$$

$$c_1 = \left(\frac{E(1 - v)}{\rho(1 + v)(1 - 2v)}\right)^{\frac{1}{2}} = \left(\frac{2(1 - v)}{(1 - 2v)}\right)^{\frac{1}{2}} c_2, \quad \text{for plane strain} \tag{550}$$

$$c_2 = \left(\frac{E}{2\rho(1 + v)}\right)^{\frac{1}{2}} \tag{551}$$

where E is the modulus of elasticity of the material, v is Poisson's ratio and ρ is the mass density of the material.

The functions $\Omega_j(z_j)$ and $\Omega_j'(z_j)$ are sectionally analytic functions of the complex variable $z_j = \xi + i\beta_j\eta$. The function $\Omega_j'(z_j)$ for $z_j = \zeta_j + a$ is given by:

$$\Omega_j'(\zeta_j) = T_{0_j}\left(1 - \frac{\zeta_j + a}{[\zeta_j(\zeta_j + 2a)]^{\frac{1}{2}}}\right) \tag{552}$$

where:

$$T_{01} = \frac{-\sigma(1 + \beta_2^2)}{2\mu R(c)}, \quad T_{02} = \frac{-\sigma\beta_1}{\mu R(c)} \qquad (553)$$

with:

$$R(c) = 4\beta_1\beta_2 - (1 + \beta_2^2)^2 \qquad (554)$$

and σ is the applied stress at infinity of the plate. By retaining the singular and the constant term of the Taylor series expansion of the function (552), it may be derived that:

$$\Omega'_j(\zeta_j) \simeq T_{0j} - T_{0j}\frac{a}{\sqrt{2a\zeta_j}} \qquad (555)$$

and from Rel. (555) is obtained:

$$\mathrm{Re}\,\Omega'_j(\zeta_j) = -T_{0j}\sqrt{\frac{a}{2r_j}}\cos\left(\frac{\theta_j}{2}\right) + T_{0j} \qquad (556)$$

$$\mathrm{Im}\,\Omega'_j(\zeta_j) = T_{0j}\sqrt{\frac{a}{2r_j}}\sin\left(\frac{\theta_j}{2}\right) \qquad (557)$$

Using now the transformation:

$$\tan\theta_j = \beta_j\tan\theta$$
$$r_j = r(\cos^2\theta + \beta_j^2\sin^2\theta)^{\frac{1}{2}} \qquad (558)$$

relations (546), (547) and (548) are written as follows:

$$\sigma_\xi = \frac{k_1}{\sqrt{rR(c)}}[(1 + \beta_2^2)(2\beta_1^2 - \beta_2^2 + 1)F_1 - 4\beta_1\beta_2F_2]$$
$$-\frac{k_1}{\sqrt{a/2\,R(c)}}[(1 + \beta_2^2)(2\beta_1^2 - \beta_2^2 + 1) - 4\beta_1\beta_2] \qquad (559)$$

$$\sigma_\eta = \frac{k_1}{\sqrt{rR(c)}}[4\beta_1\beta_2F_2 - (1 + \beta_2^2)^2F_1] - \frac{k_1}{\sqrt{a/2\,R(c)}}[4\beta_1\beta_2 - (1 + \beta_2^2)^2] \qquad (560)$$

$$\tau_{\xi\eta} = \frac{2k_1\beta_1(1 + \beta_2^2)}{\sqrt{rR(c)}}[G_2 - G_1] \qquad (561)$$

with:

$$F_j = \left(\frac{1}{2}\frac{(\cos^2\theta + \beta_j^2\sin^2\theta)^{\frac{1}{2}} + \cos\theta}{\cos^2\theta + \beta_j^2\sin^2\theta}\right)^{\frac{1}{2}} \qquad (562)$$

$$G_j = \left(\frac{1}{2}\frac{(\cos^2\theta + \beta_j^2\sin^2\theta)^{\frac{1}{2}} - \cos\theta}{\cos^2\theta + \beta_j^2\sin^2\theta}\right)^{\frac{1}{2}} \qquad (563)$$

and:

$$k_1 = \sigma\sqrt{a/2} \qquad (564)$$

3.3 Parametric Equations of the Caustics

A parallel convergent light beam impinging in the vicinity of the tip of a cracked plate and reflected from the front (f) or rear (r) face of the specimen or transmitted (t) through the specimen, is received on a reference plane, placed at a distance z_0 from the specimen (Fig. 56). On this reference screen a caustic is formed by reflection of the light rays from the face of the plate, whose parametric equations in a reference $O\xi\eta$-coordinate system, moving together with the crack tip, and corresponding with another $O\Xi H$-reference frame on the reference screen are given by [34, 164]:

$$\Xi_{r,t,f} = \lambda_m \xi + \varepsilon z_0 dc_{r,t,f} \frac{\partial(\sigma_{xx} + \sigma_{yy})}{\partial \xi}$$

$$H_{r,t,f} = \lambda_m \eta + \varepsilon z_0 dc_{r,t,f} \frac{\partial(\sigma_{xx} + \sigma_{yy})}{\partial \eta} \tag{565}$$

where d is the thickness of the specimen, λ_m is the magnification ratio, $c_{r,t,f}$ are the stress optical constants and the complex coordinate z_1 is:

$$z_1 = \xi + i\beta_1 \eta = r_1 e^{i\theta_1} = r_1(\cos\theta_1 + i\sin\theta_1) \tag{566}$$

By Using the Relation [115, 277]

$$(\sigma_{xx} + \sigma_{yy}) = -2(\beta_1^2 - \beta_2^2)\,\mathrm{Re}\, F_1''(z_1) = 2\,\mathrm{Re}\,\Omega(z_1) \tag{567}$$

From Rels (565), the equation of the caustic is given by:

$$Z = z + C_{r,t,f}^*[(1 - \beta_1)\Omega'(z_1) + (1 + \beta_1)\overline{\Omega'(z_1)}] \tag{568}$$

with:

$$Z = (\Xi + iH)/\lambda_m \quad \text{and} \quad z = \xi + i\eta \tag{569}$$

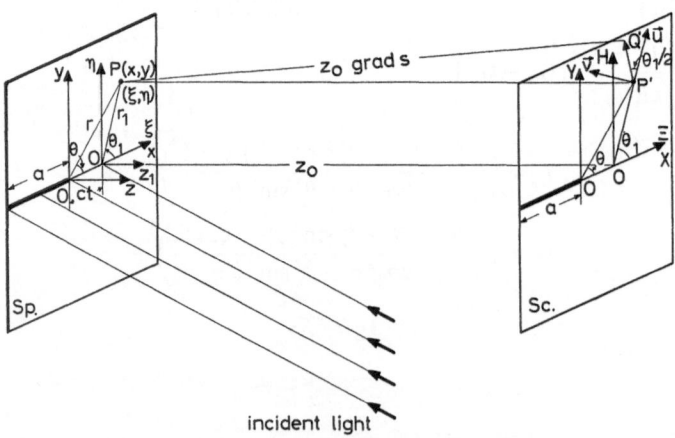

Fig. 56. The optical set-up for the method of reflected caustics. Relative position of specimen and reference screen.

and the overall constant $C^*_{r,t,f}$ expressed by:

$$C^*_{r,t,f} = \frac{\varepsilon z_0 \, dc_{r,t,f}}{\lambda_m} \qquad (570)$$

Relation (568) represents the deviations of the light rays reflected from the specimen and impinging on the reference screen, and establishes the correspondence between the points (ξ, η) of the specimen and the corresponding points (Ξ, H) on the screen. If these deviated rays form a caustic curve on the reference screen, this curve is expressed also by Rel. (568) with the constraint that points (ξ, η) of the specimen cause the vanishing of the Jacobian determinant [34]:

$$J = \frac{\partial(\Xi, H)}{\partial(\xi, \eta)} = 0 \qquad (571)$$

By taking into account Rel. (568) and the partial derivative:

$$\frac{\partial}{\partial \xi} = \frac{\partial}{\partial z_1} + \frac{\partial}{\partial \bar{z}_1}, \quad \frac{\partial}{\partial \eta} = i\beta_1 \left(\frac{\partial}{\partial z_1} - \frac{\partial}{\partial \bar{z}_1} \right) \qquad (572)$$

we derive from Rel. (571) that:

$$[2\beta_1 C^*_{r,t,f} |\Omega''(z_1)|]^2 = 1 + 2C^*_{r,t,f}(1 - \beta_1^2)\operatorname{Re}\Omega''(z_1) \qquad (573)$$

Relation (573) is the equation of the initial curve of the caustic formed on the screen.

For a specimen made of an optically inert material whose dynamic, mechanical and optical properties are known for the relevant velocity c of propagation of the crack, Rels (568) and (573) allow the complete theoretical determination of the shape of the caustic formed on the reference screen provided that the complex potential $\Omega(z_1)$ is known. It is worthwhile remarking that Eqs (568) and (573) exhibit similarities with the corresponding equations for the case of a stationary crack in an orthotropic medium [161].

The function $\Omega(z_1)$ is a function of the complex variable z_1, according to Cragg's model [171, 172], given by:

$$\Omega(z_1) = -(\beta_1^2 - \beta_2^2)F_1''(z_1) = (2\pi)^{-\frac{1}{2}}K_1^d z_1^{-\frac{1}{2}} \qquad (573a)$$

where K_1^d expresses the dynamic stress intensity factor and the complex coordinate z_1 is referred to the $O\xi\eta$-system.

From Rel. (573a) together with Eqs (568), (569) and (573) we obtain the parametric equations of the caustic as well as the equation of its initial curve. These relations are given by:

$$\Xi_{r,t,f}/\lambda_m = r_1 \cos\theta_1 + \mu_{r,t,f} r_1^{-\frac{1}{2}} \cos\frac{3\theta_1}{2}$$

$$H_{r,t,f}/\lambda_m = \frac{r_1}{\beta_1}\sin\theta_1 + \mu_{r,t,f}\beta_1 r_1^{-\frac{1}{2}}\sin\frac{3\theta_1}{2} \qquad (574)$$

and:

$$r_1^5 - \frac{9}{4}\mu_{r,t,f}^2\beta_1^2 + \frac{3}{2}\mu_{r,t,f}(\beta_1^2 - 1)r_1^{\frac{3}{2}}\cos\frac{5\theta_1}{2} = 0 \qquad (575)$$

with:

$$\mu_{r,t,f} = -\frac{\varepsilon z_0 \, dc_{r,t,f} K_1^d}{\lambda_m (2\pi)^{\frac{1}{2}}} \qquad (576)$$

and:

$$K_1^d = \frac{(\beta_1^2 - \beta_2^2)(1 + \beta_2^2)}{R(c)} k_1 \tag{577}$$

In Rels (574) r_1 represents a positive real solution of Eq. (575) expressing the initial curve of the caustic. This solution is given by:

$$r_1 = (\tfrac{3}{2}\mu_{r,t,f} R)^{\tfrac{2}{3}} \tag{578}$$

where:

$$R = \frac{1}{2}(1 - \beta_1^2)\cos\frac{5\theta_1}{2} + \frac{1}{2}\left[(1 - \beta_1^2)^2 \cos^2\frac{5\theta_1}{2} + 4\beta_1^2 \right]^{\tfrac{1}{2}} \tag{579}$$

Relations (574), by introducing Rel. (578) become:

$$\Xi_{r,t,f}/\lambda_m = \left(\frac{3}{2}\mu_{r,t,f}\right)^{\tfrac{2}{3}}\left[R^{\tfrac{2}{3}}\cos\theta_1 + \frac{2}{3}R^{-\tfrac{1}{3}}\cos\frac{3\theta_1}{2} \right]$$

$$H_{r,t,f}/\lambda_m = \left(\frac{3}{2}\mu_{r,t,f}\right)^{\tfrac{2}{3}}\left[\frac{1}{\beta_1}R^{\tfrac{2}{3}}\sin\theta_1 + \frac{2}{3}\beta_1 R^{-\tfrac{1}{3}}\sin\frac{3\theta_1}{2} \right] \tag{580}$$

Relations (580) for values of angle θ_1 and the coefficient β_1 limited by:

$$-\pi \leqq \theta_1 \leqq \pi \quad \text{and} \quad 0 < \beta_1 \leqq 1$$

yield caustics which surround the tips of cracks propagating with velocities c which may be compared to the velocities c_1 of the longitudinal waves of the material of the specimen ($c < c_2 \simeq 0.6c_1$). For β_1, which corresponds to $c = 0$, the caustics formed correspond to those obtained for stationary cracks.

Figure 57 presents the caustics formed by reflections from the front (C_f) and rear (C_r) faces of the specimens for propagating cracks with relative velocities $c/c_1 = 0$, 0.2, 0.4, 0.6 and 0.8. In the same figure the respective initial curves (I_f, I_r) derived from Eqs (578) and (580) are also plotted by the computer.

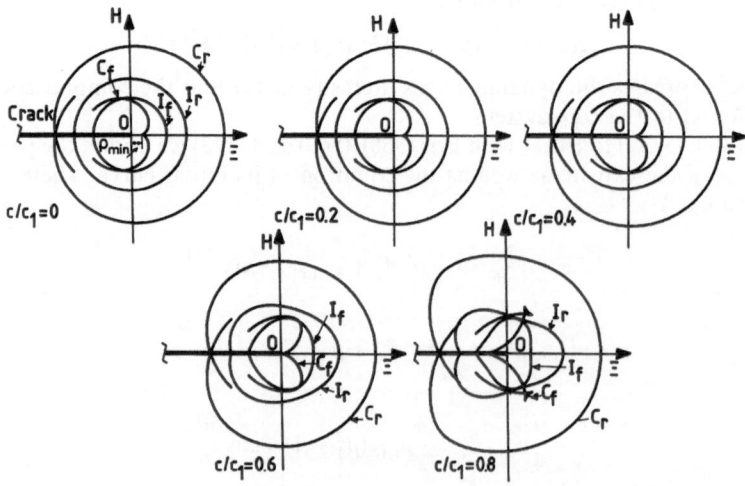

Fig. 57. Initial curves and respective caustics from reflected light rays on a cracked plate made of an isotropic elastic material for relative crack velocities $c/c_1 = 0$, 0.2, 0.4, 0.6 and 0.8 as plotted by computer, for $-\pi < \theta < +\pi$.

It can be derived from Rel. (576) that the factor $\mu_{r,t,f}$ is proportional to the optical constant $c_{r,t,f}$ which is inversely proportional to the modulus of elasticity E of the material [163]. Therefore, the factor $\mu_{r,t,f}$ is inversely proportional to the modulus of elasticity E, that is:

$$\mu_{r,t,f} = \mu_{r,t,f}\left(\frac{1}{E}\right) \tag{581}$$

In cases of dynamic loading of plates, a large increase of E is observed and a relatively small increase of Poisson's ratio v [163]. This means that in dynamic loading of cracked plates the factor $\mu_{r,t,f}$ is decreasing since E^d is always increasing. Therefore, the static value $\mu_{r,t,f}^s$ is always larger than the corresponding dynamic value $\mu_{r,t,f}^d$, i.e. it is always valid that:

$$\mu_{r,t,f}^s > \mu_{r,t,f}^d \tag{582}$$

Therefore, dynamic caustics are always smaller than their respective static ones [162] and for constant propagating velocities of the cracks the caustics formed shrink as the respective c increases. Figure 58 presents the variation in shape, size and position of the initial curves (a, c) and the caustics (b, d) derived from reflections from either the rear face (a, b), or the front face (c, d) as the relative velocity c/c_1 is increased and approaching unity.

If we plot the initial curves and caustics for different values of the relative velocity c/c_1 and varying values of the factors $\mu_{r,t,f}^d$ we can state that if we change only the values of $\mu_{r,t,f}^d$, the size of the respective initial curves and caustics is reduced considerably. Therefore, for E^d increasing, the size of the caustics is reduced considerably (figure not shown). On the other hand, if E^d remains constant we can derive from Fig. 58 that if the relative velocity c/c_1 increases, the transverse diameter

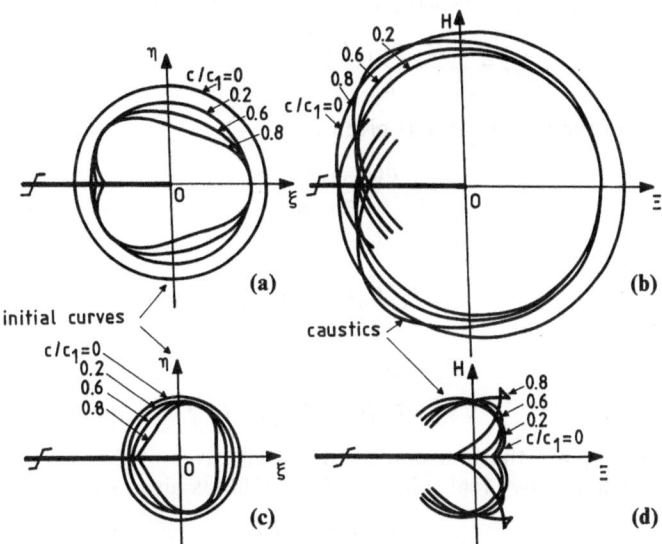

Fig. 58. The relative positions of initial curves and the respective external caustics for $\mu_r = 8$ (a, b) or internal caustics for $\mu_f = -2.8$ (c, d).

D_t of the caustic increases, whereas the longitudinal diameter D_l is reduced and the caustic becomes more and more elongated. Furthermore, the distance between the cusp of the internal caustic (C_f) and the crack tip is reduced. This is indicated in Figs 57 and 58 where $\mu_{r,t,f}^d$ was kept constant. Therefore an increase of E^d, for constant relative velocity c/c_1 of the crack, has as result the reduction of the size of the caustic only, while for E^d constant an increase of the relative velocity c/c_1 results in a variation of the size of the caustic and a displacement of the internal caustic towards the crack tip. This causes an increase of the eccentricity of the internal caustic relative to the external caustic which is insensitive to variation of c/c_1.

By measuring the various characteristic quantities of the specimen and the optical set-up we have found that [163]:

$$\mu_f^d/\mu_r^d = -0.35, \quad \mu_f^s/\mu_r^s = -0.29$$

so that:

$$\mu_r^s/\mu_r^d = 1.50 \tag{583}$$

Figures 57 and 58 were plotted by taking into consideration Rel. (583), so that the relative size of the two groups of caustics from the front and rear faces were designed on this basis.

By Using the Stress Field (559)–(561) (in the Physical Plane (r,θ)) or Using the Transformation (558) for the Polar Coordinates (r_1,θ_1)

$$\tan \theta_1 = \beta_1 \tan \theta \tag{584}$$

$$r_1 = r(\cos^2 \theta + \beta_1^2 \sin^2 \theta)^{\frac{1}{2}}$$

The parametric equations of caustics (574) in the physical plane are given by:

$$\Xi_{r,t,f}/\lambda_m = r \cos \theta + C_{r,t,f}^d r^{-\frac{3}{2}}(F_1 \cos \theta + 2F_1' \sin \theta) \tag{585}$$

$$H_{r,t,f}/\lambda_m = r \sin \theta + C_{r,t,f}^d r^{-\frac{3}{2}}(F_1 \sin \theta - 2F_1' \cos \theta) \tag{586}$$

$$C_{r,t,f}^d = -\frac{\varepsilon z_0 d c_{r,t,f} K_I^d}{\lambda_m}$$

The equation of the initial curve is given by [164]:

$$J = \frac{\partial(\Xi_{r,t,f}, H_{r,t,f})}{\partial(r,\theta)} = 0 \tag{587}$$

which becomes:

$$r^5 + \frac{3}{2}C_{r,t,f}^d R_0 r^{\frac{5}{2}} - \frac{3}{2}(C_{r,t,f}^d)^2 S_0 = 0 \tag{588}$$

with:

$$R_0 = \frac{2}{3}(F_1 - 2F_1'') - F_1 \tag{589}$$

$$S_0 = F_1(F_1 - 2F_1'') + 6F_1'^2 \tag{590}$$

where F_1' and F_1'' are the first and second derivatives of the function (562) with respect to θ. The positive real solution of Eq. (588) is given by:

$$r = (\tfrac{3}{2}C_{r,t,f}^d R)^{\frac{2}{5}} \tag{591}$$

where:

$$R = -\tfrac{1}{2}R_0 + \tfrac{1}{2}(R_0^2 + \tfrac{8}{3}S_0)^{\frac{1}{2}}$$

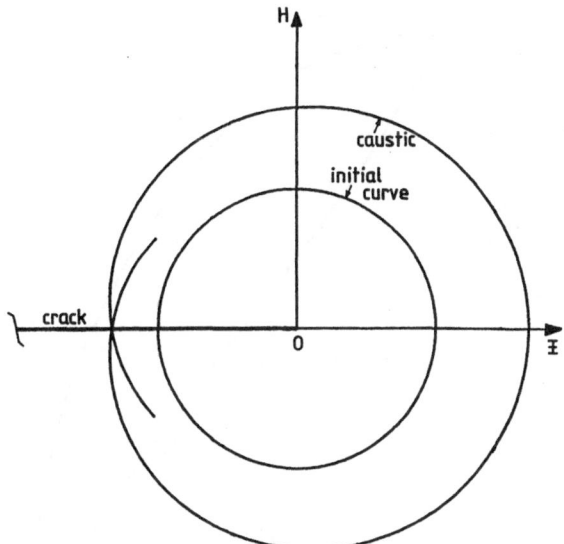

Fig. 59. The caustic and the initial curve for $v = 0.34$ and $c/c_2 = 0.2$.

Then, the parametric equations of the caustics, (585) and (586), become:

$$\Xi_{r,t,f}/\lambda_m = (\tfrac{3}{2}C^d_{r,t,f})^{\tfrac{1}{3}}[R^{\tfrac{2}{3}}\cos\theta + \tfrac{2}{3}R^{-\tfrac{1}{3}}(F_1\cos\theta + 2F'_1\sin\theta)]$$

$$H_{r,t,f}/\lambda_m = (\tfrac{3}{2}C^d_{r,t,f})^{\tfrac{1}{3}}[R^{\tfrac{2}{3}}\sin\theta + \tfrac{2}{3}R^{-\tfrac{1}{3}}(F_1\sin\theta - 2F'_1\cos\theta)]$$

(592)

Figure 59 presents the caustic formed from light rays transmitted through the specimen for a propagating crack with relative velocity $c/c_2 = 0.2$ (in the physical plane).

The application of the optical method of reflected caustics yields all the necessary information for the complete solution of the problem of the mode of propagation of cracks. Indeed, the size of caustic (external) yields the value of K^d_I, while the relative position of the internal caustic with respect to the external one yields the possibility to determine the exact instantaneous position of the crack tip and from this the velocity of propagation of the crack.

From Figs 57 and 58 we can deduce that if the relative velocity of propagation of the crack is increasing, the relative position of the two caustics defined by the distance ρ_{min} of the cusp of the internal caustic from the crack tip is diminishing. The variation of ρ_{min} in terms of the diameter D_l^{max} is presented in Fig. 60 with respect to either the relative velocity c/c_1, or the diameter ratio D_t^{max}/D_l^{max}. We can observe that the ratio ρ_{min}/D_l^{max} takes values for large velocities approaching the longitudinal stress wave velocity c_1, which means that in these cases the cusp of the internal caustic is behind the crack tip. However, it can be deduced from this figure that this variation of ρ_{min} is always small and therefore it is a good approximation to accept always that the crack tip and the cusp of the internal caustic coincide, although this is strictly true only when $c = c_2 \simeq 0.6c_1$. The error introduced by this assumption in the estimation of c/c_1 does not exceed 1%.

Any increase of c/c_1 results in a change of shape of the caustics, as shown in Figs 57, 58 and 62, by increasing the transverse diameter of the external caustic and decreasing the longitudinal diameter.

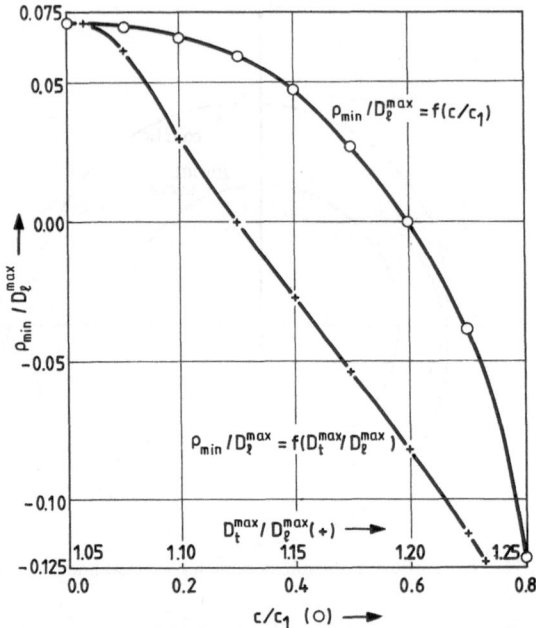

Fig. 60. The variation of the ratio of the minimum distance between the crack tip and the cusp of the internal caustics ($\rho_{\min}$) and the maximum longitudinal diameter $D_l^{\max}$ versus either the ratio of the maximum diameters $D_t^{\max}/D_l^{\max}(+)$, or the relative crack velocity $c/c_1(\circ)$

3.4 Dynamic Stress Intensity Factors K_I^d and K_{II}^d

In order to evaluate the dynamic stress intensity factor K_I^d, the evaluation of the maximum diameters of the caustic is needed. The second of Rels (580) presents extrema which may be found by zeroing the derivative $dH_{r,t,f}/d\theta_1$. Thus, we have [164]:

$$dH_{r,t,f}/d\theta_1 = \frac{2R^{-\frac{3}{5}}}{5\beta_1}R^* \sin \theta_1 + \frac{R^{\frac{2}{5}}}{\beta_1}\cos \theta_1 - \frac{2\beta_1 R^{-\frac{3}{5}}}{5}R^* \sin \frac{3\theta_1}{2} + \beta_1 R^{-\frac{3}{5}}\cos \frac{3\theta_1}{2} = 0 \tag{593}$$

with:

$$R^* = -\frac{5}{4}(1 - \beta_1^2)\sin \frac{5\theta_1}{2} - \frac{5}{8}(1 - \beta_1^2)^2 \sin 5\theta_1 \left[(1 - \beta_1^2)^2 \cos^2 \frac{5\theta_1}{2} + 4\beta_1^2 \right]^{-\frac{1}{2}} \tag{594}$$

From the solutions of Rel. (593) for $0 \leq \beta_1 \leq 1$ we can determine the positions of $\theta_{1t}^{\max}$ corresponding to maxima of the second of Rels (580). Figure 61 yields the values for $\theta_{1t}^{\max}$ in terms of the relative velocity c/c_1.

The second of Rels (580) may be put in the form:

$$D_t^{\max} = 2H_{r,t,f}^{\max} = \lambda_m (\tfrac{3}{2}\mu_{r,t,f})^{\frac{3}{2}} \delta_t^{\max}(c/c_1) \tag{595}$$

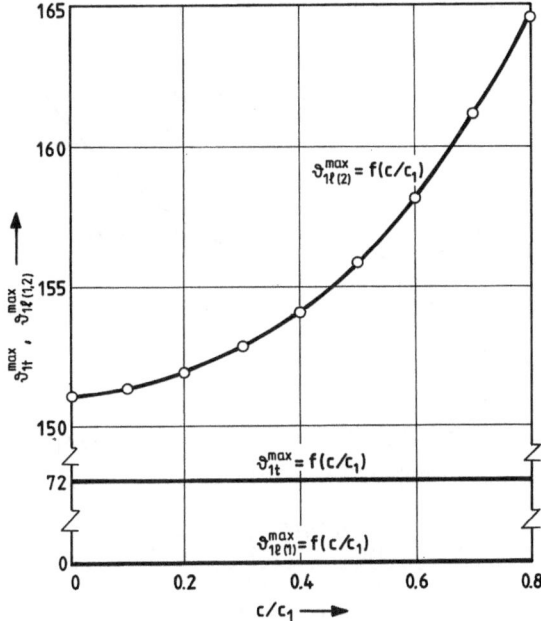

Fig. 61. The variation of the angle $\theta_{1t}^{\max}$ corresponding to a maximum of the external caustic ($D_t^{\max}$) and the angles $\theta_{1l(1,2)}^{\max}$ corresponding to maxima of the longitudinal diameter, versus the relative crack velocity c/c_1.

where $D_t^{\max}$ denotes the maximum transverse diameter of the caustic and $\delta_t^{\max}$ a correction factor given by:

$$\delta_t^{\max} = 2\left(\frac{1}{\beta_1} R^{\frac{2}{3}} \sin\theta_{1t}^{\max} + \frac{2}{3}\beta_1 R^{-\frac{2}{3}} \sin\frac{3}{2}\theta_{1t}^{\max}\right) \tag{596}$$

In the positions where $\theta_{1t}^{\max}$ exists, Rel. (596) takes its maximum values which when introduced into Eq. (595) yield the maximum diameter of the caustic.

Figure 62 presents the maximum values for the correction factors $\delta_t^{\max}$ and $\delta_t^{\max}$ as functions of the relative velocity c/c_1, as well as of the ratio of maximum transverse and longitudinal diameters of the caustic $D_t^{\max}/D_l^{\max}$. In order to determine the longitudinal diameter of the caustic $D_l^{\max}$ (which is always a maximum for the diameters) the angles $\theta_{1l(1,2)}^{\max}$ are determined for which the second of Rels (580) becomes zero. For these values, the first of Rels (580) becomes maximum and yields:

$$D_l^{\max} = \Xi_{r,t,f(1)}^{\max} + \Xi_{r,t,f(2)}^{\max} = \lambda_m (\tfrac{3}{2}\mu_{r,t,f})^{\frac{2}{3}} \delta_l^{\max}(c/c_1) \tag{597}$$

where $D_l^{\max}$ is the longitudinal maximum diameter of the caustic and $\delta_l^{\max}(c/c_1)$ a correction factor expressed by:

$$\delta_l^{\max}(c/c_1) = (R_1^{\frac{2}{3}}\cos\theta_{1l(1)}^{\max} + \tfrac{2}{3}R_1^{-\frac{2}{3}}\cos\tfrac{3}{2}\theta_{1l(1)}^{\max}) + (R_2^{\frac{2}{3}}\cos\theta_{1l(2)}^{\max} + \tfrac{2}{3}R_2^{-\frac{2}{3}}\cos\tfrac{3}{2}\theta_{1l(2)}^{\max}) \tag{598}$$

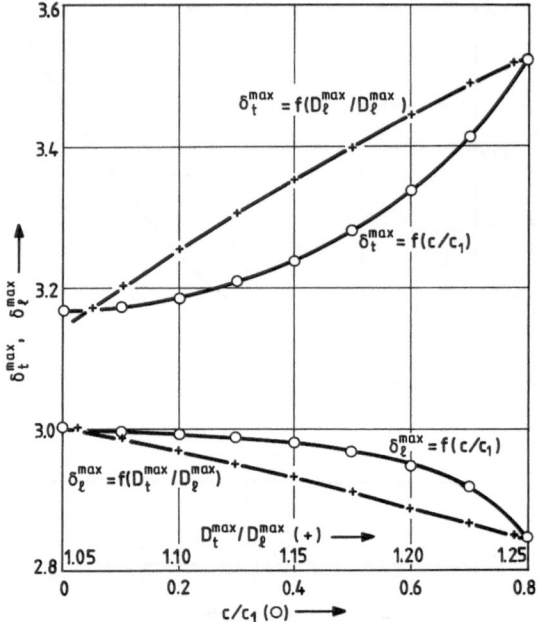

Fig. 62. The variation of the correction factors δ_t^{max} and δ_l^{max} versus either the relative crack velocity c/c_1 or the ratio of the maximum diameters of the caustic D_t^{max}/D_l^{max}.

with:

$$R_{1,2} = \tfrac{1}{2}(1 - \beta_1^2)\cos\tfrac{5}{2}\theta_{1l(1,2)}^{max} + \tfrac{1}{2}((1 - \beta_1^2)^2\cos^2\tfrac{5}{2}\theta_{1l(1,2)}^{max} + 4\beta_1^2)^{\frac{1}{2}} \qquad (599)$$

In these last three relations the indices 1 and 2 indicate the two angles θ_1 and θ_2 for which the external caustic cuts the crack-axis. One of them (θ_1) is always zero. Figure 61 presents the variation of angles $\theta_{1l(1,2)}^{max}$ in terms of the relative velocity c/c_1, whereas Fig. 62 shows the variation of the correction factor δ_l^{max} in terms of c/c_1 and D_t^{max}/D_l^{max}.

From Rels (595) and (597) we derive the ratio of the maximum diameters given by:

$$D_t^{max}/D_l^{max} = \delta_t^{max}/\delta_l^{max} \qquad (600)$$

Therefore, the ratio of the transverse and longitudinal maximum diameters of the external caustic yield the ratio of the respective correction factors. Figure 63 presents the variation of these ratios with respect to the relative velocity c/c_1.

Finally, from Rels (576), (595) and (597) the dynamic stress intensity factor K_l^d may be derived:

$$K_l^d = \frac{2(2\pi)^{\frac{1}{2}}}{3\varepsilon z_0\, d\lambda_m^{\frac{3}{2}} c_{r,t,f}} \left(\frac{D_{t,l}^{max}}{\delta_{t,l}^{max}(c/c_1)}\right)^{\frac{5}{2}} \qquad (601)$$

Relation (601) yields the mode-I dynamic stress intensity factor from the geometric elements of the caustic.

From Fig. 61 it may be derived that $\theta_{1t}^{max} = 72°$ holds for any relative velocity. This means that the position of the maximum for the transverse diameter remains constant, coincides with the position for the static case and is independent of the

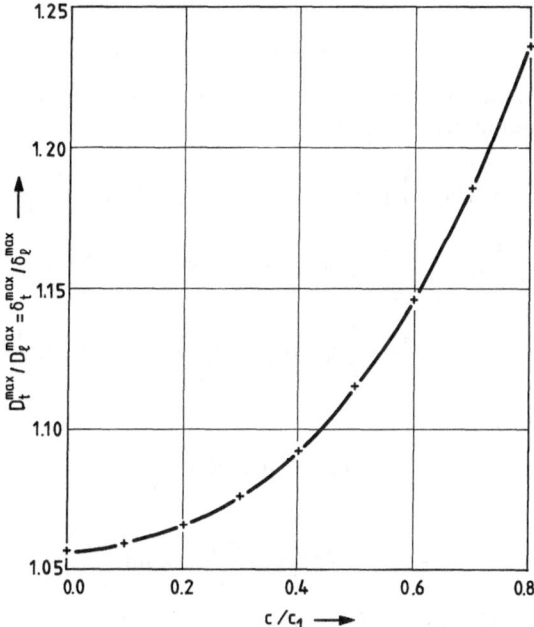

Fig. 63. The variation of the ratio $D_t^{\max}/D_l^{\max}$ of the maximum diameters of the caustics versus the relative crack velocity c/c_1.

relative velocity c/c_1. The same phenomenon occurs also for the internal caustic; $\theta_{1l(1)}^{\max}$ is always 0°. This means that for mode-I deformation the front point of the caustic remains at the same position relative to the crack tip and the dynamic caustic is always symmetric to the crack-axis. This phenomenon does not hold for the rear point of intersection of the caustic with the crack-axis. This point corresponds to angles $\theta_{1l(2)}^{\max}$ which increase as c/c_1 increases and this indicates that the rear point of intersection of caustic and crack-axis approaches the crack tip as c/c_1 increases (Fig. 58(b)).

Finally, from Fig. 63 it may be derived that for small relative velocities of the propagating crack ($c/c_1 \leq 0.4$) the relative variations of the diameters $D_t^{\max}$, $D_l^{\max}$ are rather small, while for crack velocities exceeding the above limit the curve $D_t^{\max}/D_l^{\max} = f(c/c_1)$ becomes steeper. For these velocities not only does the transverse diameter increase but the longitudinal diameter decreases significantly, so that their ratio increases rapidly.

In the case of an inclined crack ($\omega \neq 0°$, Fig. 5), the $\Omega(z_1)$-function is expressed by [165]:

$$\Omega(z_1) = (2\pi)^{-\frac{1}{2}}(K^d)^* z_1^{-\frac{1}{2}} \tag{602}$$

where $(K^d)^*$ denotes the dynamic complex stress intensity factor, given by:

$$(K^d)^* = K_I^d - i\,K_{II}^d \tag{603}$$

Equation (567) in this case becomes:

$$\sigma_{xx} + \sigma_{yy} = 2(2\pi)^{-\frac{1}{2}}K_I^d\left(\cos\frac{\theta_1}{2} - \varkappa\sin\frac{\theta_1}{2}\right) \tag{604}$$

where the factor $\varkappa$ is given by [165, 166]:

$$\varkappa = K_{II}^d / K_I^d \tag{605}$$

The parametric equations of the caustics, and the equation of the initial curve for a mixed-mode loading are given by:

$$\Xi_{r,t,f} = \lambda_m \left(\frac{3}{2} \mu_{r,t,f} \right)^{\frac{3}{2}} \left[R_d^{\frac{2}{3}} \cos \theta_1 + \frac{2}{3} R_d^{-\frac{3}{2}} \left(\cos \frac{3\theta_1}{2} - \varkappa \sin \frac{3\theta_1}{2} \right) \right] \tag{606}$$

$$H_{r,t,f} = \lambda_m \left(\frac{3}{2} \mu_{r,t,f} \right)^{\frac{3}{2}} \left[\frac{1}{\beta_1} R_d^{\frac{2}{3}} \sin \theta_1 + \frac{2}{3} \beta_1 R_d^{-\frac{3}{2}} \left(\sin \frac{3\theta_1}{2} + \varkappa \cos \frac{3\theta_1}{2} \right) \right] \tag{607}$$

$$r_1^{\frac{5}{3}} - \frac{9}{4} \beta_1^2 \mu_{r,t,f}^2 (1 + \varkappa^2) + \frac{3}{2} \mu_{r,t,f} (\beta_1^2 - 1) \left(\cos \frac{5\theta_1}{2} - \varkappa \sin \frac{5\theta_1}{2} \right) r_1^{\frac{5}{3}} = 0 \tag{608}$$

from which we take:

$$r_1 = (\tfrac{3}{2} \mu_{r,t,f} R_d)^{\frac{3}{2}} \tag{609}$$

with:

$$R_d = \frac{1}{2}(1 - \beta_1^2) \left(\cos \frac{5\theta_1}{2} - \varkappa \sin \frac{5\theta_1}{2} \right)$$

$$+ \frac{1}{2} \left[(1 - \beta_1^2)^2 \left(\cos \frac{5\theta_1}{2} - \varkappa \sin \frac{5\theta_1}{2} \right)^2 + 4\beta_1^2 (1 + \varkappa^2) \right]^{\frac{1}{2}}$$

Relations (606) and (607) for $\beta_1 = 1$ ($c = 0$) yield caustics corresponding to the limiting case of stationary cracks. For $\varkappa \neq 0$, an angular displacement of the axis of symmetry of the caustic relative to the crack-axis is observed and the angle ϕ, subtended between these two axes, is expressed by:

$$\tan \frac{\phi}{2} = K_{II}^d / K_I^d = \varkappa \tag{610}$$

Fig. 64 presents the shapes of reflected caustics for values of $\varkappa = 0$, 0.4 and 1.0 and crack velocities $c/c_1 = 0$, 0.2, 0.4 and 0.6.

Equation (607) presents extrema whose positions are defined by setting the first derivative equal to zero:

$$dH_{r,t,f}/d\theta_1 = 0 \tag{611}$$

and checking the validity of $d^2 H_{r,t,f}/d\theta_1^2 < 0$.

For the angles θ_{1t}^{max} where the transverse diameter of the caustic presents maxima, we have:

$$D_t^{max} = 2H_{r,t,f}^{max} = \lambda_m (\tfrac{3}{2} \mu_{r,t,f})^{\frac{3}{2}} \delta_t^{max}(\varkappa, c/c_1) \tag{612}$$

where $\delta_t^{max}(\varkappa, c/c_1)$ expresses a correction factor of the maximum transverse diameter of the caustic D_t^{max} taking care of the distortion of the shape of the caustic due to the dynamic effect. The correction factor δ_t^{max} is expressed by:

$$\delta_t^{max}(\varkappa, c/c_1) = 2[\beta_1^{-1} R_d^{\frac{2}{3}} \sin \theta_{1t}^{max} + \tfrac{2}{3} \beta_1 R_d^{-\frac{3}{2}} (\sin \tfrac{3}{2} \theta_{1t}^{max} + \varkappa \cos \tfrac{3}{2} \theta_{1t}^{max})] \tag{613}$$

On the other hand, the maximum longitudinal diameter D_l^{max} of the caustic may be derived from Eq. (606) for angles θ_{1lj}^{max} ($j = 1, 2$), by making the relation in Eq. (607) zero. For these places the maximum longitudinal diameter of the caustic D_l^{max} is

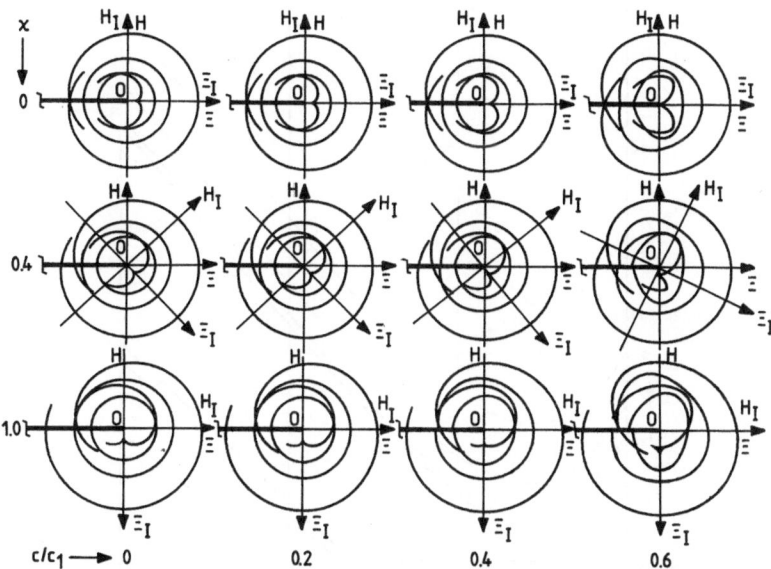

Fig. 64. Initial curves and respective caustics from reflected light rays on a cracked plate of an isotropic elastic material for relative crack velocities $c/c_1 = 0$, 0.2, 0.4 and 0.6 and for a ratio of stress intensity factors $\varkappa = K_{II}^d / K_I^d = 0, 0.4$ and 1.0, as plotted by computer for $-\pi \leqq \theta_1 \leqq \pi$.

given by:

$$D_l^{\max} = \sum_{j=1}^{2} \Xi_{r,t,f,j}^{\max} = \lambda_m (\tfrac{3}{2} \mu_{r,t,f})^{\tfrac{2}{3}} \delta_l^{\max}(\varkappa, c/c_1) \tag{614}$$

where $\delta_l^{\max}(\varkappa, c/c_1)$ is a correction factor taking care of the distortion of the caustic due to the dynamic effect along its maximum longitudinal diameter. This factor is expressed by:

$$\delta_l^{\max}(\varkappa, c/c_1) = \sum_{j=1}^{2} [R_{dj}^{\tfrac{2}{3}} \cos \theta_{11j}^{\max} + \tfrac{2}{3} R_{dj}^{-\tfrac{1}{3}}(\cos \tfrac{3}{2}\theta_{11j}^{\max} - \varkappa \sin \tfrac{3}{2}\theta_{11j}^{\max})] \tag{615}$$

with:

$$R_{dj} = \tfrac{1}{2}(1 - \beta_1^2)(\cos \tfrac{5}{2}\theta_{11j}^{\max} - \varkappa \sin \tfrac{5}{2}\theta_{11j}^{\max}) + \tfrac{1}{2}[(1 - \beta_1^2)^2$$
$$\times (\cos \tfrac{5}{2}\theta_{11j}^{\max} - \varkappa \sin \tfrac{5}{2}\theta_{11j}^{\max})^2 + 4\beta_1^2(1 + \varkappa^2)]^{\tfrac{1}{2}}, \quad j = 1, 2 \tag{616}$$

From Eqs (612) and (614) we obtain the ratio of the maximum diameters of the caustic given by:

$$D_t^{\max} / D_l^{\max} = \delta_t^{\max} / \delta_l^{\max} \tag{617}$$

Figure 65 presents the variation of the correction factors $\delta_t^{\max}$ and $\delta_l^{\max}$ versus the ratio $\varkappa = K_{II}^d / K_I^d$ with relative velocity c/c_1 of crack propagation as variable.

Finally, from Eqs (605), (576), (612) and (614), the dynamic stress intensity factors K_I^d and K_{II}^d may be derived from the following relations:

$$K_I^d = \frac{2(2\pi)^{\tfrac{1}{2}}}{3\varepsilon z_0 \, d\lambda_m^{\tfrac{3}{2}} c_{r,t,f}} \left(\frac{D_{t,l}^{\max}}{\delta_{t,l}^{\max}(\varkappa, c/c_1)} \right)^{\tfrac{5}{2}} \tag{618}$$

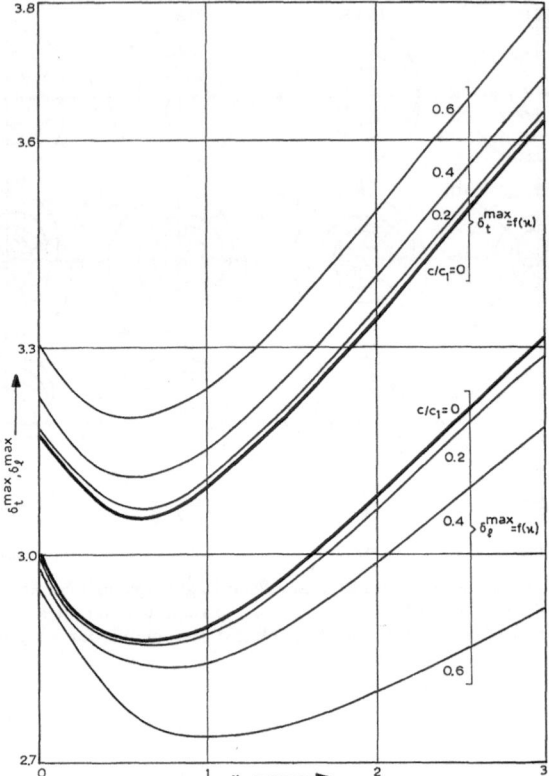

Fig. 65. The variation of the correction factors δ_t^{max} and δ_l^{max} versus the ratio $\varkappa$ for different values of the relative crack velocity c/c_1.

and:

$$K_{II}^d = \varkappa K_I^d = K_I^d \tan\frac{\phi}{2} \tag{619}$$

The theory of caustics developed here for cracks propagating in an arbitrary direction with constant velocity allows evaluation of the dynamic stress intensity factors K_I^d and K_{II}^d by an easy and accurate method. From the maximum diameters D_t^{max} and D_l^{max} of the external branch of the reflected caustic, the K_I^d stress intensity factor may first be evaluated. The internal branch of the caustic, which normally takes the form of a cusp, allows the exact determination of the instantaneous crack tip and the angle subtended between the instantaneous crack-axis and the axis of symmetry of the respective caustic. This can be achieved by tracing the common tangent to the two lobes of the cuspoid curve and constructing the mid-normal to this tangent. The intersection of the normal to the tangent with the crack-axis defines exactly the position of the crack tip. As soon as this line is defined, the angle subtended by the normal and the crack-axis yields the value of the angular displacement of the caustic. The tangent of this angle defines K_{II}^d from K_I^d. Furthermore, the definition of the exact position of the instantaneous crack tip by the internal branch of the caustic allows the correct evaluation of the velocity.

It can be deduced from Fig. 64 that as the velocity increases, the transverse diameter D_t^{max} of the respective caustic also increases, whereas the longitudinal diameter D_l^{max} diminishes correspondingly. Thus, for low values of c/c_1 the ratio of the diameters does not change significantly; but for high values of c/c_1 the distortion of the shape of the caustic is important. For values of c/c_1 between zero (stationary crack) and 0.2 the dynamic caustic resembles the stationary caustic, with no distortion of its shape. This result is important because it can then be concluded that for all normal types of fracture where crack velocities do not exceed the value $c/c_1 = 0.4$, the shape of the caustic is not significantly distorted and only its size is changed. Therefore, the simple theory for stationary cracks is still valid to a good approximation.

On the other hand, if the ratio $\varkappa = K_{II}^d/K_I^d$ increases, that is, if the contribution of shear to the deformation of the cracked plate increases, the caustic is angularly displaced without suffering significant distortions. The influence of stress intensity factor ratio $\varkappa$ is insignificant for low values of relative velocity of crack propagation, whereas for high values of c/c_1 this influence is significant. Furthermore, as the relative velocity of crack propagation increases with $\varkappa = $ constant, the cusp point of the internal branch of the respective caustic continuously approaches the instantaneous crack tip. One may assume that the cusp point coincides with the crack tip with an error not exceeding 1%. Then it is natural to assume for high velocities that the cusp of the branch of the caustic coincides with the respective crack tip.

Figure 65 yields the variation of the correction factors δ_t^{max} and δ_l^{max} with $\varkappa$ for selected relative velocities. These factors vary significantly under the influence of $\varkappa$ but are rather insensitive to the influence of c/c_1. For values of $\varkappa \leq 0.5$ ($\omega = 26.6°$), the correction factors diminish with increasing $\varkappa$, passing through a minimum value in the interval $0.5 < \varkappa < 1.1$, and then increasing rapidly with increasing velocities.

We must note here that the Freund and Clifton [64] stress field is valid for $c < c_2 \simeq 0.6\, c_1$. This means that Figs 57, 58, 60-65 are valid up to $c/c_1 \simeq 0.6$. For $c > c_2 \simeq 0.6\, c_1$ the Freund and Clifton stress field needs modification. According to Freund and Clifton [64] theory, the dynamic stress intensity factor for the mode-I deformation is defined as:

$$K_1(t) = \lim_{r \to 0} [(2\pi r)^{\frac{1}{2}} \sigma_{yy}(r, y = 0, t)]$$

and the relation expressing the dynamic stress intensity factor, in terms of the maximum transverse diameter of the caustic (Rel. (601)), is given as:

$$K_1(t) = \frac{2(2\pi)^{\frac{1}{2}}}{3\varepsilon z_0 d\lambda_m^{\frac{3}{2}} c_{r,t,f}} \left(\frac{D_{t,l}^{max}}{\delta_{t,l}^{max}(c/c_1)} \right)^{\frac{5}{2}} \frac{R(c)}{(1+\beta_2^2)(\beta_1^2 - \beta_2^2)}$$

or:

$$K_1(t) = K_I^d \frac{R(c)}{(1+\beta_2^2)(\beta_1^2 - \beta_2^2)}$$

where K_I^d is defined by Rel. (601).

3.5 Application of Caustics in Dynamic Problems

3.5.1 Mode-I Dynamic Crack Propagation

In order to show the potentialities of the method of reflected caustics we shall apply it to plexiglas specimens containing an initial edge transverse crack. Plexiglas plates $0.3 \times 0.1 \, \text{m}^2$ and of thickness $d = 0.003 \, \text{m}$ were initially cracked by saw-cuts of length $a_0 = 0.01 \, \text{m}$ and dynamically loaded by falling weights of 372.78 N from a height of 0.25 m.

Figure 66 presents a series of photographs taken during the propagation of the transverse crack by a Cranz–Schardin 24-spark high-speed camera at time intervals of 8 µs. At the tips of the cracks two caustics were formed, one from reflections from the rear face and the other from reflections from the front face. The instantaneous tip of the crack was defined by the position of the cusp of the internal caustic.

Since the velocity of propagation of these cracks was of the order of 430 m/s for all experiments, the relative velocity $c/c_1 \simeq 0.2$, since the longitudinal wave velocity for plexiglas $c_1 \simeq 2000 \, \text{m/s}$. This ratio $c/c_1 \simeq 0.2$ indicates that the shapes of the caustics must resemble the theoretical shape shown in Fig. 57. Random tests for

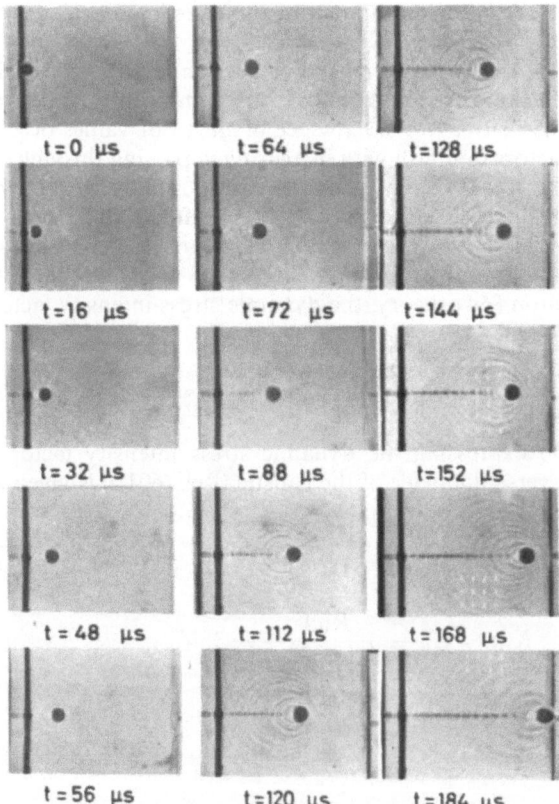

t = 0 µs	t = 64 µs	t = 128 µs
t = 16 µs	t = 72 µs	t = 144 µs
t = 32 µs	t = 88 µs	t = 152 µs
t = 48 µs	t = 112 µs	t = 168 µs
t = 56 µs	t = 120 µs	t = 184 µs

Fig. 66. A series of photographs of a propagating edge crack in a plexiglas plate under mode-I deformation with an initial length of crack $a_0 = 0.010 \, \text{m}$.

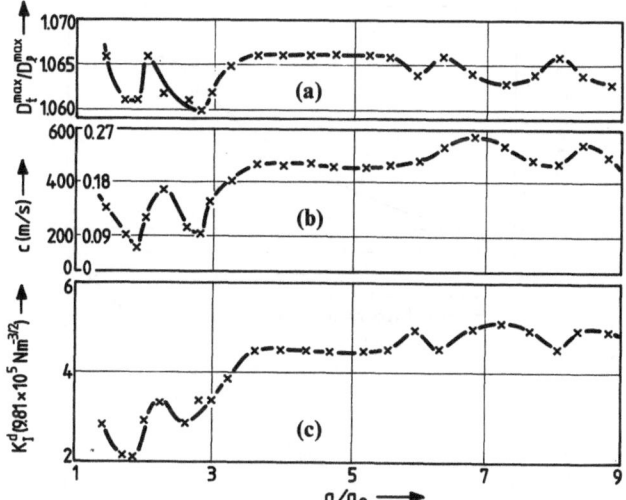

Fig. 67. (a) The ratio of maximum diameters of the caustics D_t^{max}/D_l^{max}, (b) the relative crack velocity c/c_1 or crack velocity c and (c) the dynamic stress intensity factor K_I^d; versus the crack length a normalized to its initial length a_0.

the shape of caustics in the area of constant velocity $c = 430$ m/s showed satisfactory correlation of experimental and theoretical shapes. During the propagation of the crack the caustic changes dimensions because the propagation of the crack does not take place under constant velocity. The change in dimensions of the caustic indicates the variation of K_I^d.

Figure 67(a) presents the variation of the ratio of the maximum diameters D_t^{max}/D_l^{max} with respect to the instantaneous crack length a, normalized to the initial crack length a_0. Since in all experiments the initial crack length was kept constant, its influence on the characteristics of propagation of the cracks was excluded. It can be seen from this figure that the maximum transverse diameter is always larger than the corresponding maximum longitudinal diameter. By comparing the ratios of the diameters of the dynamic caustics to the respective ratio for the static case, which is always $D_t^{max}/D_l^{max} = 1.0567$, the fact, derived from the theory, that the ratio of the maximum diameters for the dynamic case is always larger than the respective ratio for the static case can be verified.

In Fig. 67(b) the variation of the velocity c of propagation of the crack is given in terms of the normalized length of the crack a/a_0. As may be deduced from this figure, the variation of c at the initiation of propagation is large, as it is at the end of propagation when the crack tip approaches the opposite boundary of the plate and is influenced by it. In between, the crack velocity remains almost constant. A similar behaviour appears in the values of K_I^d, plotted in Fig. 67(c).

Also of interest are the fluctuations in the propagation velocity c and the dynamic stress intensity factor K_I^d during the periods of initiation and termination of the crack, when the crack is influenced mainly by the boundaries of the plate as well as by the stress waves created at the crack tip at each time instant.

Figure 68 presents the variation of the dynamic stress intensity factor K_I^d (Fig. 68(a)) and the ratio of the diameters of the caustics D_t^{max}/D_l^{max} (Fig. 68(b)) versus the relative velocity of the crack c/c_1. These values have been derived from the

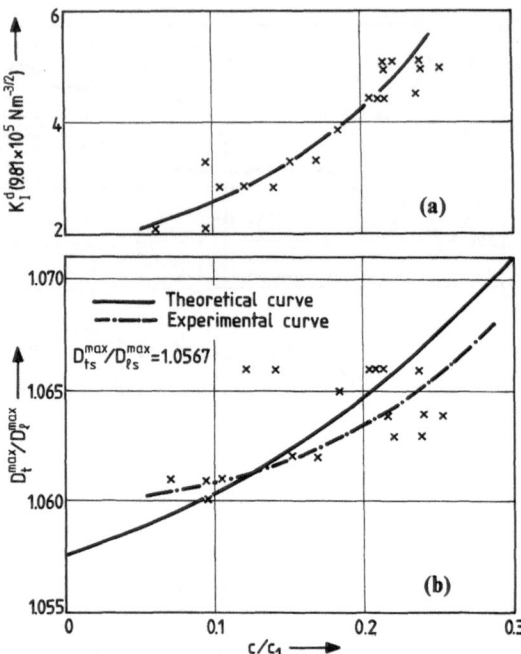

Fig. 68. The variation of the dynamic stress intensity factor K_I^d and the ratio D_t^{max}/D_l^{max} versus the relative crack velocity c/c_1.

experimental shapes of the respective caustics and compared with the dynamic theory developed in this chapter. The increase in K_I^d corresponds to change in shape of the caustics with increasing c/c_1 indicated in Fig. 68(b). Comparing the experimental values for D_t^{max}/D_l^{max} with the theoretical ones we have found good agreement, the discrepancies not exceeding 1%.

It remains to compare the approximate theory, which uses a correction factor to compensate for the error introduced by using the static theory [173], with the results given here, in order to find out if this correction factor agrees with the results of the dynamic theory.

Figure 69 presents in detail the variation in relative velocity of the running crack when this is influenced by the opposite transverse boundary of the plate. The wavy variation of c/c_1 as the crack tip approaches the boundary indicates that there is not only an interaction between the boundary and the crack, but also a strong influence of the stress waves created at each time instant from the crack tip and reflected from the boundary. Their influence as the crack approaches this boundary becomes more and more significant and results in an additive or cancelling effect in the velocity of the crack and the dynamic stress intensity factor. The velocity of these waves was found to be $c_s = 1200$ m/s.

Figure 70 presents a series of photographs taken with an initial delay of $t_d = 120$ μs, so that the last phase of the propagating crack is photographed, together with the effect following complete rupture of the specimen. It may be observed from these photographs that as the crack is propagated, stress waves are created from the tip of the crack, which propagate with a circular front having as centre the instantaneous

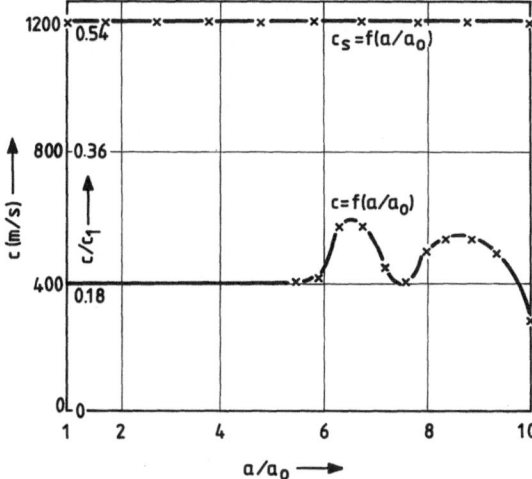

Fig. 69. The variation of the relative crack velocity c/c_1 during the exit of the crack in a plexiglas plate under mode-I deformation and the corresponding value c_s of the velocity of the reflected shear stress wave from boundary of the specimen, versus the ratio a/a_0.

position of the crack tip. The ratio of the wave-front velocity to the velocity of propagation of the crack is of the order of 2.3, which for an average velocity of propagation of the crack of the order $c = 500$ m/s yields a wave-front velocity $c_2 = 1150$ m/s. Since for plexiglas it can be found from Rel. (550) that the shear-wave velocity $c_2 = 1200$ m/s for an average longitudinal velocity $c_1 = 2200$ m/s found in the experiments [60], it can be concluded that the circular front waves created during the propagation of the crack are shear waves.

In the same figure it can be observed that after the rupture of the specimen (at time $t = 88$ μs) a residual elastic deformation remains on the lips of the crack. This residual deformation propagates back along the lips of the crack in the ruptured specimen with a velocity $c_r = 1202$ m/s which is identical with the shear-wave velocity c_2. This accords with the theory that as the crack tip disappears at the boundary of the ruptured specimen, a shear wave is created which, reflected on the transverse boundary of the plate, propagates in the opposite direction along both lips of the crack where it creates an intensive elastic deformation, forming a black dot and a faint caustic.

Another type of experiment with a different material yields the same conclusions. In this case the crack was arrested midway and the specimen was not ruptured. The specimen was made of polycarbonate which is an optically anisotropic material. Figure 71 presents the evolution of stress waves in an arrested crack propagated transversely in a polycarbonate plate. Since the crack is arrested, there is no production of new waves and only those created during the moment of arrest of the crack appear in the figure. The times indicated in the photographs correspond to the total time intervals from the initiation of the crack. The velocity of these propagating waves was found to be $c_s = 700$ m/s. For polycarbonate the velocity of the longitudinal waves was found to be $c_1 = 1500$ m/s. Thus, from Rel. (551) it can be derived that the shear waves have a velocity $c_2 = 800$ m/s, which is close to c_s, thus indicating that the waves shown in Fig. 71 are shear waves.

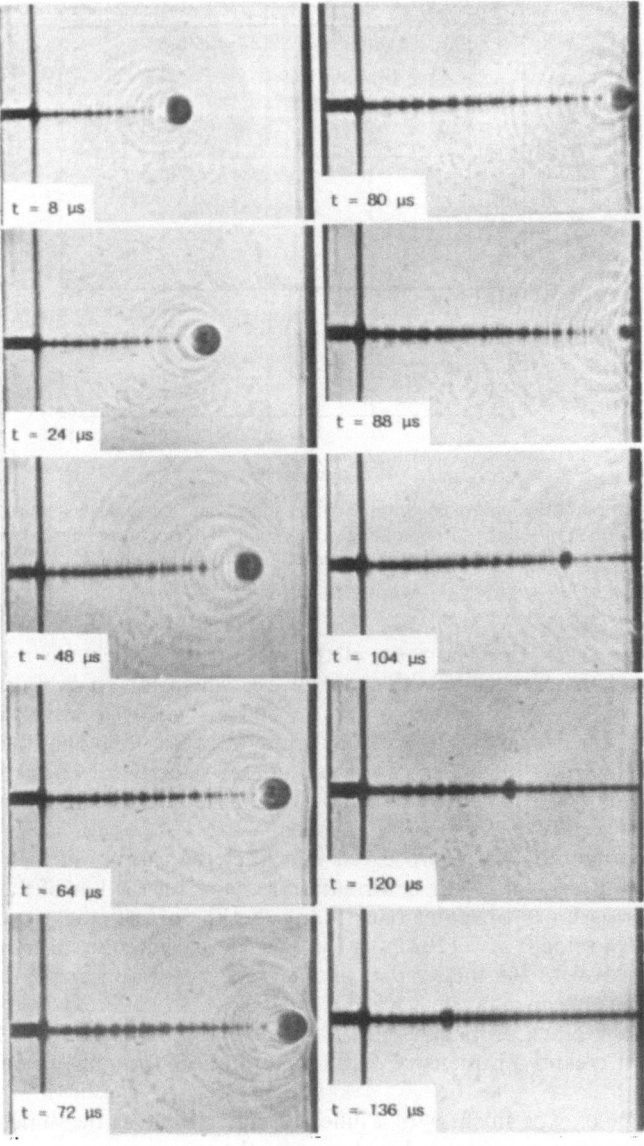

Fig. 70. A series of photographs of a cracked plexiglas plate containing an initial crack length $a_0 = 0.010\,\mathrm{m}$ and subjected to mode-I deformation, taken after a time $t_d = 120\,\mu\mathrm{s}$ so that the end of fracture phenomenon and its after-effects could be recorded. (Note the small caustic running along the lips of the crack after complete rupture of the specimen.)

It is also worthwhile indicating the phenomenon of deformation of the transverse boundary of the specimen and the lips of the crack as the fronts of the shear wave, represented by the bright fringes, interfere with these boundaries, creating wavy surfaces which are clearly shown in these photographs. An analysis of the shape of these boundaries at the intersections of the fronts may yield interesting results

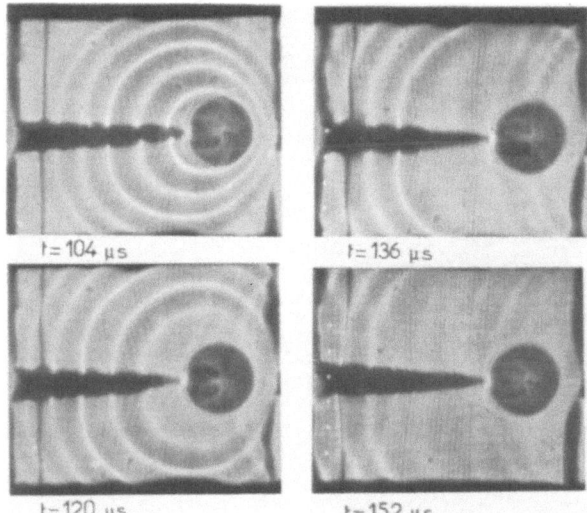

Fig. 71. A series of photographs of an arrested crack in a polycarbonate plate where the shear stress waves generated at the crack tip travel along the specimen (initial crack length $a_0 = 0.010$ m).

concerning the regional stress field at these points. However, it is clear that, besides the influence of neighbouring boundaries of the specimen on the value of the propagation velocity of the running crack, there is also an interaction between the reflected shear wave from the opposite boundary and the stress field around the tip of the propagating crack causing a wavy form to the values of instantaneous propagation velocity of the crack and the dynamic stress intensity factor. Figures 67 and 70 indicate this influence. It is worthwhile remarking here that this influence results in increased c and K_I^d above their average values at the end of the phenomenon while during the acceleration period at the beginning of the crack the fluctuating values of c and K_I^d are much smaller than the average values.

3.5.2 Crack Propagation Model

The theory developed was applied to plexiglas specimens subjected to dynamic loads and containing initially slant edge cracks. The inclination of the cracks was always $\omega = 45°$. The dimensions of the plates were 0.3×0.1 m^2 and their thickness $d = 0.003$ m.

Figure 72 presents a series of photographs of the propagating crack. The continuous angular displacement of the caustic relative to the crack-axis indicates the existence of a K_{II}^d stress intensity factor. As can be seen from the whole series of photographs, the caustic ceases to rotate after a time $t_s = 72$ μs, indicating that the direction of crack propagation has become stable and K_{II}^d has reached its asymptotic value of $K_{II}^d = 0$ [165, 174]. By measuring the angle of rotation of the caustic relative to the instantaneous crack-axis it was possible to evaluate the ratio K_{II}^d / K_I^d. The variation of this ratio with crack length a is given in Fig. 73(a). It may be seen from this figure that the initially slant crack of length a_0 develops a stationary K_{II}^s value, $K_{II}^s = 0.48\,K_I^s$ [102, 121]. As soon as the crack starts to propagate, the

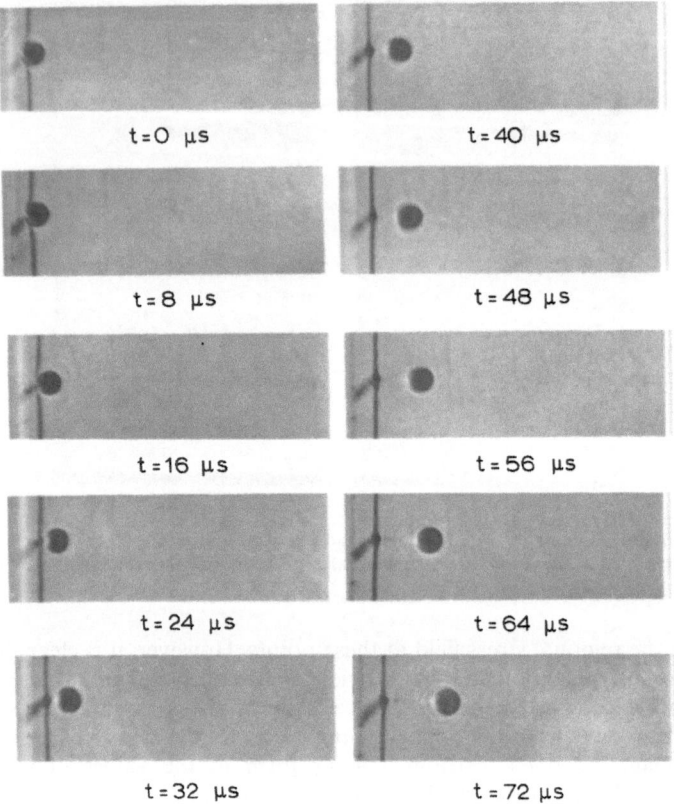

Fig. 72. A series of photographs of a propagating inclined edge crack in a plexiglas plate under mixed mode deformation with an initial length of crack $a_0 = 0.010\,\text{m}$ and angle of inclination $\omega = 45°$.

K_{II}^d factor takes the value $K_{II_0}^d = 0.40\,K_{I_0}^d$. Subsequently, the relative angular displacement of the caustic decreases and the K_{II}^d value tends to zero. This zero value occurs at $t = 32\,\mu\text{s}$, while for $t > 32\,\mu\text{s}$ the K_{II}^d factor takes negative values. This means that the angle subtended between the crack-axis and the axis of symmetry of the caustic has now changed sign. In this way the values for K_{II}^d oscillate about the $(K_{II}^d = 0)$-line and tend asymptotically to a limiting state of crack propagation, where only the K_I^d factor is operative.

This oscillating variation of K_{II}^d may be explained by a zig-zag propagation of the crack. A model for this type of propagation of initially slant cracks was introduced in Ref. [83], and was based on the elastic strain-energy density criterion. However, after a number of oscillations and zig-zags, the crack always tends to a stable state of propagation at a constant velocity dominated by a K_I^d mode of deformation. Similar behaviour was described by Cotterell and Rice [175].

From the point of initiation of the crack, which corresponds to the tip of the stationary crack, to the point of the first zero in K_{II}^d, there is a corresponding crack-step $a_1 = 0.0082\,\text{m}$, whereas for the distance between the first and second zeroes of K_{II}^d there is an interval $a_2 = 0.005\,\text{m}$. The third and fourth intervals between the second and the third zero values and the third and fourth zeroes of the ratio K_{II}^d/K_I^d are approximately equal, $a_3 \simeq a_4 \simeq 0.004\,\text{m}$. It can be observed that, while

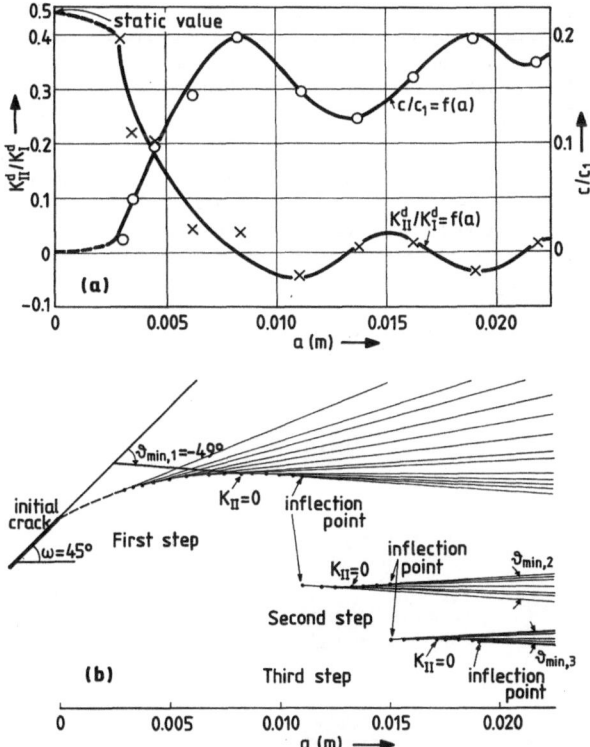

Fig. 73. (a) The variation of the crack velocity c/c_1 and the variation of the ratio of dynamic stress intensity factors K_{II}^d/K_I^d versus the crack length a from the tip of the initial crack. **(b)** The path of crack propagation as measured from the variation of the ratio K_{II}^d/K_I^d of (a).

the crack accelerates along an oblique path of propagation, the ratio of the stress intensity factors oscillates about the zero K_{II}^d value with continuously attenuating amplitudes and wave lengths until it tends asymptotically to the stable case of a transverse propagation with $K_{II}^d = 0$. The establishment of this stable crack propagation under constant velocity coincides with the cancellation of the contribution due to K_{II}^d. In order to plot the path of the propagating crack at each instantaneous position of the crack tip, the value of angle ω_i was measured from the angle of rotation of the respective caustics shown in Fig. 72. In Fig. 73(b), the path of the propagating crack is plotted. The zero points of the ratio K_{II}^d/K_I^d correspond to horizontal (self-similar) propagation ($K_I^d \neq 0$, $K_{II}^d = 0$), while the extreme values of the same ratio correspond to inflection points of the path.

3.6 Crack Propagation in Composite Materials

3.6.1 Particulate Composites

The behaviour of cracked composite specimens under dynamic load was studied by the method of caustics. The composite material used was an epoxy resin reinforced with iron particles in several volume concentrations. The main goal of this study

Table 8. Properties of composite materials

Volume fractions	Elastic modulus E (N/m^2)	Poisson's ratio v	Density ρ(kg/m^3)	Velocity of longitudinal waves c_1(m/s)	Average velocity of crack $\bar{c}$(m/s)	Ratio of velocities $\bar{c}/c_1$
3%-Fe	3.7×10^9	0.3579	1388	1749	422.1	0.2414
5%-Fe	3.9×10^9	0.3565	1521	1714	419.1	0.2445
7%-Fe	4.1×10^9	0.3551	1653	1685	408.8	0.2426

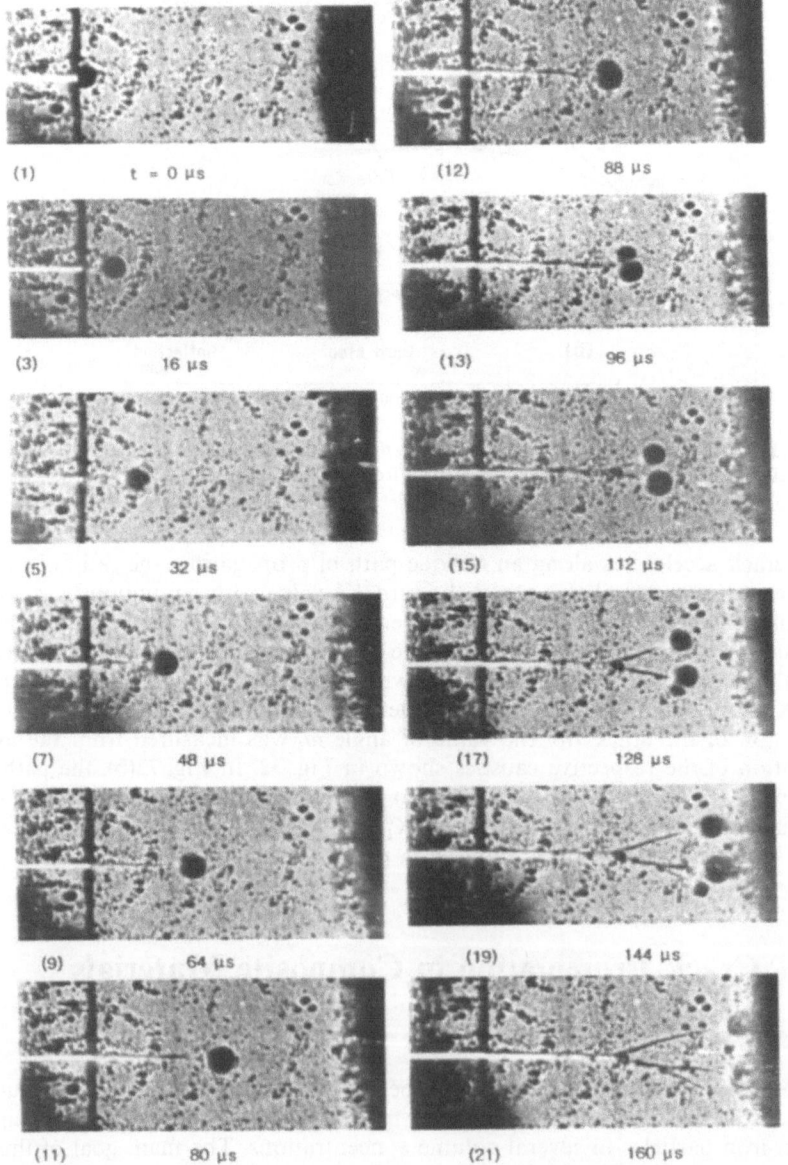

(1)   t = 0 μs   (12)   88 μs

(3)   16 μs   (13)   96 μs

(5)   32 μs   (15)   112 μs

(7)   48 μs   (17)   128 μs

(9)   64 μs   (19)   144 μs

(11)   80 μs   (21)   160 μs

was the determination of the dependence of the crack propagation velocity, as well as of the K_I^d stress intensity factor of the particulate composite upon the filler volume fraction [176, 177].

The matrix material was in all cases a cold-setting system based on a diglycidyl ether of bisphenol A resin (Epicote 828, Shell Company) cured with 8% triethylene-tetramine, which is slightly lower than stoichiometric. One particle size (150 μm) of iron powder was applied. The properties of composite materials for various volume fractions are presented in Table 8. A dynamic load with a constant strain rate of $\dot{\varepsilon} = 0.7/s$ was applied.

Figure 74 presents a series of photographs showing crack propagation through composite specimens reinforced with filler volume fraction 3%. Figures 75(a), (b), (c) present the variation of the dynamic stress intensity factor K_I^d, for the three composite specimens, normalized to its value at the initiation of the crack, $K_{I_0}^d$, as a function of crack length a normalized to its initial length a_0. From these figures, one can see that the stress intensity factor increases by steps. This mode of increase is due both to the existence of the rigid interfaces presented by the filler particles which play the role of barriers and to the different properties of the filler material. The presence of the metal particles in the polymeric matrix results in bifurcation or multifurcation effects (Fig. 74). These effects are accompanied by a reduction in the critical crack propagation velocity with increasing filler volume fraction.

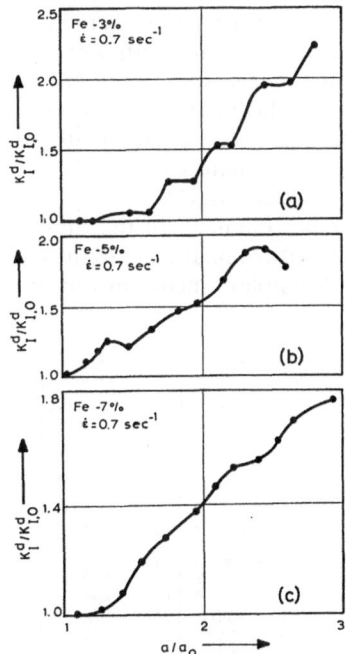

Fig. 75. Diagram of the normalized stress intensity factor versus normalized crack length for the three composite specimens.

Fig. 74. Series of photographs showing crack propagation in a 0.15 mm iron particle–epoxy composite with volume fraction 3%.

3.6.2 Rubber-Modified PMMA Models

The dynamic crack propagation behaviour of several rubber-modified composites has been studied. Results of crack propagation mode observation, fracture toughness and crack propagation velocity measurements are presented [178]. In the case of two "complex" inclusions it was found that the crack propagation mode is highly rate-dependent. At low test rates the crack growth tends to follow an almost straight crack path, while an increase in strain rate in general results in the formation of a kink in the interparticle area. In the same area a crack propagation delay, and in some cases arrest, was observed while both the crack propagation velocity c and dynamic stress intensity factor K_I^d showed an intense variation. For the sake of comparison, specimens with one and/or two press-fitting inclusions were fractured under dynamic loads. In all cases both qualitative and quantitative results were obtained [178]. Also, a study was made of the effect on the dynamic crack propagation mode of eccentricity of the initial crack path from the specimen centreline on which the complex inclusion centre is located [179]. It was shown that the eccentricity of the initial crack path largely determines both the crack propagation delay observed as the crack approaches the inclusion and the bifurcation of the crack. It was also found that the maximum crack propagation velocity, the length of the crack path along the interface and the angle of incidence where the propagating crack meets the interface, increase as the eccentricity increases.

The matrix material chosen for the present experimental investigation was in all cases plexiglas (PMMA). The specimens used were in the form of rectangular plates with dimensions $0.30 \times 0.10 \times 0.003 \, \text{m}^3$ with an edge artificial crack of length $a_0 = 0.010 \, \text{m}$. Two different models were prepared. First, an interference fit was obtained by milling holes in the matrix and press-fitting a single or a pair of PMMA circular inclusions. The hole was nominally 0.034 m in diameter. Next, in order to simulate rubber-modified materials, models with circular PMMA inclusions surrounded by concentric rubber rings were prepared. The inclusion diameter ϕ was 0.023 m whereas the outer diameter of the rubber ring Φ was 0.034 m. In all cases the inclusions were placed symmetrically to the initial crack direction (Fig. 76). Finally, a model with one "complex" inclusion was prepared to study the effect of

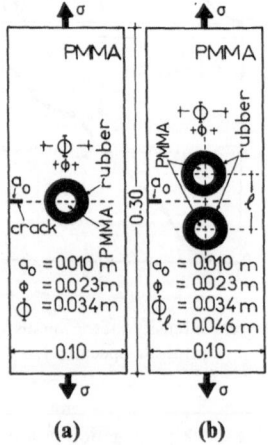

(a)   (b)

Fig. 76. Geometry of specimens $a_0 = 0.010 \, \text{m}$, $\phi = 0.023 \, \text{m}$, $\Phi = 0.034 \, \text{m}$, $l = 0.046 \, \text{m}$.

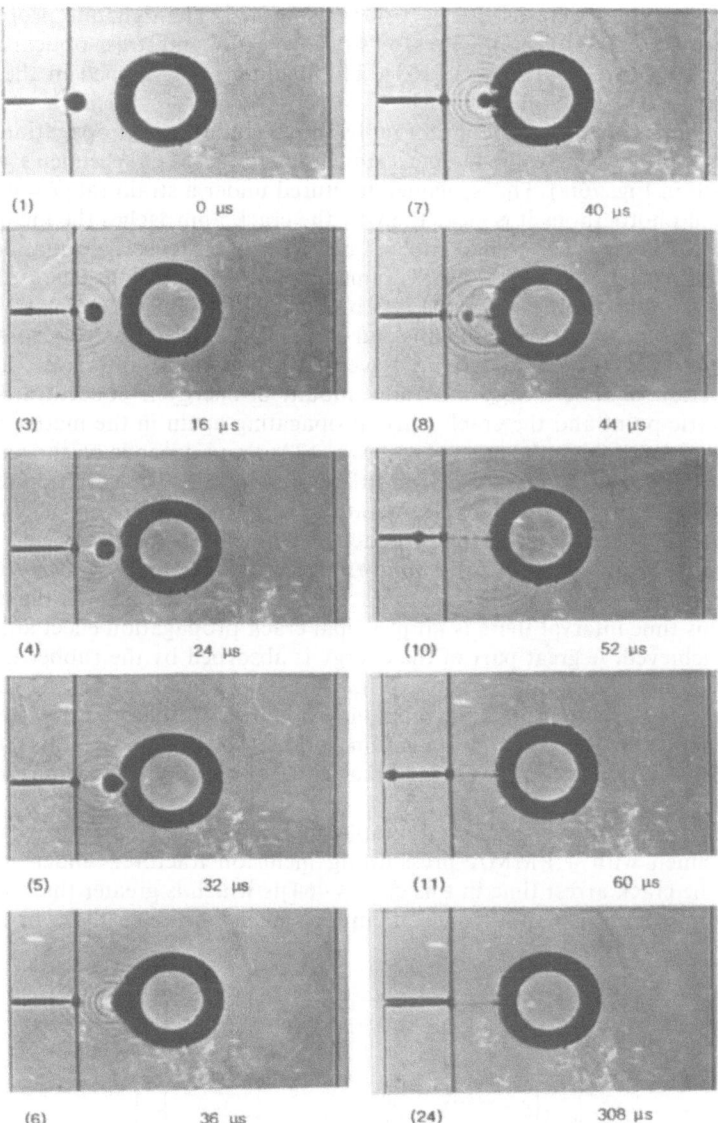

Fig. 77. Series of photographs showing crack propagation in a specimen with a single "complex" inclusion with a rubber interphase, subjected to a strain rate $\dot{\varepsilon} = 4/s$.

eccentricity of the crack path from the specimen centreline on which the "complex" inclusion centre is located, on the dynamic crack propagation mode. The inclusion diameter was 0.016 m whereas the outer diameter of the rubber ring was 0.0265 m and the eccentricities were $e = 0.005$ and 0.0125 m.

The specimens were subjected to a dynamic tensile load until fracture by a Hydropulse High-speed Testing machine of the type of Carl-Schenk Co. with a maximum available strain rate $\dot{\varepsilon} = 80/s$. In the optical set-up used in the experiments the following values were used: $z_0 = 0.80$ m, $\lambda_m = 0.75$. The dynamic properties of PMMA are $E = 4.3 \times 10^9$ N/m², Poisson's ratio $v = 0.34$ and stress-optical constant $c_t = 0.74 \times 10^{-10}$ m²/N (Table 4) [163]. The loading rates applied in the present work were $\dot{\varepsilon} = 0.8, 2, 4$ and $8/s$ [178].

Figure 77 presents a series of photographs showing the crack propagation process in a single "complex" inclusion having a rubber interphase. The specimen's geometry is presented in Fig. 76(a). This specimen fractured under a strain rate $\dot{\varepsilon} = 4/s$. From the series of photographs it is clear the way the crack approaches the inclusion. As the crack tip approaches the "complex" inclusion an intense deformation of the rubber interphase is observed (Fig. 77, frames 4 and 5). A part of the energy delivered is absorbed by the rubber while the rest is consumed by the reflected stress waves (Fig. 77, frame 8). At the same time the end points of the reflected stress fronts follow symmetrically the rubber–matrix interface and thus form an interface crack path. After a period of time (arrest time) an amount of energy is concentrated at the antidiametric point and the crack starts propagating again in the matrix material. The crack arrest time in the present case was 172 μs and depends on the nature and quality of the bond that exists between rubber and matrix. The variation of K_1^d and c versus the crack length a for strain rate $\dot{\varepsilon} = 4/s$ is given in Fig. 78. It may be observed that when the crack tip approaches the inclusion both K_1^d and c attain a maximum value and subsequently an abrupt decrease is observed. Next, the main crack stops momentarily while the interface crack propagates around the inclusion. During this time interval there is no principal crack propagation effect and a crack arrest is achieved. A great part of the energy is absorbed by the rubber interphase and there is no energy transmission to the PMMA inclusion. Thus the role of the rubber interphase is to protect the main inclusion from the effect of the stress waves developed due to the crack propagation process. The degree of this protection depends on the quality of adhesion and the degree of compatibility of the constituent materials.

Figure 79 presents a series of photographs showing the crack propagation process in a specimen with a PMMA press-fitting inclusion fractured under strain rate $\dot{\varepsilon} = 4/s$. The crack arrest time in this case is 464 μs which is greater than that in the case of the specimen with a single "complex" inclusion. From these photographs

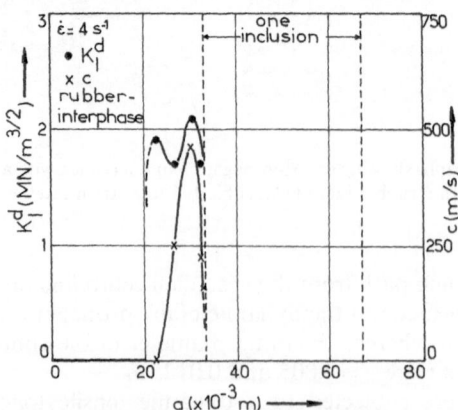

Fig. 78. Variation of the stress intensity factor K_1^d and the crack propagation velocity c, versus the crack length a, for the specimen of Fig. 77.

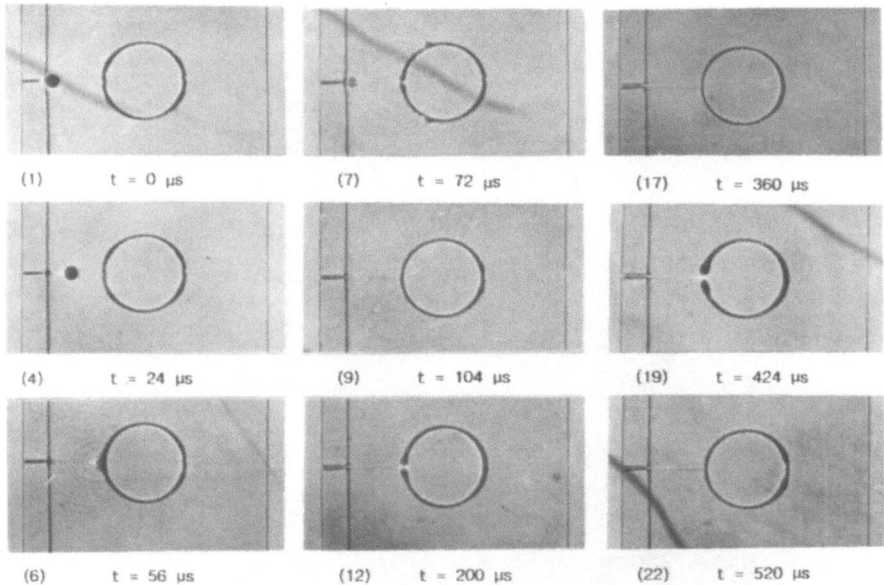

Fig. 79. Series of photographs showing crack propagation in a specimen having one press-fitting PMMA inclusion. The strain rate was $\dot{\varepsilon} = 4/s$.

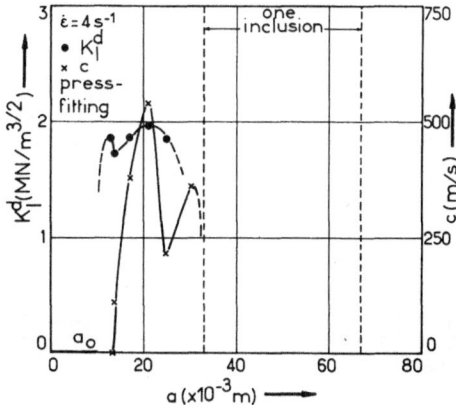

Fig. 80. Variation of the stress intensity factor K_I^d and the crack propagation velocity c, versus the crack length a, for the specimen of Fig. 79.

we may observe an oscillation of the interface area which follows the reflection of the stress waves from both the matrix–inclusion interface and the specimen borders. Indeed, this phenomenon may be observed twice: the first time in frames 9–17 having a duration of 260 μs, and the second time in frames 17–22 having the same duration of 260 μs. A characteristic phenomenon observed in frame 22 is the intense deformation of the antidiametric point. From this point the crack will start to propagate into the matrix after it has followed the matrix–inclusion interface.

Figure 80 shows the variation of crack propagation velocity c as well as that of the stress intensity factor K_I^d versus the crack length a, for strain rate $\dot{\varepsilon} = 4/s$. The

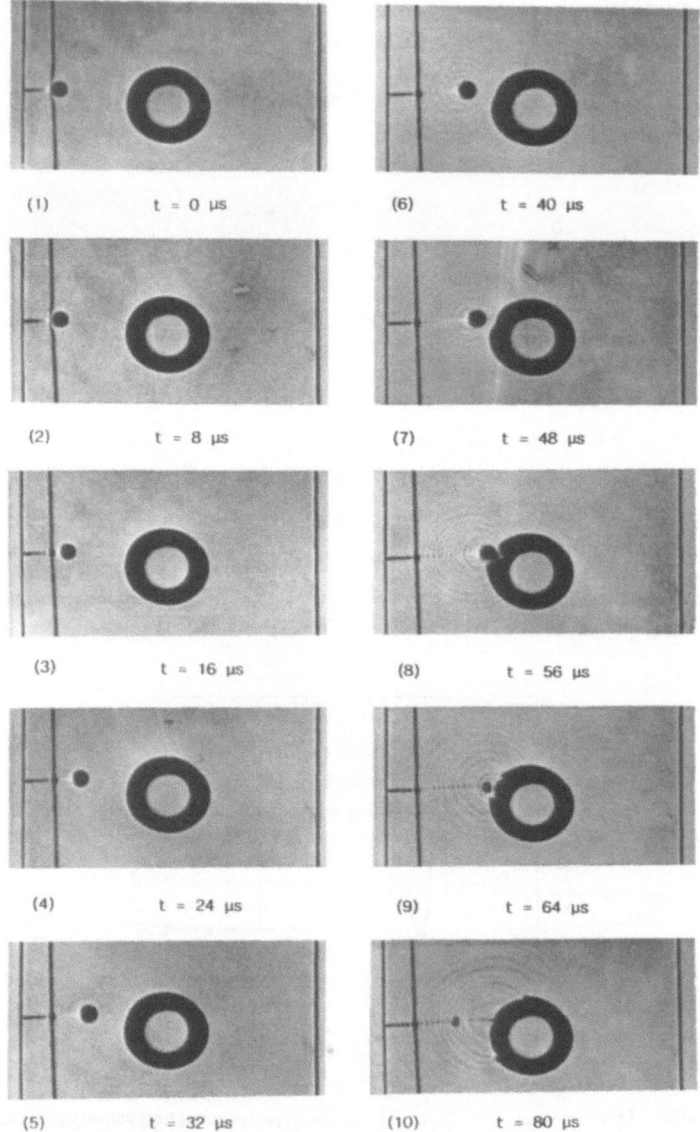

Fig. 81. Series of photographs showing crack propagation in a specimen having one "complex" inclusion. The eccentricity of the initial crack was $e = 0.005$ m and the strain rate was $\dot{\varepsilon} = 4$/s.

crack approaches the inclusion in a way similar to that observed in the case of the rubber interphase.

So, it may be summarized that: (i) In the case of a single "complex" inclusion, as the crack approaches the inclusion an intense deformation of the rubber interphase and an interface crack path were observed. Also due to the energy absorbed by the rubber interphase, crack delay and in some cases crack arrest phenomena were observed, and (ii) An oscillation of the interface area has been observed in the case of a single press-fitting inclusion while the crack arrest time was about 464 μs.

Figure 81 presents a series of photographs showing crack propagation in a specimen with one "complex" inclusion. The eccentricity of the initial crack was $e = 0.005$ m and the strain rate $\dot{\varepsilon} = 4/s$. As the crack tip approaches the inclusion the same intense phenomena are observed as in previous cases. The time needed for the crack to approach the inclusion is about 64 μs. We may also observe the initiation and propagation of the stress waves originating from the tip of the crack (frames 8, 9 and 10).

Figure 82 shows the variation of K_I^d and of c versus the crack length a. We may observe that when the crack tip is in the close vicinity of the "complex" inclusion interface, c attains a maximum value of about 521 m/s and subsequently c dramatically decreases when the crack tip arrives at the interface of the inclusion. The maximum value for K_I^d is attained earlier than the corresponding maximum for the crack propagation velocity.

Figure 83 presents a series of photographs showing the crack propagation process in a specimen with eccentricity of the initial crack $e = 0.0125$ m. The applied strain rate was $\dot{\varepsilon} = 4/s$. We may observe the way that the crack approaches the inclusion. The measured time required for the crack to approach the inclusion was about 80 μs. Next, we may observe that the point of initiation of the final crack is very close to the point of incidence of the initial main crack on the circumference of the inclusion. The variation of K_I^d as well as of c versus the crack length a for the specimen of Fig. 83 is shown in Fig. 84. The maximum values of c and K_I^d attained in the close vicinity of the "complex" inclusion, are 580 m/s and 1.98 MN/m$^{\frac{3}{2}}$ respectively. The measured crack arrest time was about 256 μs and after this arrest time a new crack propagates. The final crack is characterized by higher values of velocity and K_I^d in comparison with the respective values of the initial crack.

The effect of eccentricity of the initial edge crack on the dynamic crack propagation of several rubber-modified composite models may be summarized as follows: (i) In all cases the crack propagation velocity c as well as K_I^d attain a maximum value when the crack tip is in the close vicinity and before the inclusion. This is followed by an abrupt decrease of the respective c and K_I^d values. (ii) In all cases as the crack

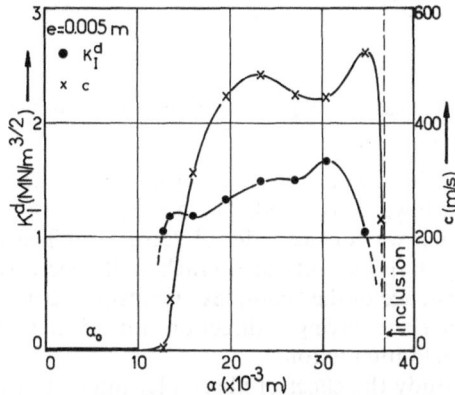

Fig. 82. Variation of the stress intensity factor K_I^d and the crack propagation velocity c, versus the crack length a, for the specimen of Fig. 81.

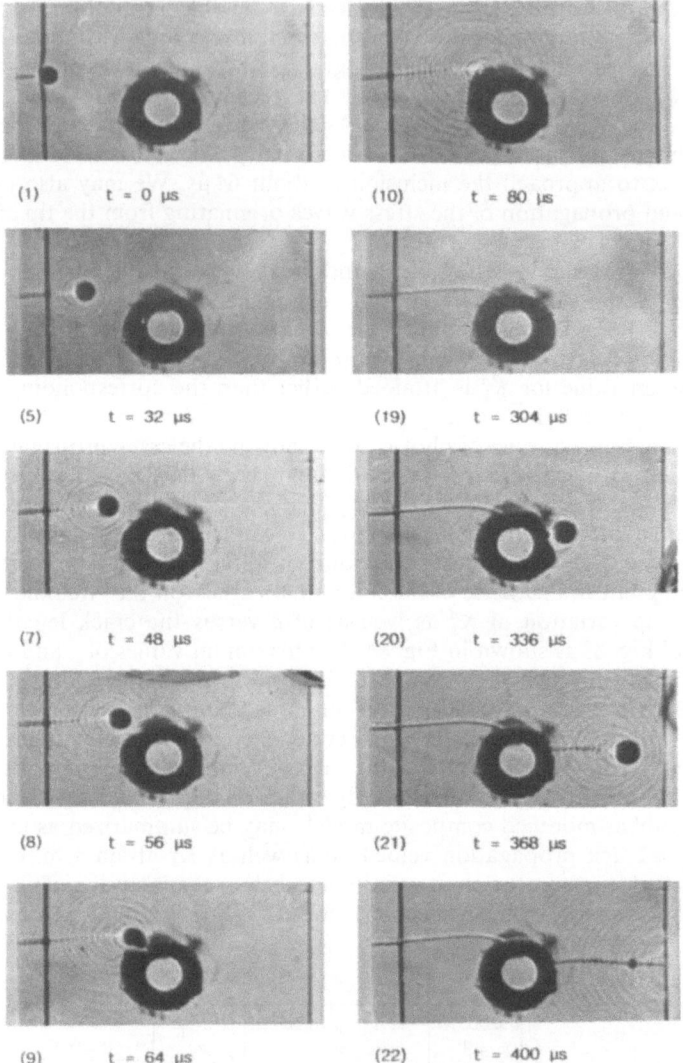

Fig. 83. As in Fig. 81, for eccentricity $e = 0.0125\,\mathrm{m}$.

tip approaches the inclusion an intense deformation of the rubber interphase and an interface crack path were observed. These effects were accompanied by a crack delay arrest time which was of the order of $256\,\mu s$. (iii) The velocity c_{max} increases with the eccentricity. (iv) The interface crack path decreases as the eccentricity increases. (v) The existence of the "complex" inclusion "attracts" the crack which is diverted onto the interface having a direction not normal to the inclusion; this is attributed to stress field interaction.

Next, in order to study the effect of filler–filler interactions as well as strain rate effects, specimens with two "complex" inclusions at constant inter-inclusion separation (Fig. 76(b)) were fractured under four different strain rates [178].

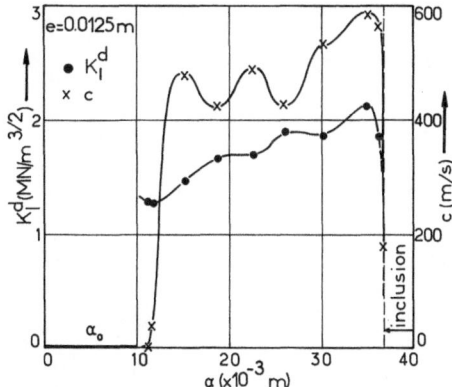

Fig. 84. As in Fig. 82, for eccentricity $e = 0.0125$ m.

The detailed crack propagation process for the case of two inclusions and for strain rate $\dot{\varepsilon} = 4/s$ is shown in the series of photographs presented in Fig. 85. From these photographs evaluation of the instantaneous velocities of crack propagation was obtained, while from the geometric characteristics of the respective caustics the dynamic stress intensity factor K_I^d was evaluated by using Rel. (601). Also in these photographs the whole crack propagation process in the area between the two inclusions may be observed in detail, as well as the effect of the presence of the inclusions on the crack propagation velocity and the stress field developed at the tip of the crack. In the interparticle area the observed crack propagation delay is of the order of 112 μs (frames 9–23 in Fig. 85). Specifically in frames 18–20 a more pronounced effect of the inclusions on the crack propagation process is observed which results in the development of a kink.

The variation of c and of K_I^d versus the crack length a for the specimen of Fig. 85 is shown in Fig. 86. From this figure it may be observed that the crack propagation velocity is strongly affected by the presence of the two inclusions and this is more pronounced in the interparticle area.

The attraction of the crack by one of the inclusions and the development of the kink (Fig. 85) strongly depends on the relative eccentricity of the initial notch. The crack propagation process in the interparticle area is characterized by the attraction of the crack by one of the inclusions and the absorption of a great part of the crack propagation energy by the rubber interphase. Because of this absorption the crack is not permitted to propagate through the inclusion.

The same effects, but less pronounced, are observed in the case of press-fitting inclusions. Figure 87 presents a series of photographs showing the detailed crack propagation process for the case of two press-fitting inclusions. As the crack propagates into the inter-inclusion area it shows a crack delay of about 88 μs (Fig. 87, frames 12–19). This crack delay is less than that observed in the case of two inclusions with a rubber interphase (Fig. 85). Also, the kink developed in this case is much more abrupt than in the case of the rubber interphase, since the energy absorbed by the PMMA inclusion is much lower than that absorbed by the rubber interphase.

The variation of c and of K_I^d versus the crack length a for the specimen of Fig. 87 is shown in Fig. 88. These variations are more intense than in the case of the rubber interphase, while at the point where the kink developed a crack arrest of about 20 μs (Fig. 87, frames 16 and 17) is observed, which was not apparent in the

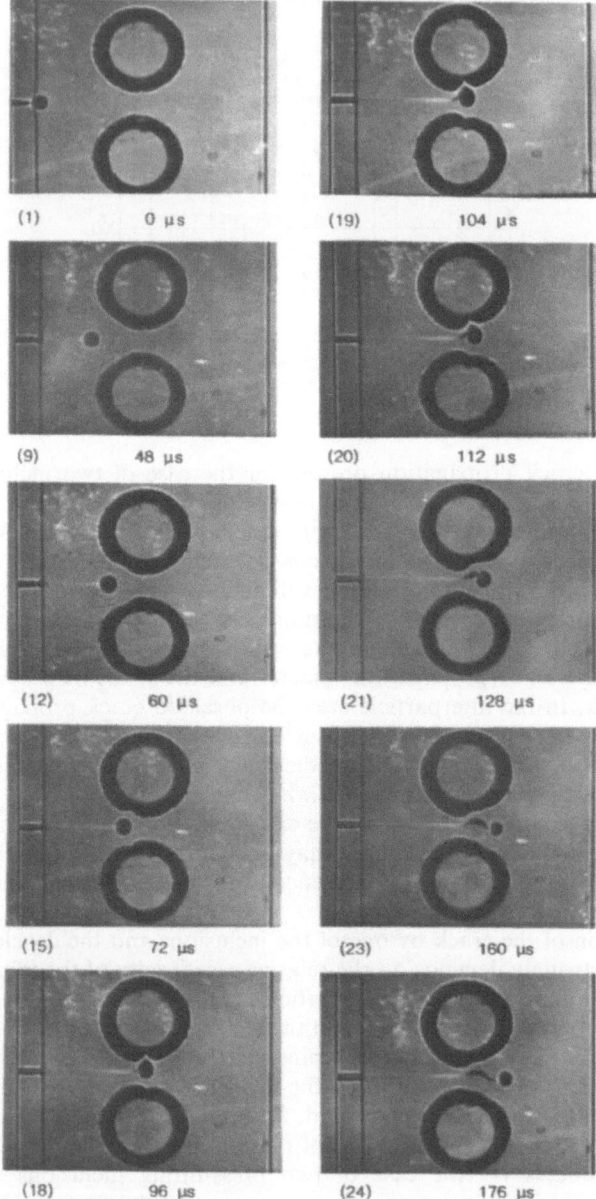

Fig. 85. Series of photographs showing crack propagation in a specimen having two "complex" inclusions. The strain rate was $\dot{\varepsilon} = 4/s$.

case of the rubber interphase. Immediately after the crack momentarily stops at the matrix–inclusion interface (Fig. 87, frame 17) a reinitiation of the crack is observed while the crack propagation velocity increases again (Fig. 88).

So, it may be summarized that: (i) In the case of specimens with two "complex" inclusions it was found that at low strain rates the crack path is almost straight

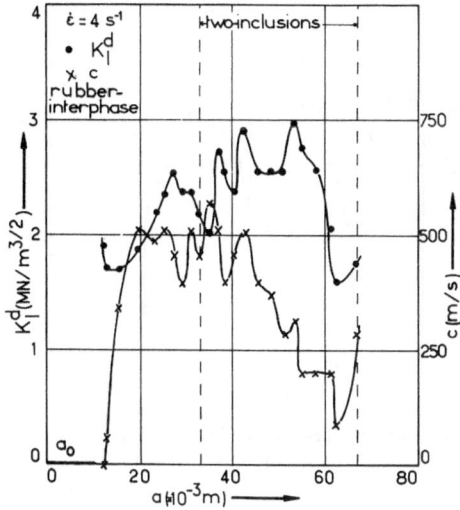

Fig. 86. Variation of the stress intensity factor K_I^d and the crack propagation velocity c, versus the crack length a, for the specimen of Fig. 85.

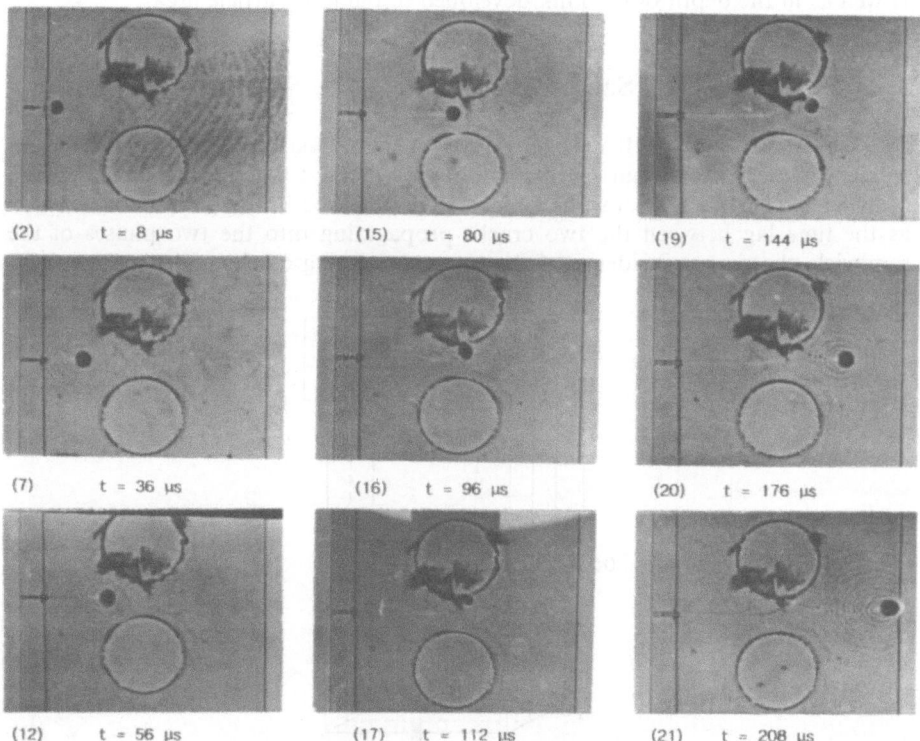

(2) t = 8 μs   (15) t = 80 μs   (19) t = 144 μs

(7) t = 36 μs   (16) t = 96 μs   (20) t = 176 μs

(12) t = 56 μs   (17) t = 112 μs   (21) t = 208 μs

Fig. 87. Series of photographs showing crack propagation in a specimen having two press-fitting inclusions. The strain rate was $\dot{\varepsilon} = 4/s$.

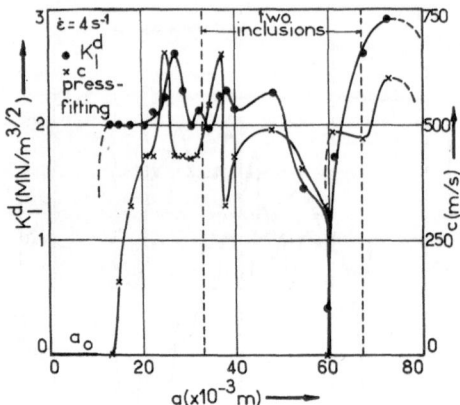

Fig. 88. Variation of the stress intensity factor K_I^d and the crack propagation velocity c, versus the crack length a, for the specimen of Fig. 87.

while at higher strain rates a kink was observed in the interparticle area. Also, in the same area, there is an intense variation of both c and K_I^d, the degree of which depends on the strain rate. (ii) Similar effects have been observed in the case of two press-fitting inclusions, the difference being in the degree of variation of c and K_I^d, as well as in the depth of the kink developed in the interparticle area.

3.6.3 PCBA–PMMA Sandwich Plates

Crack propagation in PCBA (Lexan)–PMMA (plexiglas) sandwich plates has been studied by means of a high-speed photography method together with the method of dynamic caustics. Various phenomena were observed in these experiments, such as the time lag between the two cracks propagating into the two phases of the sandwich, the time coincidence of the two propagating cracks and phenomena of

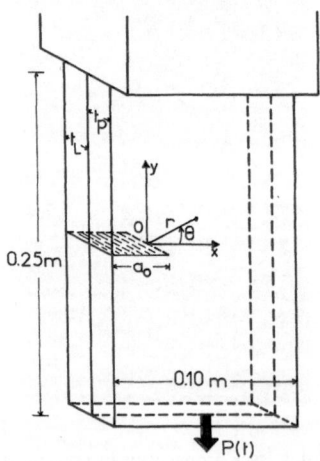

Fig. 89. Geometry of the sandwich specimens.

acceleration, deceleration and bifurcation of the propagating cracks. More precisely, the initial crack starts to propagate first into the brittle (PMMA) phase while a second crack starts to propagate later into the ductile (PCBA) phase of the sandwich plate. The time lag and the time coincidence depend on the nature and the degree of compatibility of the two phases of the sandwich plate [180].

Notched Lexan–plexiglas sandwich plates of area $0.3 \times 0.1\,\text{m}^2$ were used for the experimental investigation. The thickness t_L of Lexan used was $0.0016\,\text{m}$ and $0.002\,\text{m}$ and the thickness t_P of the plexiglas plate was $0.001\,\text{m}$. The specimens initially contained an edge transverse artificial crack $a_0 = 0.01\,\text{m}$ as shown in Fig. 89. The specimens were tested to fracture under a dynamic tensile load with strain rates $\dot{\varepsilon} = 24$ and $40/\text{s}$ [180]. The optical constant for the plexiglas was (see Table 4) $c_t^{(P)} = 0.74 \times 10^{-10}\,\text{m}^2/\text{N}$ and for the Lexan (see Table 4) $c_t^{(L)} = 1.55 \times 10^{-10}\,\text{m}^2/\text{N}$ [163].

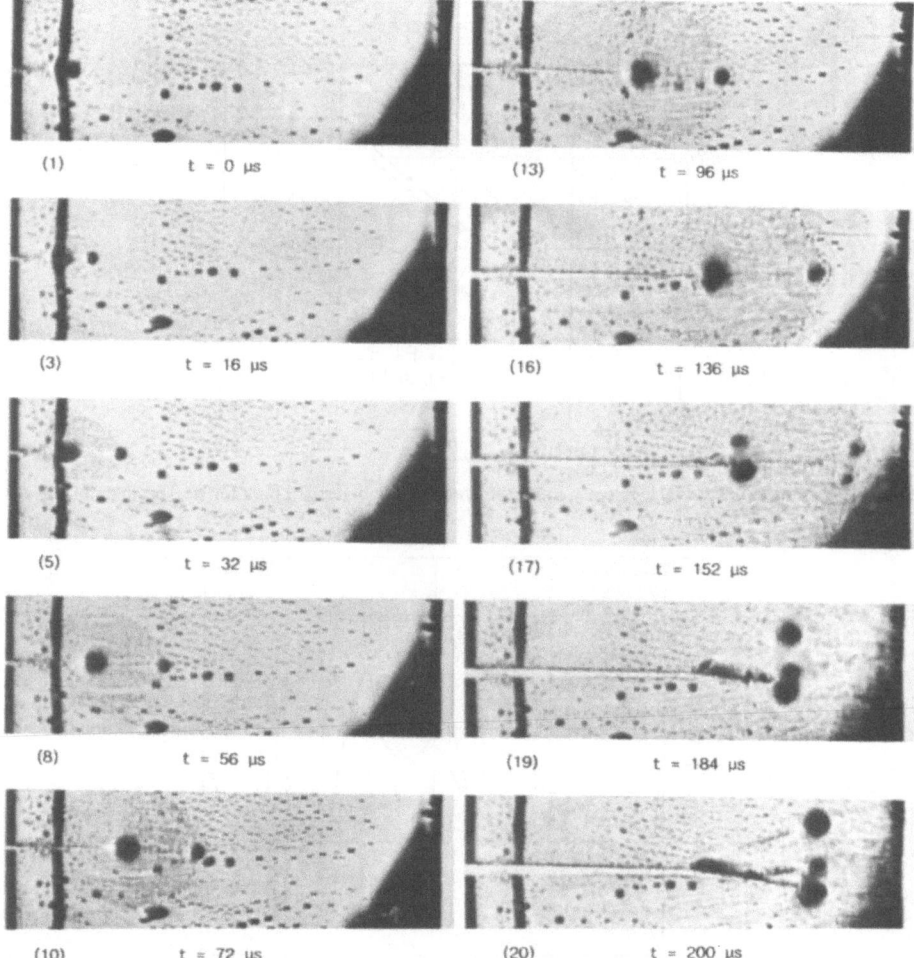

Fig. 90. Series of photographs showing crack propagation in a sandwich specimen with $t_P = 0.001\,\text{m}$ and $t_L = 0.0016\,\text{m}$ subjected to a strain rate $\dot{\varepsilon} = 24/\text{s}$. The two plates were bonded with epoxy resin.

In order to study the behaviour of dynamic crack propagation in sandwich plates a number of specimens consisting of two materials bonded together were fractured at various strain rates. Figure 89 shows the geometry of the specimens. The specimens consisted of a plexiglas and a Lexan plate which were bonded using either epoxy resin or the adhesive cement trichloroethylene–dichloromethane (2/1).

Figure 90 presents a series of photographs showing crack propagation in a sandwich specimen consisting of a plexiglas plate ($t_P = 0.001$ m) and a Lexan plate ($t_L = 0.0016$ m) bonded with epoxy resin. This specimen was subjected to a strain rate $\dot{\varepsilon} = 24/s$. The photographs show clearly that crack propagation started in the plexiglas plate and was followed about 30 µs later (time lag) by propagation in the Lexan plate because debonding takes place around the propagating crack. This implies that the adhesion between the plates and the epoxy resin phase was poor. The time lag of 30 µs depends on the nature and degree of compatibility of the two plates. The crack velocities c_P and c_L and the crack lengths a_P and a_L in the plexiglas and Lexan plates are plotted against time t in Fig. 91. Acceleration, deceleration

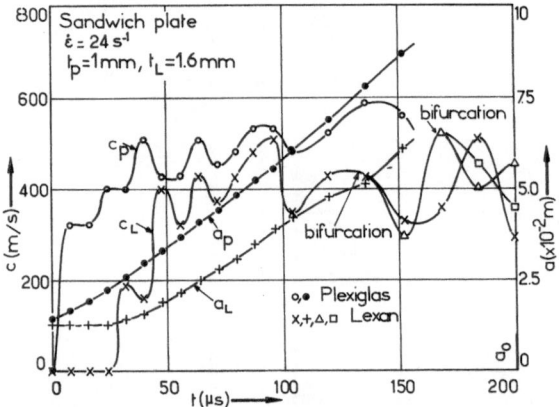

Fig. 91. Variation of the crack velocity c and the crack length a versus time t for the specimen of Fig. 90.

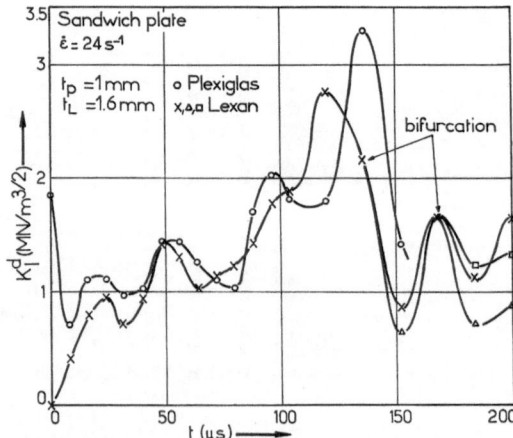

Fig. 92. Variation of the stress intensity factor K_I^d versus time t for the specimen of Fig. 90.

and crack bifurcation phenomena can be observed in both plates in this experiment (Fig. 90, frames 16 and 17). The dynamic stress intensity factor K_I^d in the plexiglas and Lexan plates is plotted versus time t in Fig. 92.

Figure 93 presents a series of photographs showing crack propagation in a sandwich specimen consisting of a plexiglas plate ($t_P = 0.001$ m) and Lexan plate ($t_L = 0.002$ m) bonded with adhesive cement trichloroethylene–dichloromethane (2/1) so that there was no other phase between the two plates. This specimen was subjected to a strain rate $\dot{\varepsilon} = 40$/s. It appears that the adhesion was very good and the degree of compatibility was high, and no debonding of the two plates during crack propagation was observed. As can be seen in this figure, the crack initiated in both plates but with different velocities and with a time lag shorter than 4 μs (because the time between frames is 4 μs). After about 40 μs (Fig. 93, frame 12) the two

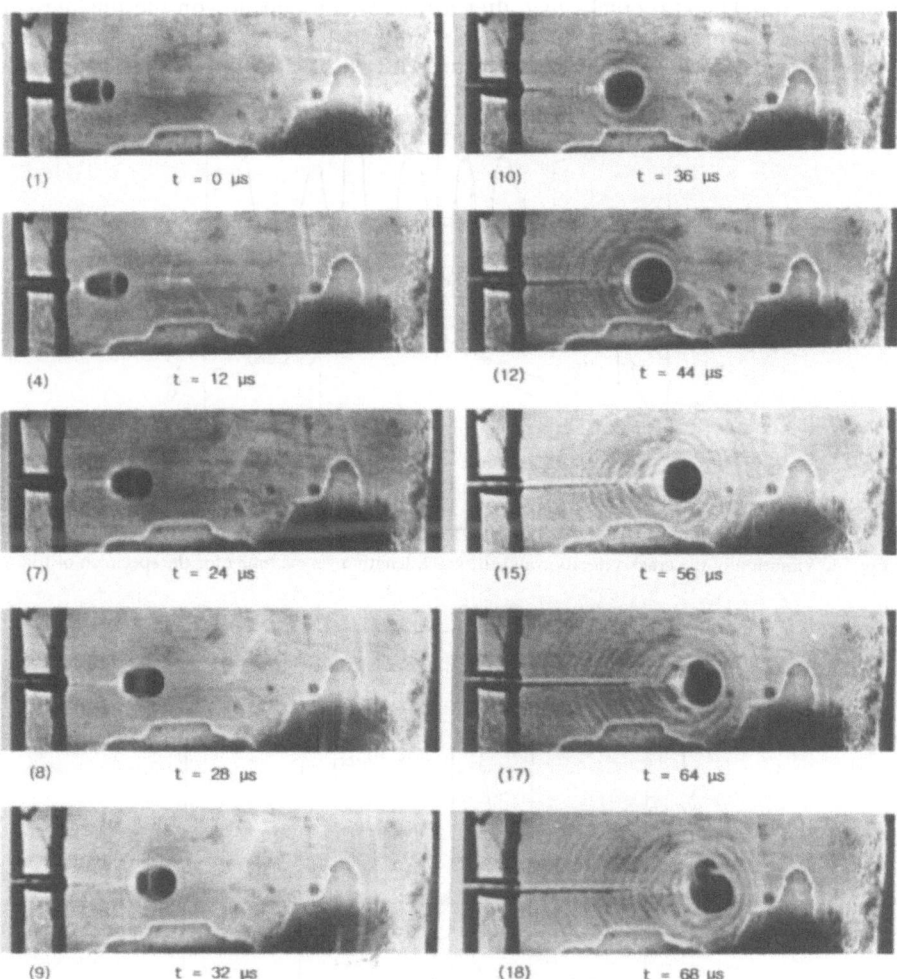

Fig. 93. Series of photographs showing crack propagation in a sandwich specimen with $t_P = 0.001$ m and $t_L = 0.002$ m subjected to a strain rate $\dot{\varepsilon} = 40$/s. The two plates were bonded with the adhesive cement trichloroethylene–dichloromethane.

propagating cracks met and then propagated as a single crack. After a time of about 60 μs a bifurcation was observed. The crack velocities c_P (in plexiglas) and c_L (in Lexan) and the crack lengths a_P and a_L are plotted against time t in Fig. 94. It can be seen that the crack in the Lexan plate accelerates and the crack in the plexiglas plate decelerates until the two cracks coincide after about 40 μs. The variation of the dynamic stress intensity factors K_1^d in plexiglas and Lexan plates is plotted against time t in Fig. 95. It can be observed that the values of the stress intensity factor in the plexiglas plate were strongly influenced by the Lexan plate.

From this research the following observations can be made: (i) Debonding of the phases took place. The degree of debonding depends on the adhesion between the two phases, which in turn depends on the adhesive used. (ii) There is a time lag between crack initiation in the two phases which depends on the nature of the phases and their degree of compatibility. This time lag decreases as the strain rate increases. (iii) The two cracks met after a time which depends on the thickness of the phase and on the strain rates. So, the coincidence time t_{coin}, decreases as the strain rate increases. Also, the time t_{coin} decreases as the thickness, t_L, of the ductile

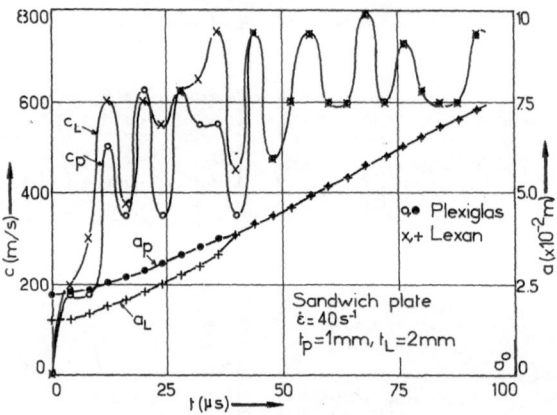

Fig. 94. Variation of the crack velocity c and the crack length a versus time t for the specimen of Fig. 93.

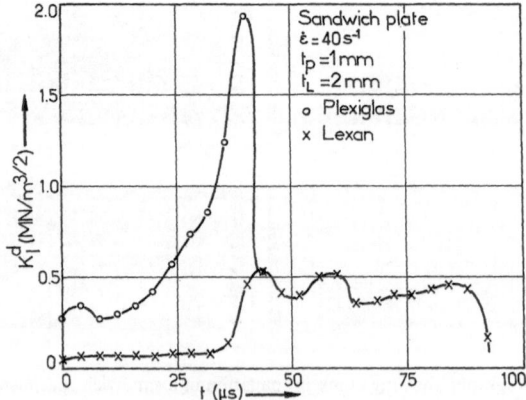

Fig. 95. Variation of the stress intensity factor K_1^d versus time t for the specimen of Fig. 93.

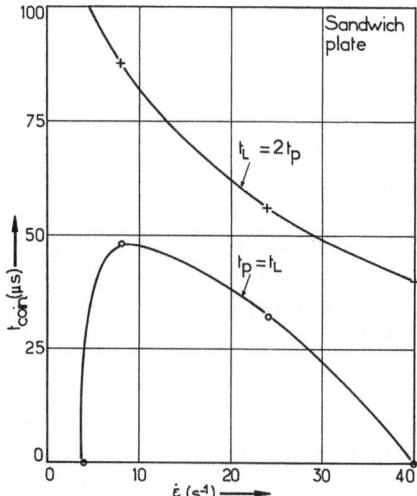

Fig. 96. Variation in the time of coincidence t_{coin} of the two cracks versus strain rate $\dot{\varepsilon}$.

phase, decreases. (iv) The velocities of the two propagating cracks and the stress intensity factor for the sandwich with $t_P = t_L$ up to the time of meeting of the two cracks, remain smaller than the corresponding values for the sandwich with $t_L = 2t_P$.

Figure 96 shows the variation of coincidence time, t_{coin}, with the strain rate, $\dot{\varepsilon}$, for sandwich specimens with thickness of the plates $t_L = t_P$ and $t_L = 2t_P$. It can be observed that for sandwich specimens fractured with the same strain rate, t_{coin} is longer where $t_L = 2t_P$ than where $t_P = t_L$.

3.7 Crack Propagation in Polystyrene

Dynamic crack propagation in thin, edge-notched polystyrene specimens was studied by the method of dynamic caustics [181]. During crack propagation, an intensive zone of crazing surrounds and precedes the propagating crack. Therefore, an active zone ahead of the crack tip is developed. This active zone is related to the velocity of crack propagation and the strain rate of loading. The velocity of the crack and the stress intensity factor, K_I, or the energy release rate, G_I, were strongly influenced by the development of the active zone at the crack tip.

Fatigue crack propagation and damage in thin, single edge-notched polystyrene specimens was studied by Botsis et al. [182–184]. According to their studies, an intensive zone of crazing surrounds and precedes the propagating crack. The system of the crack and craze zone constitute a crack layer, part of which, ahead of the crack tip where crazing accumulates prior to crack layer growth, is called the active zone. A schematic drawing of a rectilinear crack layer is shown in Fig. 97 [183]. The area ahead of the crack tip, limited by the trailing and leading edges, is defined as the active zone. Within this zone the rate of damage growth is always positive. That part of a crack layer complementary to the active zone is the wake zone.

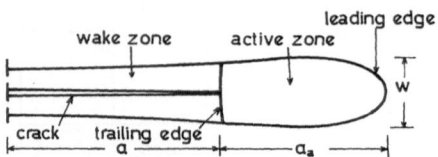

Fig. 97. Schematic drawing of a crack layer.

The width of the active zone was found to increase during slow crack layer propagation. Analysis of craze distribution within the active zone reveals that the craze density decreases away from the crack. Also, a considerable difference in the critical energy release rates was observed in specimens fatigued under different loading levels. This is attributed to the difference in the density of crazes at the critical crack tip.

Recently, experimental studies demonstrated that a zone in the vicinity of the crack tip precedes the propagating crack and this zone is usually termed a plastic zone [185], process zone [186, 187], dissipation zone [188] or deformation zone [189].

The objective of this study was to examine crack propagation in polystyrene and the influence of the active zone and the strain rate on the crack velocity, on the stress intensity factor and on the energy release rate, using the method of dynamic caustics [164, 165, 167]. In addition, the energy release rate was calculated from experimental values of the stress intensity factor.

The method of dynamic caustics was used to study the variation of crack propagation velocity and stress intensity factor K_1. The dynamic stress intensity factor K_1 was evaluated by using Rel. (601).

The elastic energy release rate, G_1, is calculated for plane stress by Rel. (121). The value of G_1 at the crack initiation is defined as G_{1c}, the critical strain energy release rate. The fracture surface energy, γ, as defined in the Griffith theory [190] equals:

$$\gamma = G_1/2 = K_1^2/2E \tag{620}$$

The stress intensity factor K_1, is experimentally calculated from the caustics by Rel. (601). The critical stress intensity factor, K_{1c}, or the corresponding fracture surface energy, γ_c, is a useful index of the ultimate properties of the material. For the critical K_{1c}, γ_c is given by:

$$\gamma_c = K_{1c}^2/2E \tag{621}$$

Notched polystyrene specimens of dimensions $0.220\,\text{m} \times 0.050\,\text{m} \times 0.00025\,\text{m}$ were used for the experimental investigation. The specimens initially contained an edge transverse V-notch of $25°$ and length $a_0 = 0.006\,\text{m}$ and were subjected to a dynamic tensile load to fracture using a Hydropulse high-speed testing machine. A Cranz–Schardin high-speed camera was used to record the dynamic crack propagation. In the optical set-up used in the experiments, $z_0 = 0.80\,\text{m}$ and $\lambda_m = 0.77$. The properties of polystyrene are: $E = 2.2\,\text{GPa}$, $v = 0.3$ and the stress-optical constant $c_t = 0.74 \times 10^{-10}\,\text{m}^2/\text{N}$. The strain rate $\dot\varepsilon$ in the present experiments were $10/\text{s}$ and $20/\text{s}$.

The series of photographs in Fig. 98 shows detail of the crack propagation process in a polystyrene plate with a strain rate $\dot\varepsilon = 10/\text{s}$. From these photographs we may see that the form of the caustics is not influenced by the crack layer which is developed at the crack tip. This means that the magnitude of the crack layer is

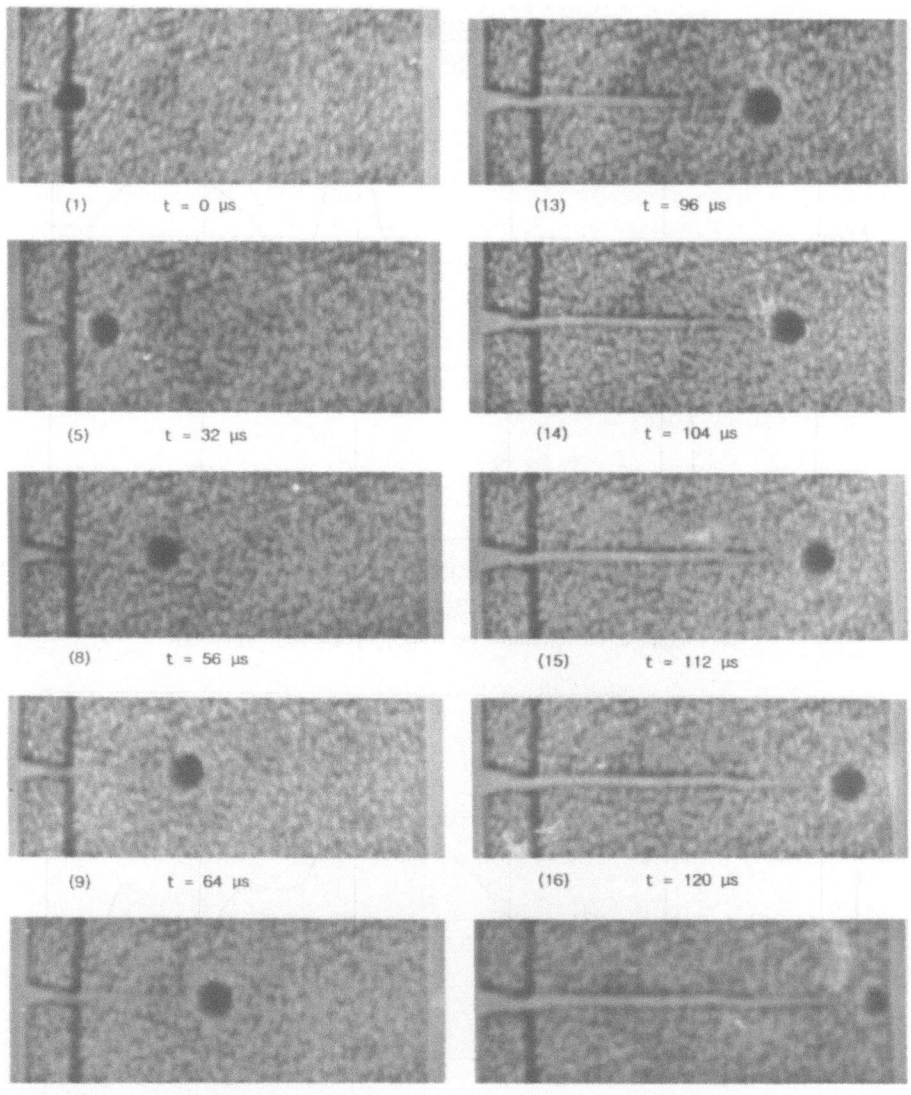

(1)   t = 0 µs

(13)   t = 96 µs

(5)   t = 32 µs

(14)   t = 104 µs

(8)   t = 56 µs

(15)   t = 112 µs

(9)   t = 64 µs

(16)   t = 120 µs

(10)   t = 72 µs

(17)   t = 128 µs

Fig. 98. A series of photographs showing crack propagation in a polystyrene plate with a strain rate $\dot{\varepsilon} = 10/s$.

smaller than the magnitude of initial curve of the caustics. The magnitude of the initial curve is given by [165]:

$$r_0 = D_t^{\max}/\lambda_m \delta_t^{\max}(c) = 0.0016 \text{ m} \tag{622}$$

where $D_t^{\max}$ is the mean value of the maximum diameter of the caustic ($= 0.0039$ m), λ_m is the magnification ratio of the optical set-up ($= 0.77$) and $\delta_t^{\max}(c)$ is the correction factor, which depends on the velocity of the crack ($= 3.175$). Therefore, the magnitude of the crack layer must be smaller than 0.0016 m. The crack layer at the crack tip is shown in Figs 99 and 102. The crack layer, a region of crazes around the crack

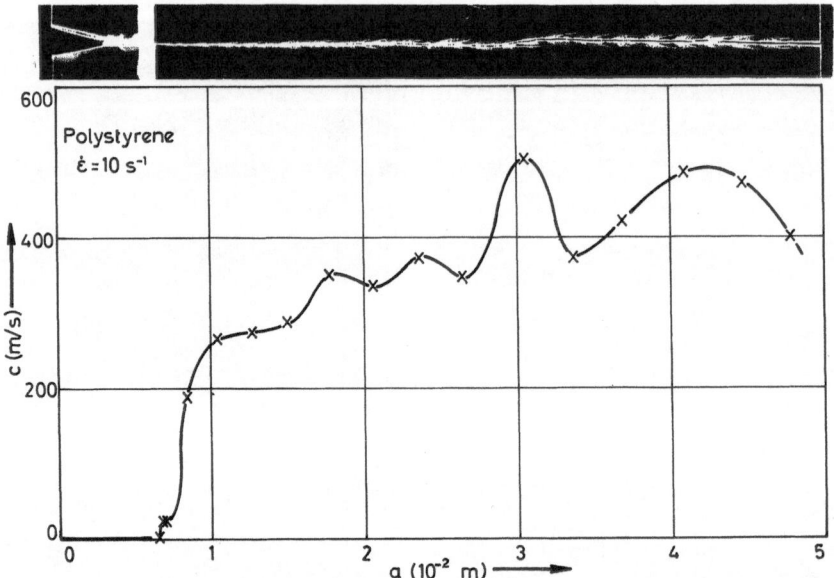

Fig. 99. Variation of the crack propagation velocity c, versus crack length a, for strain rate $\dot\varepsilon = 10/s$. The photograph at the top shows the crack propagation morphology.

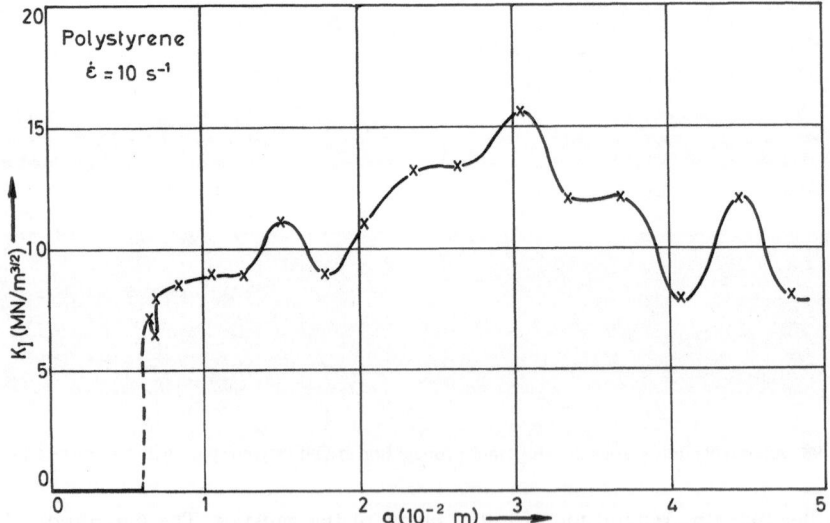

Fig. 100. Variation of the stress intensity factor K_1, versus crack length a, for strain rate $\dot\varepsilon = 10/s$.

tip, increases as the strain rate increases. As the crack layer is developed, the crack propagation velocity decreases and thus the magnitude of the crack layer decreases. After that, a new crack acceleration is observed and a new crack layer is developed. This phenomenon is presented in Figs 99 and 102. In Fig. 99 the crack propagation velocity c, versus the crack length a, is presented for strain rate $\dot\varepsilon = 10/s$. The photograph at the top of the figure shows the crack propagation morphology with

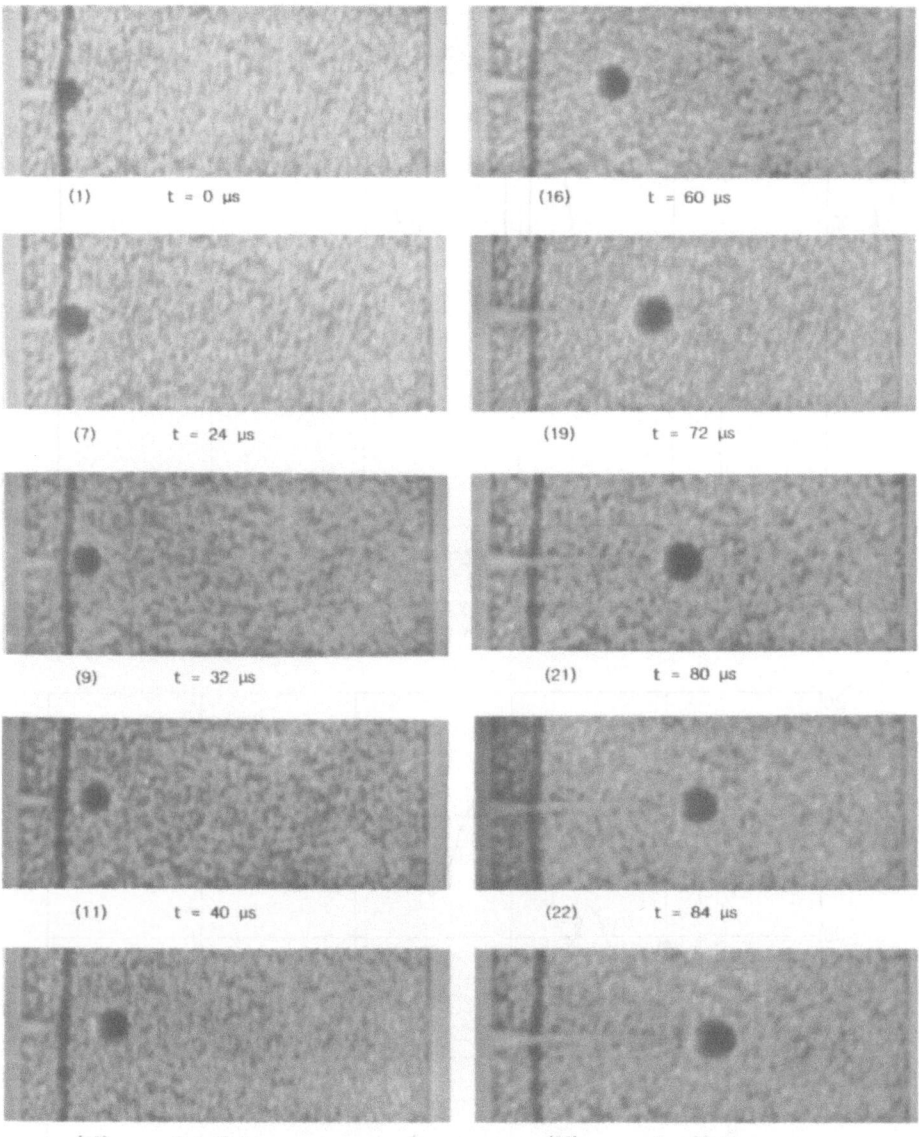

(1)   t = 0 μs
(16)   t = 60 μs
(7)   t = 24 μs
(19)   t = 72 μs
(9)   t = 32 μs
(21)   t = 80 μs
(11)   t = 40 μs
(22)   t = 84 μs
(13)   t = 48 μs
(23)   t = 88 μs

Fig. 101. A series of photographs showing the crack propagation in a polystyrene plate with a strain rate $\dot{\varepsilon} = 20/s$.

the craze regions. In this figure we can see a correspondence of the minima of velocity with the crack layer, while the maxima of velocity correspond to regions of the crack path without crack layers. It is concluded from this that the crack velocity decreases with the crack layer.

The stress intensity factor, K_I, is calculated from the diameter of the caustic by using Rel. (601). The variation of K_I with crack length a for strain rate $\dot{\varepsilon} = 10/s$ is presented in Fig. 100. From this figure it may be observed that K_I varies approximately in synchronism with the velocity.

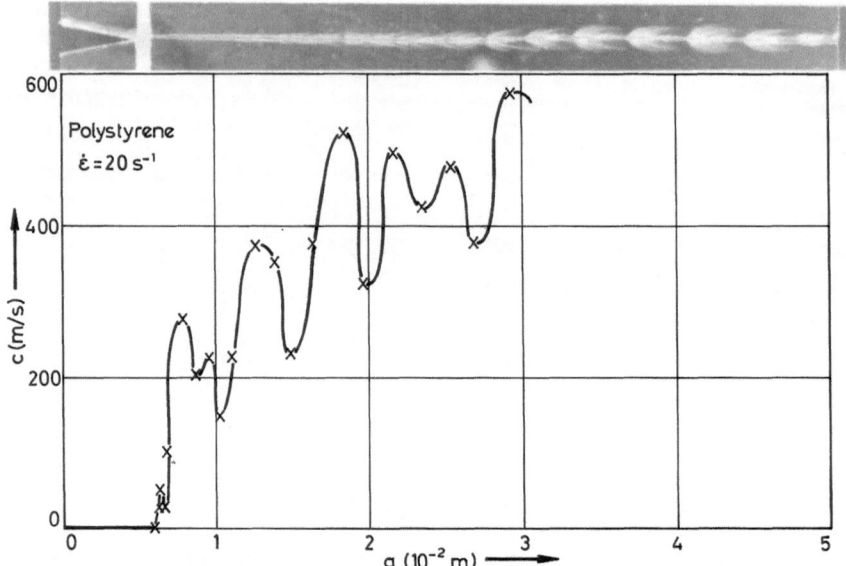

Fig. 102. Variation of the crack propagation velocity c, versus crack length a, for strain rate $\dot{\varepsilon} = 20/s$. The photograph at the top shows the crack propagation morphology.

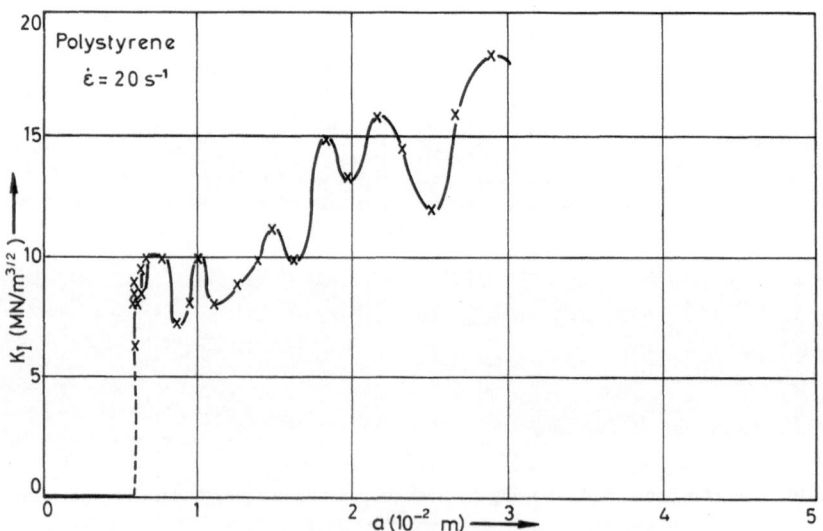

Fig. 103. Variation of the stress intensity factor K_1 versus crack length a, for strain rate $\dot{\varepsilon} = 20/s$.

A series of photographs in Fig. 101 shows the crack propagation process in a polystyrene plate with a strain rate $\dot{\varepsilon} = 20/s$. From these photographs we may see more intense examples of the phenomena in Fig. 98.

Figure 102 shows the variation in the crack propagation velocity, c, with the crack length, a, the strain rate $\dot{\varepsilon} = 20/s$. The photograph at the top of the figure shows the crack propagation morphology with the craze regions. We can see that the crack layers are wider than those of the crack path in Fig. 99 and also that the crack velocities are greater than those in Fig. 99 for strain rate $\dot{\varepsilon} = 10/s$. In addition, in Fig. 102 it may be observed that the minima of the crack velocity correspond to the crack layers, while the maxima of the velocity correspond to the path regions without crack layers (regions between crack layers). Figure 103 shows the variation of the stress intensity factor, K_1, with the crack length, a, for strain rate $\dot{\varepsilon} = 20/s$.

By comparison of Figs 99 and 102 it may be observed that the crack velocity increases as the strain rate increases. The crack velocity increases smoothly up to a maximum value of 520 m/s at the position of crack path $a = 0.031$ m for $\dot{\varepsilon} = 10/s$, while for $\dot{\varepsilon} = 20/s$ the crack velocity rapidly increases up to a maximum value of 580 m/s at $a = 0.029$ m. The same variation can be observed for the stress intensity factor, K_1.

Figure 104 shows the variation of energy release rate, G_1, which was calculated using Rel. (121), versus the crack length, a, for strain rate $\dot{\varepsilon} = 10/s$ and 20/s. The energy release rate G_1, for strain rate $\dot{\varepsilon} = 20/s$ is greater than that for $\dot{\varepsilon} = 10/s$. The critical fracture surface energy γ_c for $\dot{\varepsilon} = 10/s$ is $\gamma_c = 11.65$ KJ/m^2 and for $\dot{\varepsilon} = 20/s$ is $\gamma_c = 18.33$ KJ/m^2. It may be observed that in the crack layer, where there is a great craze density, the energy release rate G_1, reduces considerably.

The effect of the crack layer on dynamic crack propagation in polystyrene plates has been studied. The results may be summarized as follows: i) The shape and magnitude of the active zone (crack layer) are strongly dependent on the strain rate. ii) During crack propagation, the active zone is discontinuously developed at the crack tip. iii) The crack velocity and the stress intensity factor decrease where an active zone is developed, while they increase in regions where no active zone is developed. iv) The active zone is the area with intense damage. Therefore, the modulus of elasticity E decreases and then a reduction in the crack velocity is observed. v) The active zone is formed ahead of the crack propagation tip but when the crack velocity is great, the active zone behind the crack tip remains until a new active zone is formed ahead of the crack tip.

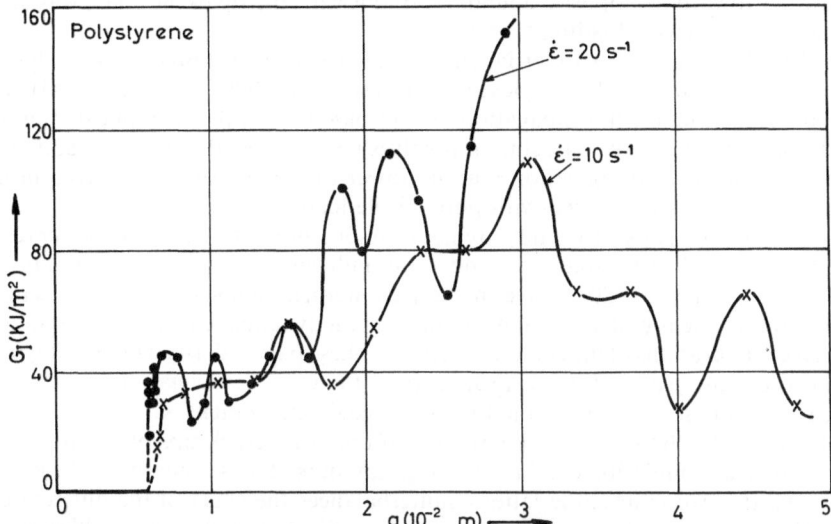

Fig. 104. Variation of the energy release rate G_1, versus crack length a, for strain rate $\dot{\varepsilon} = 10/s$ and 20/s.

3.8 Crack Propagation Under Impact Bending Load

The method of caustics has been applied to investigate crack propagation and interaction between a main propagating edge-crack and a collinear stationary one in plane specimens subjected to impact loading applied in a three-point bending mode [191]. The high sensitivity of the experimental method used, combined with variable collinear crack configurations, has disclosed interesting results concerning the continuously variable dynamic stress and strain distributions around the propagating main crack. By this arrangement of sensitive cracks it was possible to verify the existence and relative motion of the neutral zone appearing ahead of the moving crack tip which separates regions of tensile and compressive transient stress in bent bars.

The process of crack propagation in specimens subjected to impact under bending is of significant interest, and has been a subject of investigation by many authors. Indeed, three-point bending was applied to the process of stress wave emission during fracture [192–194], and for the study of stress conditions in the time intervals just before crack initiation and during the initial steps of crack propagation [195, 196].

Crack propagation under three-point bending is also of interest because the stress field is not a uniform one, as in most cases of tensile specimens, but a mixed one, where regions of tensile and compressive stresses coexist. This phenomenon was extensively discussed by Dao and Hermann [194] and by Theocaris and Andrianopoulos [197, 198]. They have studied the propagation of transverse edge-cracks in bent plates and concluded that the moving crack tip faces a diminishing stress field, which becomes zero when the crack reaches the boundary of the specimen. A neutral zone was always detected in bent bars, located along the axis of the crack, ahead of the crack tip, and separating the remaining ligament of the bar into two parts, one tensile and the other compressive. This zero-stress area was pushed forward, always staying in front of the crack tip, as the crack propagated in the bar toward its opposite boundary. In this way, the tip of the crack was always subjected to a tensile loading.

In Refs [199] and [200] inertial effects and stress wave propagation in dynamic bending were reported. From these reports (especially Refs [197] and [198]), it can be concluded that crack propagation in bent plates is a rather complicated process, and also a very sensitive one, since it involves tensile stress distributions and energy levels during crack propagation that are lower than the energies involved in most fracture processes associated with pure tensile loads.

We therefore chose to study dynamic interaction of two collinear cracks in specimens which were fractured under the influence of an impact bending load. Theocaris et al. [201–203] have in the past studied collinear crack interaction in PMMA specimens under the influence of different dynamic tensile loads, for different initial configurations of the coalescing cracks. These papers derived interesting results concerning the paths of crack propagation, the crack propagation velocities and the stress intensity factors at the tips of the coalescing cracks.

These results show that the interaction of opposing crack tips of collinear cracks leads to an accumulation of relatively large amounts of tensile energy at the ligament between the two cracks; the latter counterbalances the effect of the different crack lengths on the mode of crack propagation. Skew–parallel and collinear crack coalescence in tensile specimens is thus accompanied by such phenomena as the sudden propagation of the initially stationary crack, strong deviations of crack propagation paths, abrupt changes of the K_I-factor and velocity values, etc. These phenomena can be diminished or even cancelled by undertaking dynamic three-point bending experiments, where much lower dynamic tensile stress levels due to external loading, are involved.

Our main aim was, therefore, to extend the study of collinear crack propagation to some processes and phenomena which could not be detected under tensile loading. First, we examined the case of crack interaction between a moving crack and a permanently stationary crack. A useful comparison between tensile and bending processes could therefore be derived because the stationary crack can be used as a "witness crack" for the propagating one, its tips giving the mode and amount of stressing in its vicinity by the form and size of the caustics.

The location of the neutral zone ahead of the moving crack tip in the bent bar could be investigated by means of dynamic photoelasticity. Previous investigations using photoelasticity gave very interesting results [204, 205], but also proved that the method is not sensitive enough to follow the abrupt changes in stress distribution which take place immediately after crack initiation. The method of caustics [164], has proved to be much more sensitive in detecting such abrupt changes in stress concentration, as may occur in our specimen configurations.

The length of our witness crack was always equal to the initial length of the edge-crack, which ensured that propagation always started from the edge-crack. As the edge-crack was propagating, a change in the stress distribution at the tip of the witness crack developed, and this led to useful conclusions concerning the size and the position of the neutral zone of the fracturing bar.

Although the investigation is limited to PMMA specimens, the results can give an insight into the dynamic crack propagation processes that take place in any brittle material under impact bending loads. They can therefore be useful to many applications of modern technology, such as transformation or storage of liquified gases under very low temperatures (and high pressures), where embrittlement of the metallic materials of the containers usually takes place. On the other hand, impact processes are also connected with embrittlement, even of materials which show a ductile behaviour under static or quasi-static loading (e.g. metals).

It is also known that the process of crack creation and propagation in most engineering materials is connected with micro-crack creation and coalescence before the abrupt fracture process is initiated. Since it is known that the most severe cases of crack coalescence are those involving collinear cracks, this study is again justified. Indeed, the model of an unstable crack coalescing with a stationary collinear crack corresponds to this general problem of practical crack initiation and propagation, while, at the same time, the stationary crack can be used as a witness crack for the propagating one and its interaction with the neutral zone of the bent specimen.

For the study of the fracture process in bent bars it is necessary to determine the variation of the stress intensity factor at the tip of the moving crack and to evaluate the crack propagation velocity. From the geometric characteristics of the caustic we can calculate, with high accuracy, the stress intensity factor at the tip of the crack. Further, we can accurately deduce the position of the extremities of the moving crack front at any instant of time from the respective positions of the caustics and thus derive accurately the instantaneous value of the crack propagation velocity. The stress intensity factor K_I for PMMA plates is given by Rel. (601).

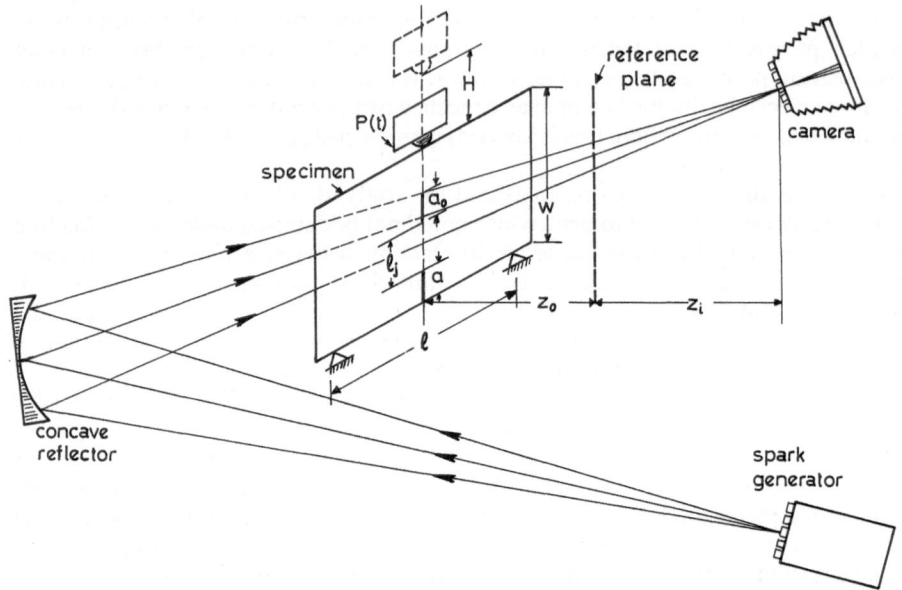

Fig. 105. Optical arrangement for transmitted caustics and geometry of specimen.

The relative instantaneous distance of the neutral point from the instantaneous position of the edge-crack tip is given by (Fig. 105):

$$h_j^r = l_j \frac{\lambda_j}{\lambda_j - 1} \tag{623}$$

where:

$$\lambda_j = \frac{1}{f_{1j}(a)} \left(\frac{D_{te}^{max}}{D_{ti}^{max}} \right)_j^{\frac{5}{2}} \left(\frac{\delta_{ti}^{max}}{\delta_{te}^{max}} \right)^{\frac{5}{2}} \tag{624}$$

where l_j denotes the distance between the edge-crack tip and the internal crack tip (near tip), $f_{1j}(a)$ is the correction factor of stress intensity factor K_I [102, 121] in terms of edge-crack length a and D_{te}^{max} and D_{ti}^{max} are the diameters of the caustics at the edge-crack tip and the internal crack tip (near tip) respectively. Quantities δ_{te}^{max} and δ_{ti}^{max} are correction factors to diameters D_{te}^{max} and D_{ti}^{max} respectively. The value of δ_{ti}^{max} is 3.1702 (value for stationary cracks). The absolute instantaneous position of the neutral point is then given by:

$$h_j^a = h_j^r + a$$

where a is the instantaneous edge-crack length.

For the study of crack propagation in a bar bent under impact a Cranz–Schardin high-speed camera was utilized capable of 24 exposures, with a maximum frequency of 10^6 frames per second. The exposure frequency was regulated in groups of 4, 5, 5, 5 and 5 intervals. This arrangement yielded flexibility for improving observation of the crack propagation phenomena. Figure 105 shows the optical part of the experimental set-up. It consists of the spark generator assembly, a concave reflector of diameter of 0.50 m and a focal distance $f = 3.50$ m, and the multiple camera for recording the phenomenon. The bending tests were executed in an impact testing machine with a semi-cylindrically notched hammer of weight $P(t) = 17.17\,\text{N}$, falling from a height $H = 0.30\,\text{m}$.

For the experimental study of crack propagation in bent bars a number of PMMA specimens were used, all having the same dimensions $L = 0.21\,\text{m}$, $w = 0.075\,\text{m}$ and $d = 0.003\,\text{m}$. The specimens contained a transverse edge-crack of initial length $a_0 = 0.01\,\text{m}$, which was sawn at mid-length of the long boundary of the bar and placed opposite to the impact point of the hammer. Along the axis of this edge-crack an internal crack of equal length ($a_0 = 0.01\,\text{m}$) was sawn. The slits were made with a very fine saw-disc, so that the opening between the lips of the artificial cracks was less than $0.0003\,\text{m}$. The distance l_j between the adjacent lips of the initial collinear cracks was varied by $0.003\,\text{m}$ for every pair of specimens, starting from a minimum distance of $0.01\,\text{m}$ up to a maximum distance of $0.039\,\text{m}$.

Figure 106 presents a series of photographs showing crack propagation and interaction in a specimen, for which the initial distance between the edge-crack tip and the nearest tip of the witness crack was $l_1 = 0.016\,\text{m}$. During the whole crack propagation process both tips of the internal crack are under the influence of an overall tensile-stress field, as can be immediately inferred from the form of the caustics, all of which are smooth and circular without any cusps. Therefore, the neutral point of the bent plate lies beyond the furthest crack tip of the witness crack, that is at a distance larger than $0.026\,\text{m}$.

It is worthwhile noticing that in frame 14 of Fig. 106 the moving crack tip reaches the nearest tip of the stationary witness crack and thus one unified singular stress concentration is created at this point. In frame 16 all singular stresses are concentrated at the furthest tip of the witness crack, which is now the unique tip of the unified crack. Nevertheless, crack propagation from the furthest point is suspended for a short time interval by a crack-arrest process, as we have already mentioned. In frame 18 the crack re-initiates with a crack propagation velocity and a value for the stress intensity factor, different from those before arrest, as may be derived from the diagrams of Figs 107 and 108, where the variation of K_I versus time and crack length are presented.

In Fig. 109, the distance between the internal crack tips of the two collinear cracks is $l_2 = 0.032\,\text{m}$. Now, both crack tips of the witness crack lie in the compressive stress field of the bar since both caustics at the tips are cuspoid caustics, which

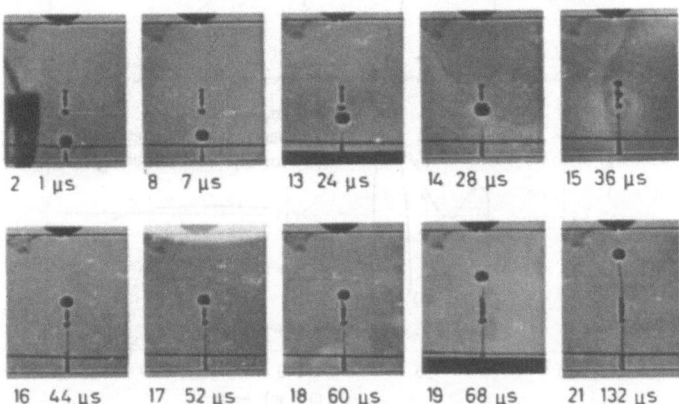

Fig. 106. Series of photographs showing crack propagation and interaction in a specimen where the initial distance l_j between the collinear cracks was $l_1 = 0.016\,\text{m}$ (internal crack under all-tensile field).

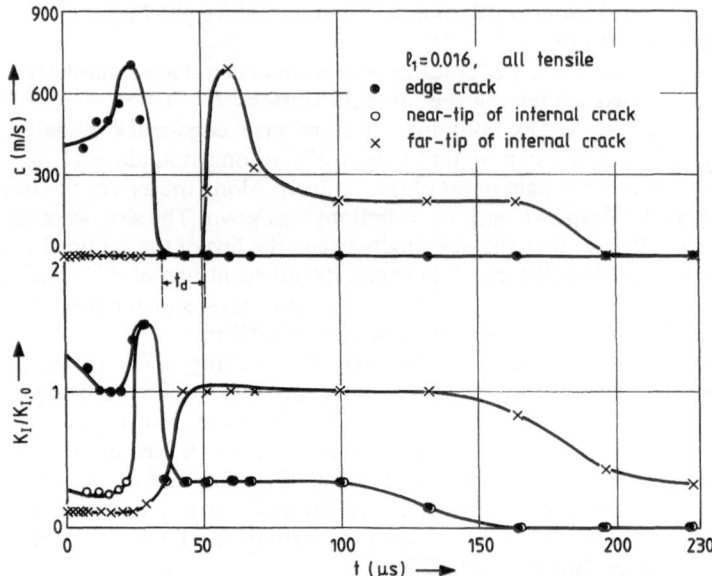

Fig. 107. Variation of crack propagation velocity c and stress intensity factor K_1 normalized to its initial value $K_{1,0}$ versus time for a specimen with $l_1 = 0.016$ m.

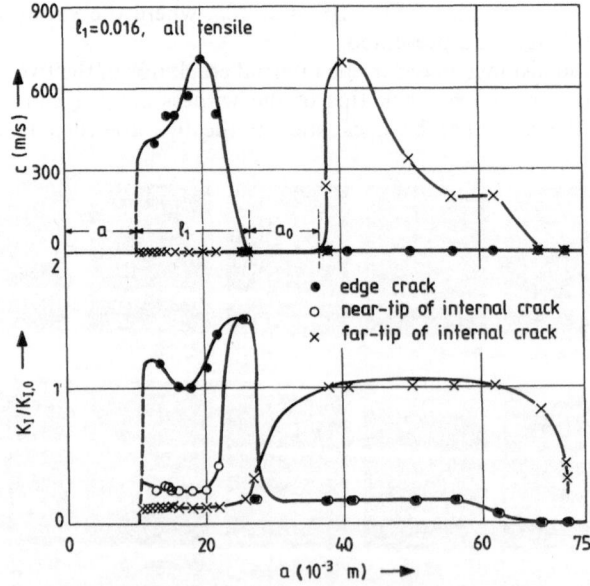

Fig. 108. Variation of the crack propagation velocity c and the stress intensity factor K_1 normalized to its initial value $K_{1,0}$ versus crack length for a specimen with $l_1 = 0.016$ m.

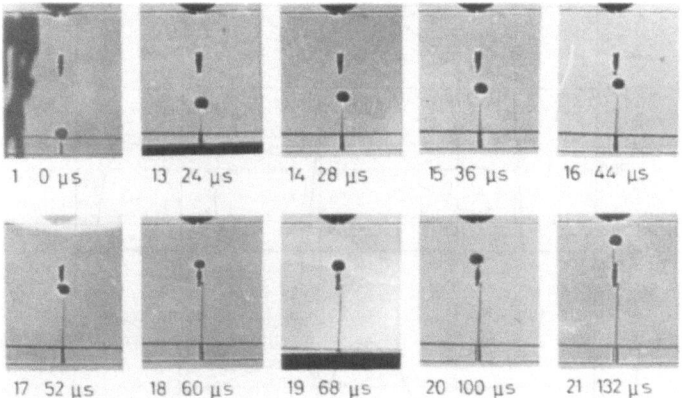

Fig. 109. Series of photographs showing crack propagation and interaction in a specimen with an initial distance between the collinear cracks $l_2 = 0.032$ m (internal crack under all-compressive field).

indicates an overall compression [60]. This state of stress lasted until the moving crack tip coalesced with the witness crack (frame 16), where the stress configuration at the first tip of the stationary witness crack become tensile. After complete crack coalescence and transfer of energy to the remotest crack tip, a tensile singular stress configuration is established in this furthest tip of the witness crack, which causes, again after a short period of crack arrest, a new crack initiation. According to the clear indications of the caustics, a neutral point should exist, as the edge-crack initiates, at a distance shorter than 0.032 m from the tip of the edge-crack. Figs 110 and 111 show associated parametric variables.

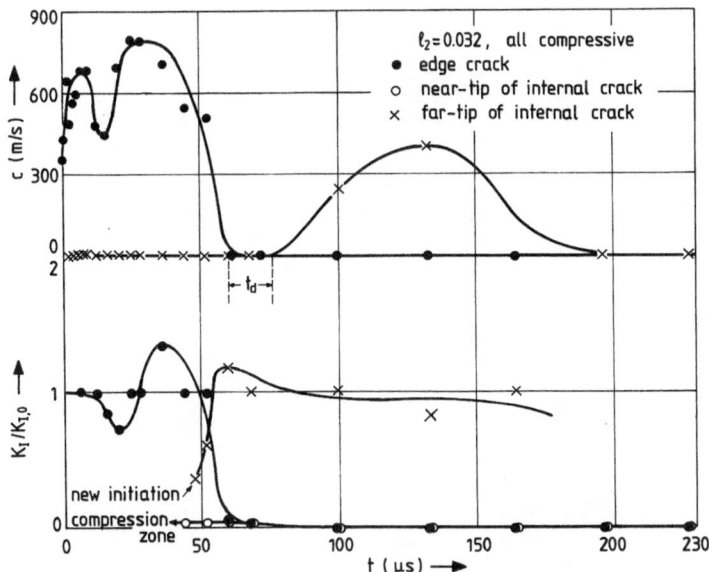

Fig. 110. Variation of the crack propagation velocity c and the stress intensity factor K_I normalized to its initial value $K_{I,0}$ versus time for a specimen with $l_2 = 0.032$ m.

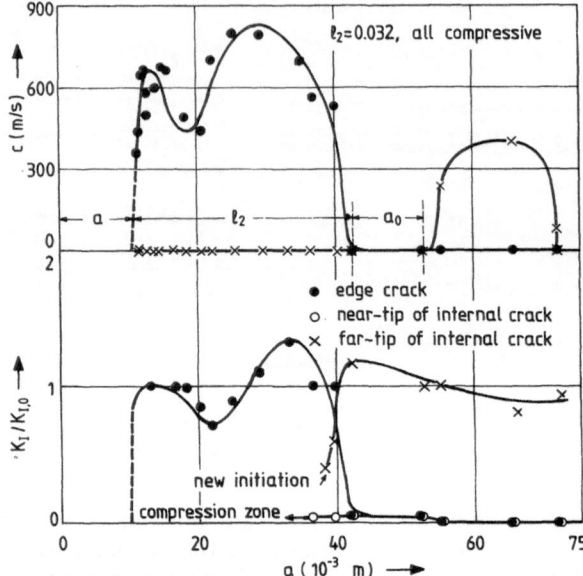

Fig. 111. Variation of the crack propagation velocity c and the stress intensity factor K_I normalized to its initial value $K_{I,0}$ versus crack length for a specimen with $l_2 = 0.032$ m.

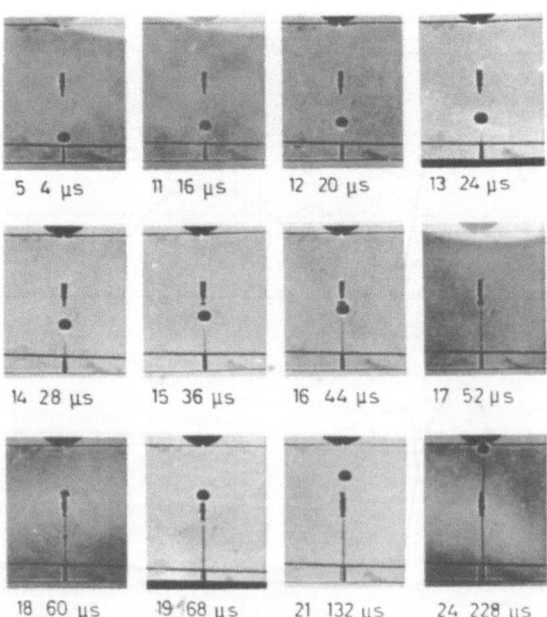

Fig. 112. Series of photographs showing crack propagation and interaction in a specimen with an initial distance between the collinear cracks $l_3 = 0.026$ m (internal crack under tensile–compressive field).

Figure 112 presents the intermediate case, where the witness crack initially has one tip in the tensile and the other in the compressive zone of the bar. This is clearly indicated by the caustics of the witness crack, the one being cuspoid (compression) and the other circular (tension). In this case, the distance between the tip of the edge crack and the nearest tip of the witness crack was $l_3 = 0.026$ m.

As the edge-crack propagated, this configuration of caustics remained practically unchanged until frame 11, when the tensile stress distribution at the near tip of the witness crack diminished, 4 µs later it vanished (frames 12 and 13), to become instantly compressive (frame 14). It returned again 18 µs later to a tensile state, (frame 15), when the moving crack approached close to the near tip of the witness crack. Subsequent steps in the crack propagation process exhibit similar phenomena to those in previous examples.

An interesting observation derived from the results of the tests with caustics is that the neutral point of the bent bar, which at the start of propagation of the edge-crack lay far away from its tip, was first attracted by the propagating tip during initiation of propagation and was then stabilized at some short distance ahead of the moving crack tip, always remaining as a precursor to the propagating crack tip. The same phenomenon appeared when the crack jumped nearly instantaneously along the witness crack, with the neutral point again remaining in a precursor position. This interesting phenomenon could be detected only because of the sensitivity of the method of caustics.

This redistribution of stresses in the bent specimen shows some similarities with the phenomena detected in the case of collinear crack interaction in tensile specimens [203]. Indeed, when there is an interaction of collinear cracks a diminution of the stress concentration at the near tip of stationary cracks is observed during a period after the initiation of the edge-crack; when the moving crack tip approaches the neighbouring tip there is a strong interaction phenomenon accompanied by abrupt increase of the K_I-value at the tip of the stationary crack. This first phase of stress redistribution is strong in the present case of a bent specimen, so that initially tensile stresses at the near tip of the stationary crack are reversed into compressive stresses.

Experimental evidence shows, therefore, that 4 µs after the initiation of propagation of the edge-crack, the neutral point of the bar lies ahead of the moving crack tip. Furthermore, 24–28 µs after crack initiation, as the moving crack tip propagates into the interior of the bent bar, the neutral point moves towards the moving crack tip, thus leaving the nearest tip of the witness crack under a compressive load; from this moment on, it follows the motion of the propagating crack, as a precursor to the propagating crack tip. The exact location of the neutral point at any instant can be calculated from the data taken from the caustics.

The variation of the crack propagation velocity c and the respective stress intensity factor K_I versus time and crack length a were derived and plotted in Figs 107, 108, 110, 111, 113 and 114 from the series of frames of all tests and for the three different positions of the stationary crack.

Notice that there is an abrupt increase in the velocity and stress intensity factors when the moving cracks arrive at a certain distance from the stationary crack. This phenomenon, which is mainly due to the coalescence of the singular stress fields at the crack tips as they oppose and approach each other, was also observed in the case of collinear crack interaction in tensile specimens. The result is high velocity and high stress intensity factor values in all cases. It may also be deduced from these figures that, as soon as the crack tip arrived at the near tip of the stationary

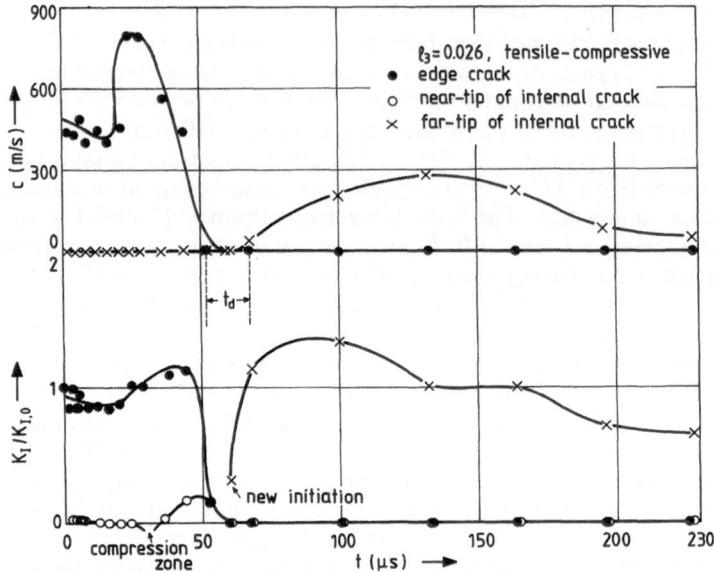

Fig. 113. Variation of the crack propagation velocity c and the stress intensity factor K_I normalized to its initial value $K_{I,0}$ versus time for a specimen with $l_3 = 0.026$ m.

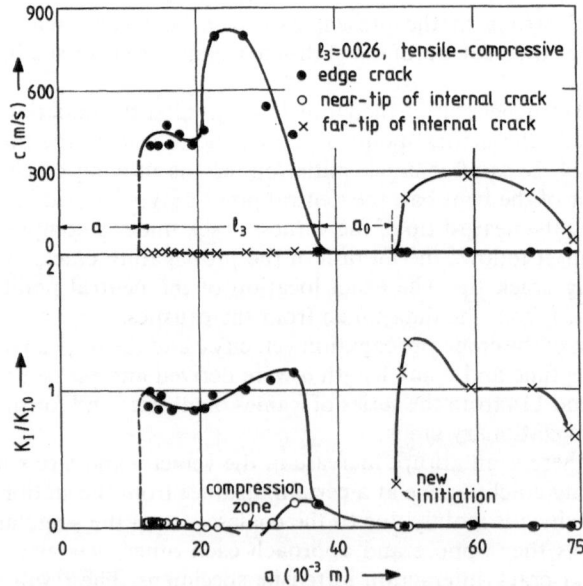

Fig. 114. Variation of the crack propagation velocity c and the stress intensity factor K_I normalized to its initial value $K_{I,0}$ versus crack length for a specimen with $l_3 = 0.026$ m.

crack, the propagating crack stopped, after which the tip of the unified crack advanced almost abruptly to the furthest tip of the witness crack.

After a period of crack arrest – time t_d – at the furthest tip of the witness crack, the crack continued with a lower crack propagation velocity in most cases. This lower value is obviously due to the fact that the tensile-stress field at the point of re-initiation is much lower than it is at first initiation, since it always lies close to but behind, the neutral point of the bent bar. The re-initiation velocity was also lower than the velocity of first initiation. This initiation velocity diminished as the initial distance between the two collinear cracks increased.

The fluctuations of crack propagation velocities detected resemble the corresponding phenomena that appear in the case of a single edge-crack propagating in a bent plate [197, 198]. The much stronger effect of collinear crack coalescence becomes obvious as the moving crack tip approaches close to the stationary crack (increase of the K_1 and c values).

This coalescence effect is weaker than in the case of collinear crack coalescence in tension specimens, and is almost insignificant during the first period of edge-crack initiation, when the stationary crack lies far away from the propagating crack in the compressive region.

Even in the case where the stationary crack lies in the tensile region, the singular stress distribution at the tip remains significantly smaller than that at the tip of the moving crack, and so the coalescence phenomena are weaker. For example, in the cases presented in Figs 107 and 108 the stress intensity factor at the tip of the witness crack is only 19 per cent of the corresponding value of K_1 for the moving crack, a stress configuration which could not appear in a pure tensile specimen.

In Figs 108, 111 and 114, the variation of the stress intensity factor with crack length, a, is presented for the tip of the moving edge crack, normalized to its initial value $K_{1,0}$ at crack initiation. The variation of the K_1-value shows typical fluctuations, which are related to the crack propagation in a bent specimen [203]. The effect of coalescence is also obvious here; the moving crack approaches the witness crack and results in an increase of the K_1-value at this region, and in an increased re-initiation value of the stress intensity factor when the period of the crack arrest is completed.

Figure 115 presents the variations of the length of the edge crack a, the relative distance of the neutral point from the moving crack tip h_j^r and of the absolute distance of this point from the longitudinal boundary of the bar h_j^a versus time for the first period of crack propagation (36 μs) in a tension–compression specimen. The measurements of these distances were executed on the photographs of Fig. 112. It may be observed from Fig. 115 that there is always a slight continuous attraction of the neutral point for the tip of the propagating crack. This attraction becomes maximum when the minimum distance appears between the neutral point and the crack tip, i.e. $h_{j\min}^r = 3 \times 10^{-3}$ m at the time instant $t = 27.5$ μs. The distance between these two points then increases continuously until complete fracture ensues.

The variation of the relative and absolute velocities of the neutral point c_p^r and c_p^a is given in Fig. 116 in terms of the instantaneous crack length a. In the same figure the variation of the velocity of crack propagation is given. It may be observed from these curves that the following relation is valid:

$$c \simeq c_j^a - c_p^r$$

The variations of these velocities are given up to $t = 36$ μs; beyond this time, the influence of the witness crack on the velocity of the moving crack becomes significant.

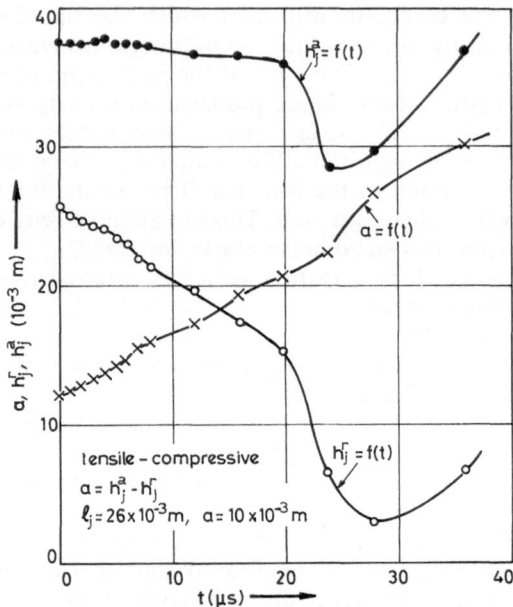

Fig. 115. Variation of the edge-crack length a, the relative neutral point distance h_j^r and the absolute neutral point distance h_j^a versus time t.

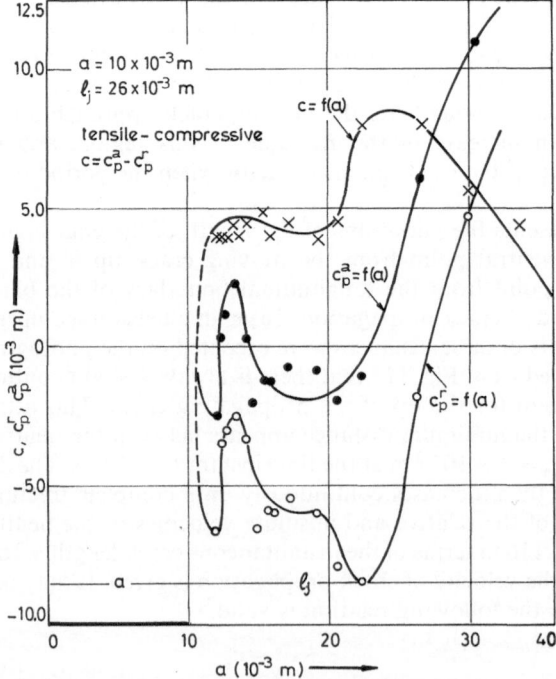

Fig. 116. Variation of the edge-crack propagation velocity c, the neutral point propagation relative velocity c_p^r and the neutral point propagation absolute velocity c_p^a versus edge-crack length.

It is clear from Fig. 116 that the velocities of the neutral point at the beginning remain small, whereas after time $t = 25\,\mu s$ (with $a = 21 \times 10^{-3}\,m$) a significant increase of these velocities is observed. This indicates the significant amount of recession of the neutral point from the crack tip as the crack propagates. The region of negative velocities may be explained as an attraction of the neutral point to the propagating crack tip, whereas in regions of positive velocity a recession phenomenon takes place.

It is easy from the last two figures to locate exactly the position of the neutral point of the bar at every time instant as the crack propagates and to evaluate its velocity of movement. In the photographs of Fig. 112 it may be observed that the caustic formed at the near tip of the witness crack presents some amount of angular displacement, implying that in this zone some shear is operative.

From this study it is concluded that: The use of a small stationary crack parallel to the main crack and placed at various positions around the initial neutral zone of the bar, allowed the study of the variation of stress intensity factor at the main crack tip and of the continuous variation of stress distribution ahead of the propagating crack.

It was shown that the moving crack tip forces the neutral zone to move forward, always remaining ahead of the moving crack tip.

It was established that at the initiation period of propagation of the crack an attraction of the neutral point towards the moving crack tip occurs, followed by a repulsion. By using the sensitive method of caustics the relative velocities of the moving crack tip and the neutral point were completely established.

Moreover, the interaction of the moving crack with the witness crack results in a momentary crack arrest, during which a redistribution of energies takes place between the arrested and the stationary crack. This indicates that the witness collinear stationary crack works as an arrester and an energy absorber to the propagating crack, thus enhancing the overall toughness of the bent bar to impact.

All fracture modes for different relative distances between collinear cracks showed some common general characteristics independent of the relative position of the cracks:

1. The propagating edge-crack, under the influence of the impact load, reached straight ahead to the stationary crack at its near tip without making deviation.

2. The path followed by the propagating crack in bent bars was completely different from the path generally followed by collinear cracks in tensile strips. Indeed, in plates with collinear cracks under simple tension, the propagating crack approaches the stationary one, interacts with it and initiates its propagation. As both cracks approach each other, they repel, but as soon as their tips overlap they start to attract and approach each other, which results in curved paths in the vicinity of their region of approach where a K_{II} component becomes apparent [203].

Moreover, at the point where one of the tips is close to the other crack path it stops, whilst the other one may continue until complete failure of the specimen occurs. It is seldom that a failure of the plate occurs by the coalescence of the secondary crack with the path of the already-passed primary crack. This mode of propagation is typical of both skew-parallel [201, 202], and collinear cracks moving in opposite directions [203].

This instability of the crack propagation paths during collinear crack coalescence was theoretically treated by Melin [206] (see also [175]), who showed that the straight path is unstable, even when dynamic effects are not taken into consideration, so that tip to tip coalescence of originally collinear cracks will not take place.

In the present experimental investigation, tip to tip coalescence was eventually possible, due to the bluntness of the stationary crack. Then, the advantage of the selected bluntness of the one crack was that, by increasing the stability of the collinear crack configuration, it allowed us to compare the stability of mode-I crack propagation under tensile loading or bending.

Contrary to the case of tensile strips, for collinear cracks in bending, the propagating crack always approaches the near tip of the stationary crack; the latter near tip is not sensitized and remains stationary as an obstacle to the path of the propagating crack. The crack which approaches it coalesces without any deviation of its path.

It seems therefore that crack propagation in bars under bending is much steadier than in tensile fields, because the tendency of the crack path to deviate from a straight path, dominated by the K_I mode, is much weaker. These weak coalescence phenomena are due to the fact that in bending, the tensile stresses at the tips of the internal crack are weak, being inside the bent bar, though sometimes they change to compressive stresses if the tip of the crack lies on the compressive side of the bar. In a uniform overall tensile field the stress concentrations at all tips are similar.

3. This last remark is also supported by the high repeatability of the tests in respect of not only the paths of the propagating cracks, but also the crack propagation velocities and the stress intensity factors.

4. The existence of the witness crack did not influence significantly the mode of fracture and the crack propagation velocity of the edge-crack since the witness crack was always placed in the neutral region of the bent bar. Only when the propagating tip was approaching the near tip of the witness crack, was some influence detected on both the stress intensity factors and the crack propagation velocities.

5. Another interesting phenomenon was the short crack-arrest period, which took place after the coalescence of the moving crack with the near tip of the witness crack. A short interval of time elapsed whilst the stresses at the tip of the edge-crack moved to the witness crack and from its near to its far tip. This phenomenon may be explained as follows: (a) The energy stored at the propagating crack tip needed a certain time for its coalescence with the near tip of the witness crack, in order to transfer energy from the near tip to the far tip of the internal crack. (b) The new and abruptly longer crack, created by the coalescence of the two cracks, necessitated another time interval, during which the energy of the tip of the new unified crack is built up and the caustic progressively enlarged in size, so as to attain the critical value of energy necessary to initiate a new and longer crack. (c) Finally, this time interval is also increased by the bluntness of the artificial witness crack, but this fact is not sufficient to explain completely the crack-arrest phenomenon. Indeed, if one examines frames 13–18 of Fig. 106, it may be observed that the successive processes of crack interaction, as exemplified in frames 13–15, are connected with the building up of the caustic at the furthest tip of the stationary crack; it is stationary in frames 15–18, and only in frame 19 is the crack again propagating.

The details of such energy redistributions would be very difficult to follow in the case of tensile fracture tests [203], where, in general, large amounts of energy are involved, which cause the energy redistribution phenomena to shorten their duration to such a point that they can hardly be detected experimentally. The duration of the crack-arrest time interval was evaluated at $t_d = 16\,\mu s$ in Figs 106, 109 and 112. This time interval depends on the initial crack length a_0 of the stationary crack.

This momentary crack-arrest process is therefore mainly a result of the bending stress configuration in the specimen. In a bending configuration, even after opposite crack coalescence and unification, the stress concentration at the distant crack tip is low and insufficient to cause crack re-initiation. A redistribution of stresses has to develop before any re-initiation of the propagation of the long crack.

3.9 Fracture Behaviour Under Stress Pulse

When a stress wave (tensile or compressive) impinges on a crack or on a V-notch existing in an elastic medium, reflection, refraction and diffraction phenomena take place. A result of diffraction is the loading of the crack or notch. While compressive stress waves do not create any stress concentration at the tip of an existing crack or notch, tensile stress waves develop stresses at the tip which may cause a propagation of the crack. If the tensile pulse is weak the crack may be propagated by steps under the action only of successive tensile stress pulses, whereas intermediate compressive stress pulses have no influence.

A complete study of the phenomena of incubation, initiation and propagation of cracks in thin plates subjected to a compressive pulse, which is subsequently reflected from the free boundaries of the plate and changed to complicated wave-trains, was undertaken in a study based on the method of caustics [133, 207].

Stress wave propagation in infinite plates due to various types of stress pulses have been extensively studied. Mathematical analyses of these phenomena have been presented by Kolsky [208], Sherwood [209], Cagniard [210], Ewing et al. [211] and Davids [212]. On the other hand, phenomena of fracture by stress waves were presented by Miklowitz [213], Thiruvenkatachar [214], Eichelberger [215], Broberg [216], Rinehart [217] and others. A mathematical analysis of the crack propagation in an elastic solid by an incident stress pulse was given by Freund [218, 219]. These references, however, constitute only a small part from a long list of papers concerned with dynamic fracture by impact.

Dynamic crack propagation studies under impact loading were advanced analytically and experimentally to yield information concerning crack initiation, bifurcation and stress arrest. In this context the early and important work by Schardin should be mentioned [220]. Moreover, Kalthoff and Shockey [221] investigated the crack propagation behaviour in a circular plate made of PC and subjected to impact, while Kobayashi et al. [222] made a similar study in a plate made of another birefringent polymer. On the other hand a mathematical analysis of crack growth under the influence of a stress wave was presented by Homma et al. [223].

Experimental studies of the dynamic behaviour of artificially cracked plexiglas strips under plane stress pulses created by projectiles shot from an air-gun have been presented in Refs [60, 224] by applying the method of caustics. The case of oblique edge-cracks in long strips under impact was studied in Refs [133, 164, 207].

Throughout these studies it was perceived that the initial edge-crack was propagated by steps under the action only of the tensile stress pulses while the compressive pulses did not create any crack propagation. In all these studies it has been shown that any fluctuations of crack velocities and variations of the values of stress intensity factors may be accurately detected by the positions, shapes, and sizes of the caustics.

During crack propagation, different types of waves may be generated at the moving crack tip. These are the longitudinal, Rayleigh and flexural waves. These phenomena of wave-emission during fracture were studied by Theocaris and Georgiadis [225, 272].

3.9.1 Dynamic Behaviour of an Oblique Edge-Crack Under Stress Pulse

When a plane dilatational (L) or distortional (T) wave impinges on a crack in an elastic medium (Fig. 117), reflections take place because the initial crack behaves as a free boundary. The amplitudes of reflected (dilation and distortion) waves are less than the amplitude of the incident wave [208], for angles of incidence between 0 and $\pi/2$. When the angle of incidence is equal to $\pi/2$ the amplitude of the reflected wave is equal to the amplitude of the impinging wave, but of opposite sign. Moreover, during normal incidence of the stress wave the reflected wave is always of the same type as the incident wave, whereas for angles of incidence smaller than $\pi/2$ two types of reflected waves (dilatational and distortional) are created.

Figure 118 presents various cases of reflection of a dilatational wave at an oblique crack. In this figure the initial crack makes an angle ω with the longitudinal free boundary of the plate (Ox-axis in the Cartesian coordinate Oxy-system, attached to the crack tip). Then, the angle of incidence of the dilatational wave is $\hat{\alpha} = (\pi/2 - \omega)$. Figure 118(a) presents the case of an initial crack subtending an angle $\omega < \pi/4$ and the angle of incidence of the dilatational wave $\hat{\alpha} > \pi/4$. Two component-waves are reflected from the leading edge of the crack, a dilatational wave at an angle $\hat{\alpha}$, and a distortional wave at an angle $\hat{\beta}$. The amplitudes of the reflected waves, A_2 and A_3 respectively, are smaller than the amplitude A_1 of the incident dilatational wave.

Subsequently, the reflected waves A_2 and A_3 impinge on the longitudinal free boundary of the specimen and four component-waves are reflected from it, these again impinge on the crack. During these successive reflections of the waves on the free boundary of the specimen and on the crack-lip, the amplitudes A_i of the

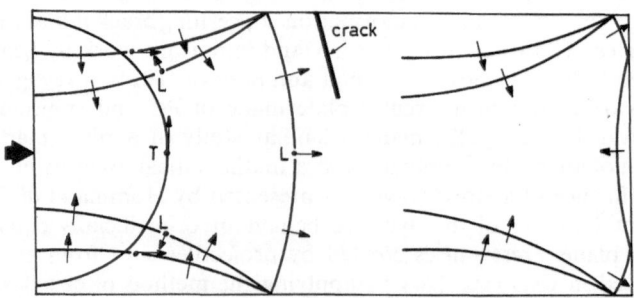

Fig. 117. Stress waves in plates: Dilatational or longitudinal (L) and distortional, transverse or shear (T).

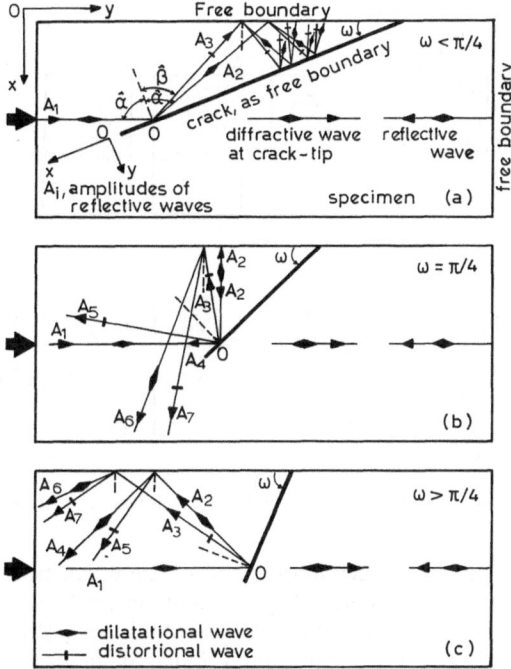

Fig. 118. Reflection of dilatational wave at crack tip of specimen for **(a)** $\omega < \pi/4$, **(b)** $\omega = \pi/4$ and **(c)** $\omega > \pi/4$.

component-waves are continuously attenuated and tend to zero. Thus, for angle ω smaller than $\pi/4$ a *cage-phenomenon* of the waves is observed.

For the limiting case of angle $\omega = \pi/4$ (Fig. 118(b)) the reflected dilatational wave A_2 impinges normal to the free boundary of the specimen and is reflected again at an angle $\hat{\alpha} = 0°$. The reflected distortional waves on the other hand impinge obliquely upon the free boundary of the specimen, generating two component-waves (dilatational and distortional), which then propagate outside the domain of the crack.

For $\omega > \pi/4$ (Fig. 118(c)) the reflected component-waves impinge obliquely on the free boundary of the specimen, generating four component-waves, which propagate outside the domain of the crack. The cage-phenomenon therefore disappears.

For obtuse angles ω between the crack-axis and the longitudinal boundary of the strip (Oy-axis) this *caging-in* phenomenon disappears [207].

The simple case of $\omega = \pi/2$ where the position of the transverse crack does not create subsequent alternating reflections on the free boundary of the strip and the lips of the crack was studied by Theocaris and Katsamanis [60]. In this case the complicated phenomenon of reinforcement or attenuation of the reflected waves from the crack lips is avoided.

When a propagating plane longitudinal stress pulse impinges on one lip of a stationary crack in a plane-stress field it is partially reflected and partially diffracted around the tip of the crack. Then, in the region around the crack tip, a transient stress field is developed because of the refraction of the stress pulse. A mathematical analysis of the problem of diffraction of the flat front of a propagating stress pulse in the region surrounding the crack tip was given by Freund [218, 219].

According to this theory of diffraction, if a plane longitudinal compressive pulse impinges along one lip of a slanted crack, it diffracts about the tip, and creates a

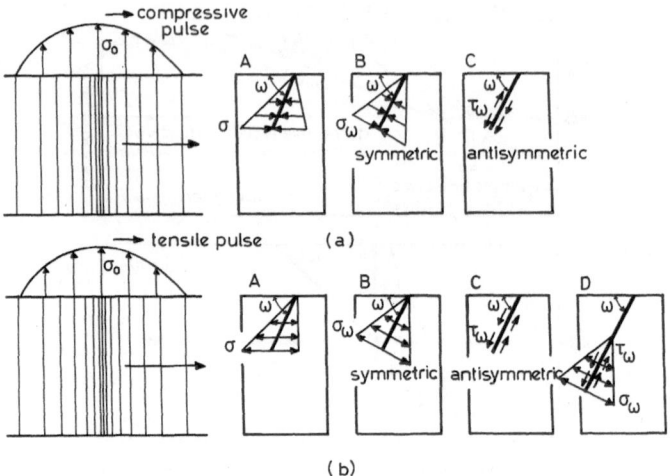

Fig. 119. Diffraction of the pulse at the crack-lip (a) compressive pulse and (b) tensile pulse.

compressive stress field in the neighbourhood of the crack tip. The mechanism of creation of such a field is indicated in Fig. 119(a) diagram A. The compressive stress field introduces a normal σ_ω-stress along the lips of the crack (Fig. 119(a) diagram B) and a shear τ_ω-stress (Fig. 119(a) diagram C). Thus, the crack is under the influence of a simple symmetric and antisymmetric state of stress.

If the impinging stress pulse is tensile (Fig. 119(b)), a symmetric (diagram B) and an antisymmetric (diagram C) state of stress are developed, and thus the crack tip is under the influence of a tensile stress field.

The tensile stress field which radiates in front of the crack tip because of the diffraction of the tensile stress pulse at the crack tip, is the cause initiating the propagation of the crack after some time-lapse (Fig. 119(b) diagram D). The delay time for the initiation of propagation of the crack is due to the fact that enough energy must be stored at the stationary crack tip, for the stress intensity factor, K_I, to reach its critical K_{Ic}-value for propagation of the crack.

The values of the σ_ω and τ_ω components of stress depend on the value of the respective velocities of the longitudinal and shear waves, the amplitude of the pulse and the angle ω subtended between the crack axis and the direction of propagation of the pulse.

The diffracted pulse travels in the part of the specimen behind the crack, it impinges normally along the transverse extremity of the strip and then reflects backwards as a pulse of opposite sign, since the transverse boundary of the strip is a free boundary. The amplitude of the reflected pulse is equal to the amplitude of the impinging pulse [208] (Fig. 118). The new pulse, of opposite sign to the primary pulse, impinges again along the other lip of the crack and creates similar phenomena, but of opposite sign.

The passage of a compressive pulse through a slant edge-crack and the influence of its consecutive reflections on the lips of the crack are depicted in Fig. 120. As the stress pulse of amplitude A_1 impinges on the lip of the crack it creates a normal compressive σ-stress along the crack and a shear τ-stress, which, because of the diffraction phenomena at the tip of the crack, appears also at the other lip of the crack. Thus, the tip of the crack is under the influence of a compressive normal

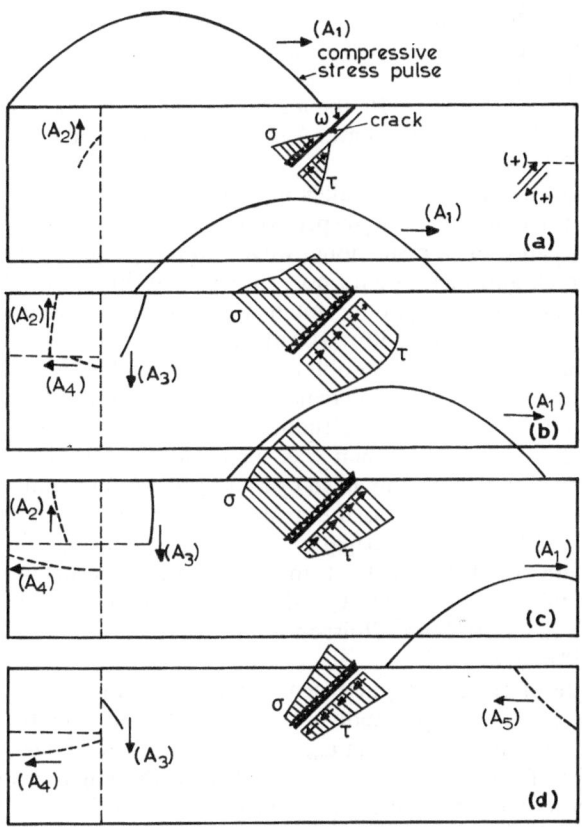

Fig. 120. Influence of a compressive stress pulse and its reflections on an oblique crack with $\omega = \pi/4$.

stress field (σ) and a shear stress field (τ) of intensities depending on the amplitude A_1 of the impinging stress pulse, the ω-angle, and the velocities of the longitudinal stress wave (c_1) and the shear stress wave (c_2) [218, 219] (Fig. 120(a)).

Furthermore, when the stress pulse impinges on the one lip of the crack, it is partially reflected from it. The reflected tensile pulse, of amplitude $A_2 = 0.55\,A_1$, travels towards the longitudinal boundary of the strip which is closer to the crack. The normally impinging pulse along this boundary is again reflected as a compressive pulse, with the same amplitude $A_3 = A_2$, and impinges once again along the lip of crack. Thus, in the already-loaded crack, a new compressive stress and a new shear stress component are added, which are derived from the diffraction of the compressive stress pulse passing near the crack tip. The new loading state at the crack tip, after the superposition of the principal stress pulse A_1 and of the reflected pulse A_3, appears in Fig. 120(b).

The compressive pulse, derived from reflection, is again partially reflected along the lip of the crack, as a tensile pulse with amplitude $A_4 = 0.55\,A_3 = 0.30\,A_1$. When the principal compressive pulse A_1 passes from the crack, the reflected pulse A_3 continues to load the crack with a compressive stress and a shear stress of opposite sign to the previous one (Fig. 120(d)). The main compressive pulse A_1 is also reflected at the transverse boundary of the strip and returns as a tensile pulse with amplitude

$A_5 = A_1$ and thus it loads the crack accordingly. The result of the loading of the tensile pulse is the initiation of propagation of the crack, due to diffraction, as shown in Fig. 119(b) diagram D.

As may be deduced from Fig. 120, the loading of the crack starts from its tip and progressively extends over all its length. A similar loading mode appears also when the angle $\omega > \pi/2$ [207].

The above detailed study has not taken into account the influence of distortional waves, which are derived from the respective reflections of the dilatational waves; this is justified because these distortional waves do not influence the normal stress loading of the crack and therefore do not contribute to initiation of propagation.

For the experimental investigation, notched PMMA plates were employed. The dimensions of specimens were $0.30 \times 0.04 \, \text{m}^2$ and their thickness was $d = 0.004 \, \text{m}$. The specimens had initially an edge-oblique artificial crack of length a_a, which was further extended by about $0.01 \, \text{m}$ to create a really natural initial crack. The angles ω between the initial crack and longitudinal free boundary of the specimen were $\omega = 15°$, $45°$ and $105°$. The specimen was suspended in a horizontal position by means of thin filaments and was impacted by a steel sphere at its transverse firing-end, remote from the crack zone, to assure a plane stress wave arriving at this zone. Dynamic crack loading and propagation was recorded using a Cranz–Schardin high-speed camera. In the optical set-up used in the experiments, the following quantities were taken: $z_0 = 0.80 \, \text{m}$, $\lambda_m = 0.75$. The dynamic properties of PMMA are (Table 4): $E = 4.3 \times 10^9 \, \text{N/m}^2$, Poisson's ratio $v = 0.34$ and stress optical constant $c_t = 0.74 \times 10^{-10} \, \text{m}^2/\text{N}$.

The magnitude of the stress pulse generated by the impact was about $\varepsilon_{max} = 3.6 \times 10^3 \, \mu\text{strain}$ [60]. The magnitude of the reflected pulse from the transverse free boundary of the specimen was again $\varepsilon_{max} = 3.6 \times 10^3 \, \mu\text{strain}$, while the magnitudes of reflected pulses from the crack lips depended on the angle of incidence [208]. The stress intensity factors K_I and K_{II} are given by Rels (618) and (619).

To exemplify the sequence of phenomena described above concerning the reinforcement and attenuation of the propagating stress pulses in an obliquely cracked strip (due to successive reflections along the boundaries of the strip and the lips of the crack, and also diffraction of stress-waves at the crack tip) a series of tests were undertaken where the loading of the crack was followed, by applying the method of caustics.

Figure 121 presents a series of photographs for a PMMA strip, containing a slant edge-crack, artificially sawn with a very fine sawing disc, having a thickness less than $0.0003 \, \text{m}$. The total initial length of the crack was $a_0 = 0.0416 \, \text{m}$ and its angle of inclination was $\omega = 15°$. The end of the artificial slit was always terminated by a natural crack created by appropriate indentations at the extremities of the lips of the initial slit. The slotted strips were subjected to a plane compressive stress pulse created by an air-gun with the bullet under an air pressure of 2.8 bars.

When the compressive stress pulse arrives at the tip of the stationary crack, compressive normal and shear components of stresses are developed at the tip of the crack because of diffraction phenomena, as explained above. Because of the small angle ω subtended by the crack-axis and the longitudinal boundary, the shear component is much stronger than the compressive normal component of stress. Therefore, the caustic developed at the tip of the crack presents a strong angular displacement relative to the crack-axis (see frame 5 of Fig. 121). Simultaneously, the stress pulse is subjected to reflections at one lip of the crack, these are caged-in to the interior of the acute angle ω, as explained in Fig. 118(a).

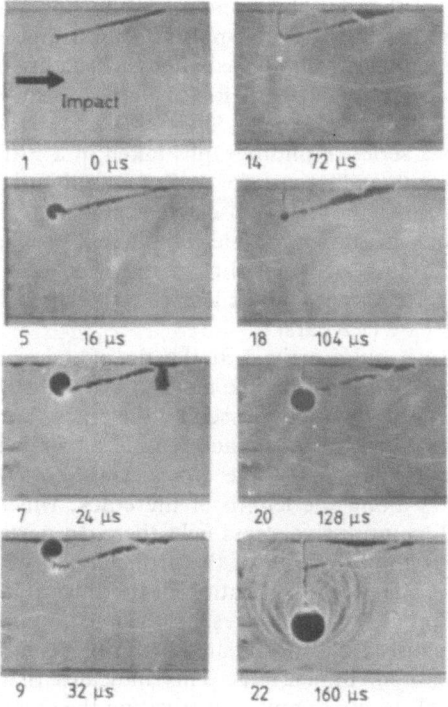

Fig. 121. Series of photographs obtained with high-speed camera in a notched PMMA plate with $\omega = 15°$ and $a_0 = 0.0416$ m. The air pressure was 2.8 bars.

A result of the caging-in phenomenon is an accumulation of energy at the crack tip, up to the final incubation and initiation of propagation of the crack towards the closer boundary of the strip (frame 9 of Fig. 121). As a result, a complete fracture of the triangular part of the strip, included between the crack and this boundary, is produced (frame 14 of Fig. 121).

However, the compressive stress pulse, reflected along this transverse boundary, returns after a while, as a tensile pulse, at the newly created acute 75° corner of the missing triangle and loads it in tension because of diffraction, as described by Figs 119 and 120. Thus, a caustic is formed (frame 18 of Fig. 121), due to the singularity existing at the obtuse angle of the corner of 285°. This caustic increases in size, as the strain energy, piled up at the apex, increases (frame 20 of Fig. 121) and when K_I attains its critical value K_{Ic}, a new crack initiates, propagating along the rest of the width of the strip with high velocity (frame 22 of Fig. 121). Simultaneously, an emission of stress waves from the tip of the crack develops; these are of mixed type, that is, longitudinal and Rayleigh waves. The distinction and separation of such waves emitted from a propagating crack tip in a strain-rate sensitive material were thoroughly studied by Theocaris and Georgiadis [225].

Photographs of the test presented in Fig. 121 indicate the creation of the cage-in phenomenon of the compressive stress pulse, inside the acute angle formed by the oblique crack and the boundary of the strip. This phenomenon contributes significantly to the loading of the crack with an intense shear stress (indicated by the orientation and size of the respective caustics) and results in the final fracture of the triangular part of the plate between crack and boundary. The tensile pulse created from the reflection of the initial pulse on the transverse free boundary of the strip, loads in tension the notch formed after the fracture of the triangular part, and results in the initiation and propagation of a crack emanating from the bottom of the notch.

Figure 122 presents a series of photographs taken in a PMMA strip containing an oblique edge-crack, subtending $\omega = 45°$ and of a total initial length $a_0 = 0.018$ m. The reflection phenomena of the stress pulse differ significantly in this case from those in the previous test, thus creating a completely different propagation pattern for the crack. In this case the initial artificial slit was sawn at a length $a_a = 0.008$ m and then extended by steps to a natural crack length $l = 0.010$ m. The air pressure in the gun was lower in this case (1.2 bars).

When the compressive pulse arrived at the tip of the stationary crack, it was loaded by diffraction with a compressive normal and a shear component of stress (frame 5 of Fig. 122). Then, the pulse passed by the crack and thus the loading was extended progressively along all the crack length, as shown in Fig. 120. Indeed, frame 8 of Fig. 122 presents, besides the typical caustic at the tip of the crack, a series of small caustics along the length of the crack, which are cuspoid curves, because of the overall compressive stress field there. In particular, such a cuspoid caustic is developed at the extremity of the artificial slit, where an irregularity in the form of the lips and an abrupt variation of the thickness of the crack-opening occurs. Frame 15 of Fig. 122 evidently presents the maximum of loading of the crack along all its length, because all caustics created there are increased in size considerably and attain their maximum dimensions. This indicates that, at this instant, the maximum amplitude of the stress pulse is activating the crack. After this maximum is attained, the crack is gradually unloaded.

This unloading procedure proceeds gradually towards the tip of the crack (frame 17 of Fig. 122). This occurs because the reflections of the stress pulse from the lip of the crack impinge along the neighbouring longitudinal boundary of the strip as described by Fig. 118(b). The new reflections impinge again on the same lip and load the crack accordingly, to create this gradual unloading moving towards

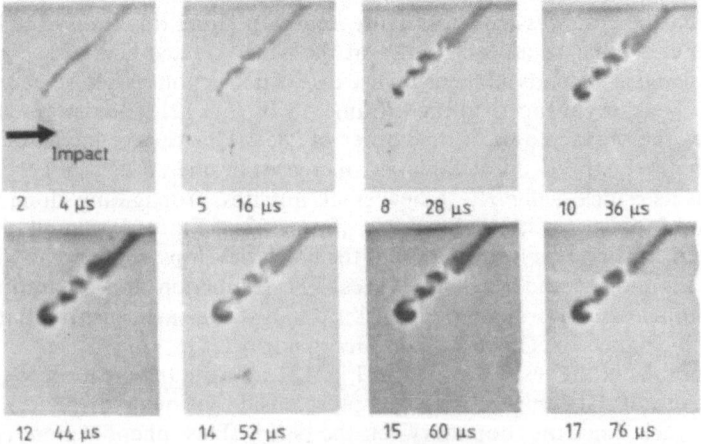

Fig. 122. Series of photographs obtained with a high-speed camera in a notched PMMA plate with $\omega = 45°$ and $a_0 = 0.018$ m (0.008 m artificial crack and 0.010 m natural crack). The air pressure was 1.2 bars.

the crack tip. This unloading tendency from the tails to the tip of the crack may be explained by the fact that the paths of reflections are increasing as we move from the extremities to the tip, and the duration of the phenomenon of reflection is accordingly extended in time.

Another result of this successive reflection process and continuous superposition of reflected pulses is that the amount of shear stress diminishes progressively, changes sign and then progressively increases in the opposite direction, as appears in Fig. 120. This results in a rotation of the caustic, which angularly oscillates about the unique position where only K_I is operative. The photographs of Fig. 122 demonstrate the fact, described by Figs 118, 119 and 120, that the stress pulse loads the crack along the whole length of its lips, whereas the reflected pulses change the sign of the shear components of stress with which the crack is loaded.

The angular oscillation of the caustic because of the continuous change of sign in the shear component of stress appears in a more clear manner in Fig. 123 [207], where a series of photographs is shown, taken during the propagation of an edge-crack, now subtending an obtuse angle $\omega = 105°$. The total initial crack length was $a_0 = 0.018$ m.

As was previously described in Fig. 120, the loading of the crack region commenced at the tip of the crack and extended afterwards along its entire length. This loading period was succeeded by a progressive unloading period. Indeed, in frame 2 of Fig. 123 an unloading process has already started from the tip of the crack. The unloading is clearly apparent in the cuspoid caustic at the extremity of the artificially sawn part of the crack, which is continuously diminishing and, by the end, is annulled (frame 4 of Fig. 123).

Afterwards, the tip of the natural crack is loaded intensively in compression, as well as with some insignificant shear-loading, creating the nucleus of a cuspoid caustic at the tip, with an infinitesimal angular displacement due to shear. The lack of appearance of a caustic in compression, although the tip of the crack is intensively loaded, has already been explained in a previous publication [60]. It is due to the fact that, since there is no rejection of the material along the lips of the physical crack, the lips of the crack close perfectly, when they are subjected to a simple compressional stress field without shear, thus eliminating completely the influence of the discontinuity of the crack. Only when shear stresses exist, do these stresses create some relative displacement between the two lips of the crack, no longer allowing a perfect fit between opposite lips, so that some deformation appears along the lips and at the tip of the crack.

As the process of the dynamic loading of the strip evolves, a reflected compressive stress pulse from the bottom longitudinal boundary of the strip approaches the crack region and introduces a loading state in the vicinity of the crack tip (frame 5 of Fig. 123). Simultaneously, a tensile stress pulse, derived from reflection from the transverse boundary of the strip, approaches and these two opposing pulses interfere and create a resultant loading on the crack, which is rather weak and not sufficient to initiate the propagation of the crack. Simultaneously, a shear component of the stress pulse is acting on the crack tip, and changes sign as the one contribution overshoots the other in the process of loading (see frames 5 to 16 of Fig. 123).

The reduction in size of the caustics is thus due to this decrease of the tensile component of normal stress, because of the influence of the corresponding compressive stress. When the compressive stress pulse passes by the region of the crack, the remaining tensile stress pulse, which follows the previous one, results in an increase in the tensile loading of the crack tip, and thus the caustic increases

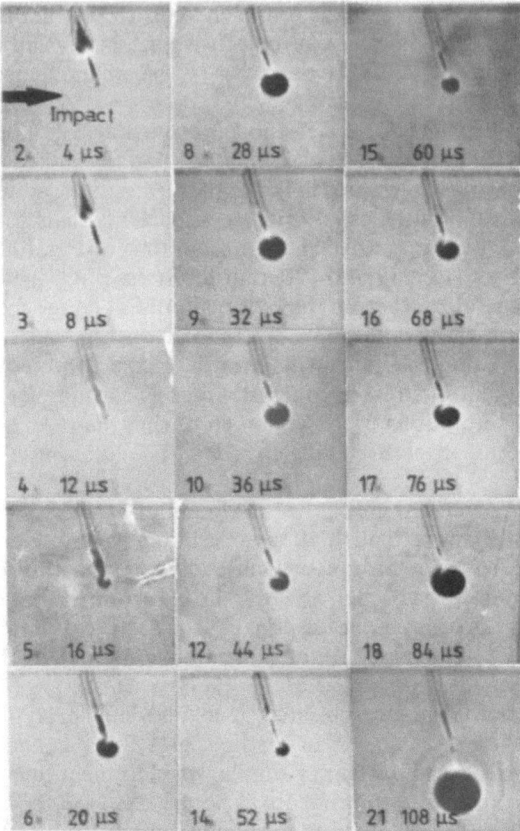

Fig. 123. Series of photographs obtained with a high-speed camera in a notched PMMA plate with $\omega = 105°$ and $a_0 = 0.018\,\mathrm{m}$ (0.008 m artificial crack and 0.010 m natural crack). The air pressure was 1.2 bars.

continuously, up to the moment when the loading field attains its critical value with $K_I = K_{Ic}$ and the crack starts to propagate along the strip (frame 18 of Fig. 123). The stress intensity factors K_I and K_{II} at this moment are $K_I = 8.78 \times 10^5\,\mathrm{N/m^{\frac{3}{2}}}$ and $K_{II} = 3.91 \times 10^5\,\mathrm{N/m^{\frac{3}{2}}}$, as derived from the magnitude and orientation of the respective caustics. The propagating crack maintains its velocity and also creates stress waves, emanating from the crack tip (frame 21 of Fig. 123). The stress intensity factors at this time are $K_I = 19.47 \times 10^5\,\mathrm{N/m^{\frac{3}{2}}}$ and $K_{II} = 0$ (crack propagates under mode-I).

Figure 124 presents the variation of stress intensity factors K_I and K_{II} versus time t, as derived from the caustics of the photographs in Fig. 123. It is clear from Figs 123 and 124 that, after a time-lapse of 150 μs, when the principal compressive pulse has passed the region of the crack, reflected pulses arrive at the vicinity of the crack from the longitudinal free boundaries of the strip and these pulses load the crack with tensile and shear stresses, as has been shown schematically in Figs 117 and 120.

A result of this successive loading is a considerable increase of the K_I-factor and an insignificant increase of the K_{II}-factor. Subsequently, the K_I-factor, after passing through a maximum value, diminishes progressively, whereas the K_{II}-factor diminishes

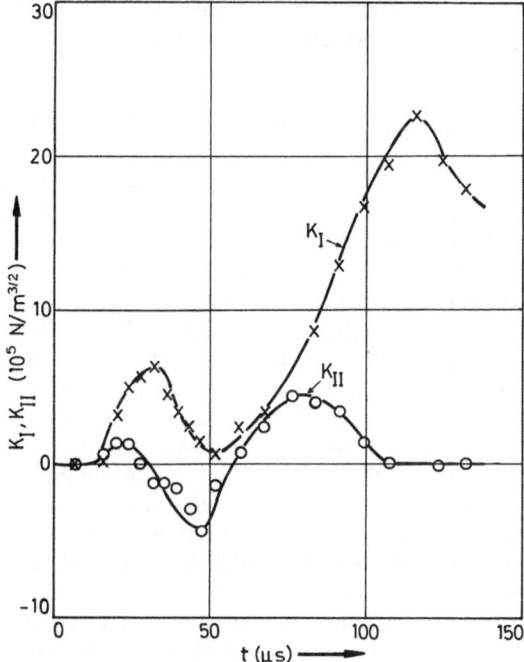

Fig. 124. Variation of stress intensity factors K_I and K_{II} versus time t.

also, passing through a zero value and subsequently taking negative values. This fact may be explained by the passage of a reflected stress pulse of opposite sign to the previous one.

During the instant of the passage of this pulse, the reflected tensile stress pulse from the transverse free boundary of the strip also arrives at the crack tip and loads the crack with additional components of tensile and shear stresses. This superposition of stresses results in an increase of the K_I-factor, which attains its critical value for crack initiation ($t = 84\,\mu s$, frame 18 of Fig. 123), whereas the K_{II}-factor increases continuously and attains its maximum value at the same instant ($t = 84\,\mu s$). Subsequently, the K_{II}-factor diminishes progressively, as the crack starts to propagate and is annulled at time $t = 108\,\mu s$, after which the crack propagates with $K_{II} = 0$, under the influence of K_I alone.

The K_I-factor passes through a maximum at $t = 116\,\mu s$ and then starts to diminish as the crack approaches the opposite boundary of the strip, because of the influence of this boundary and the reflection of the waves emanating from the instantaneous positions of the crack tip. This factor is annulled only when the crack reaches the longitudinal boundary.

The superposition of stress pulses at the tip of the crack, and the subsequent formation of a complicated mode of loading at the crack tip, is due to the particular dimensions of the strips tested, since the timing of the various reflected pulses and their interference depends mainly on the dimensions of the specimen. This super-position of pulses may be avoided by increasing the width and other dimensions of the test pieces. However, since in practice there are always parts of structures with limited dimensions, the simplicity which may be achieved by increasing the dimensions of the test pieces is not always desirable.

3.9.2 Dynamic Behaviour of a Sharp V-Notch Under Stress Pulse

Figure 125 presents the modes of propagation and reflection of longitudinal (L) and transverse (T) stress waves. It is well known [208] that when a wave impinges on the notch lips, reflections take place along the free boundaries of its lips. The amplitudes of the reflected L or T waves are inferior to the respective amplitudes of the incident waves for angles of incidence between zero and $\pi/2$. At zero angle of incidence the amplitude of the reflected wave equals the amplitude of the impinging wave, but reverses its sign. Moreover, for normal incidence, the reflected wave is always of the same type as the incident wave, whereas for angles of incidence between zero and $\pi/2$ two types of waves (dilatational (L) and distortional (T)) are created. If the normally incident stress wave is compressive, the reflected wave is a tensile wave with amplitude equal to the amplitude of the impinging wave.

Figure 125(a) presents the reflections of a dilatational wave on the lips of the notch. The V-notch of angle φ has lips subtending angles φ_1 and φ_2 with the longitudinal boundary of the strip, for which it is valid that $\varphi_1 > \varphi_2 > \pi/2$. The angle of incidence of the dilatational wave is $\hat{a} = (\varphi_1 - \pi/2)$. Two component waves are reflected from the lip of the notch, a dilatational wave at an angle $\hat{a}$, and a distortional wave at an angle $\hat{\beta} < \hat{a}$. The amplitudes of the reflected waves A_2 and A_3 are smaller than the amplitude A_1 of the incident dilatational wave. Subsequently, the two waves reflected from the closer lip impinge on the opposite longitudinal free boundary of the strip and secondary reflections take place.

In Fig. 125(b) the V-notch of angle φ has lips subtending angles φ_1 and φ_2 with the longitudinal boundary of the strip where $\varphi_1 < \pi/4$ and $\varphi_2 > \pi/2$. The angle of incidence of the dilatational wave is $\hat{a} = (\pi/2 - \varphi_1)$. As in the case of Fig. 125(a) two component waves are reflected from the closest lip to the arriving pulse, a dilatational wave at an angle $\hat{a}$ and a distortional wave at an angle $\hat{\beta}$. The amplitudes of the reflected waves A_2 and A_3 are again smaller than the amplitude A_1 of the incident

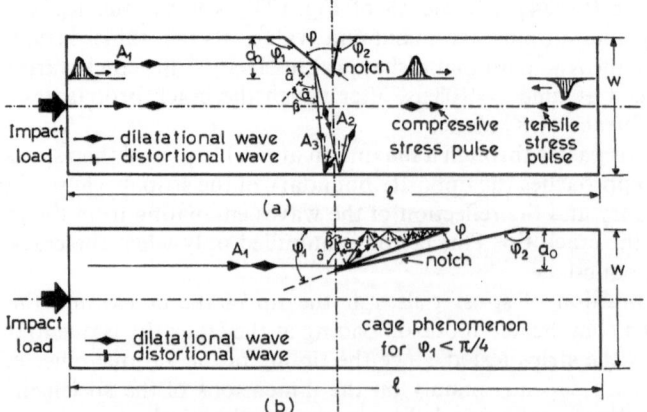

Fig. 125. Reflection of dilatational waves on the notch lips of the specimen.

wave. Subsequently, the two reflected waves impinge on the closer longitudinal free boundary of the strip and four component waves are reflected from it, each of them impinging again on the same lip of the notch. The new four component waves give eight new component waves from the reflections on the lip of the notch and so on. However, the amplitudes of the reflected components waves tend to zero, but all reflected waves remain inside the region included between the longitudinal boundary and the lip of the notch, which is of an angle φ_1.

Thus, for angles $\varphi_1 < \pi/4$ a *cage-in phenomenon* of the waves is observed, while for angles $\varphi_1 \geq \pi/4$ such a cage-in phenomenon disappears [207].

When a plane longitudinal stress pulse impinges on one lip of a notch in a plane-stress field, it is partially reflected and partially diffracted around the bottom of the notch. Then, a transient stress field is developed around the bottom, because of the refraction of the stress field. A mathematical analysis of the problem of diffraction of the flat front of a stress pulse in the region surrounding the crack tip was given in Refs [218, 219]. According to this theory of diffraction, if a plane longitudinal compressive or tensile pulse impinges along the bottom of the notch, it diffracts about this bottom and creates a compressive or tensile stress field, respectively, in the neighbourhood of the apex of the notch.

The mechanism of creation of such fields is indicated in Figs 126 and 127 for two types of sharp V-notches. The compressive or the tensile stress field introduces normal σ_{φ_1} and σ_{φ_2} stresses along the notch lips and shear τ_{φ_1} and τ_{φ_2} stresses, which depend on angles φ_1 and φ_2 respectively, whereas the singularity of the stress field at the apex of the notch depends on the angle of the notch φ [132, 139]. Thus, the notch is under the influence of simple symmetric $(\sigma_{\varphi_1}, \sigma_{\varphi_2})$ and antisymmetric $(\tau_{\varphi_1}, \tau_{\varphi_2})$ states of stress.

The tensile stress field, which radiates around the apex because of the diffraction of the tensile stress pulse at this apex and in front of it, causes the incubation and nucleation of a crack and may initiate its eventual propagation after some time lapse. The delay time for the incubation and initiation of propagation of a crack is

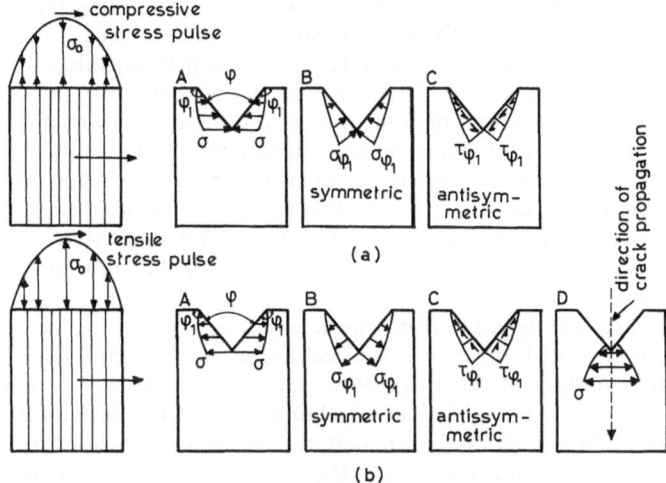

Fig. 126. Diffraction of the stress pulses at symmetric V-notches of an initial depth a_0: (a) compressive stress pulse, (b) tensile stress pulse, $\varphi_1 > 90°$.

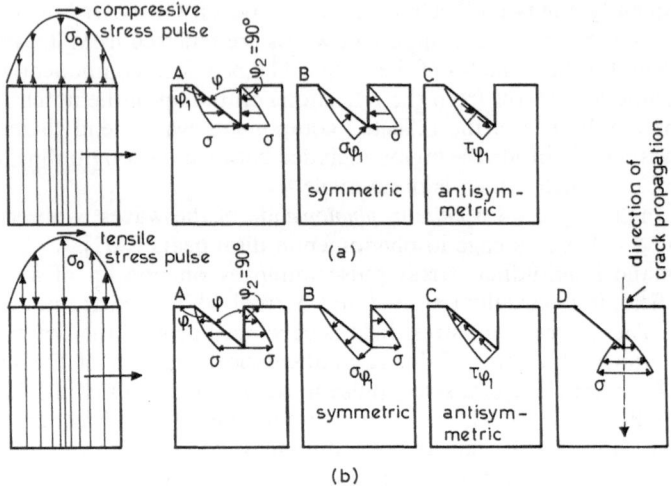

Fig. 127. Diffraction of a stress pulse at an oblique V-notch of initial length depth a_0: (a) compressive stress pulse, (b) tensile stress pulse, $\varphi_1 > 90°$ and $\varphi_2 = 90°$.

due to the fact that enough energy must be stored at the bottom of the notch, so that the stress intensity factor K_1 reaches its critical value K_{1c} for propagation of the crack.

The values of σ_{φ_1}, σ_{φ_2} and τ_{φ_1}, τ_{φ_2} components of stress depend on the respective velocities of the longitudinal and shear waves, the amplitude of the pulse and the angles φ_1 and φ_2. The diffracted pulse travels along that part of the strip behind the notch, it impinges normally along its transverse boundary, and then reflects backwards as a pulse of opposite sign (Fig. 125), since the transverse boundary of the strip is a free boundary. The reflected pulse impinges again on the notch lips and creates similar phenomena, but of opposite sign.

Figure 126 presents the diffraction phenomenon of the stress pulse at a notch symmetrically placed to the direction of pulse propagation for (a) a compressive and (b) a tensile stress pulse. The compressive stress field introduces a normal σ_{φ_1} stress along the notch lips (Fig. 126(a) diagram B) and a shear τ_{φ_1} stress (Fig. 126(a) diagram C). The tensile stress field introduces a normal σ_{φ_1} stress along the notch lips (Fig. 126(b) diagram B) and a shear τ_{φ_1} stress (Fig. 126(b) diagram C). Thus, the apex of the notch is under the influence of a tensile stress field, σ, which is propagating radially in front of the apex of the notch. This tensile stress field presents a singularity depending on the angle φ, and causes the incubation and eventual initiation of propagation of a crack along a direction normal to the direction of propagation of the stress pulse.

Figure 127 presents the same phenomena as in Fig. 126, but for the case when the V-notch is oblique, with lips which subtend angles $\varphi_1 > \pi/2$ and $\varphi_2 = \pi/2$ with the longitudinal boundary of the strip. Now, the compressive stress field introduces normal σ_{φ_1} and σ_{φ_2} stresses along the notch lips (Fig. 127(a) diagram B) and a shear τ_{φ_1} stress (Fig. 127(a) diagram C), whereas, the tensile stress field introduces normal and shear stresses and thus the apex of the notch is under the influence of a tensile stress field, σ, which causes the eventual initiation and propagation of a crack. The stress field σ varies between 0 and σ_0. If $\sigma_0 \geqq \sigma_{cr}$, where σ_{cr} is the critical value of stress for the propagation of the crack, phenomena of incubation and propagation will take place, while if $\sigma_0 < \sigma_{cr}$, such phenomena will not occur. So, the propagation phenomena depend on the amplitude of the initial stress pulse.

It is worthwhile noticing that the u and v displacements around the apex of the notch have different values and they depend on the values of stresses there. However, the stress intensities at the apex of the notch, as well as on the tips of the propagating cracks may be conveniently and accurately determined by the methods of caustics [226] and pseudocaustics [227]. On the other hand, the stress fields which were presented in Figs 126 and 127, are additionally influenced by the stress waves of the secondary reflections on the free boundaries of the plate. But the amplitudes of these stress waves are much less than the amplitude of the stress pulse first impinging on the notch, and so it is assumed that the instantaneous stress fields at the apex and the crack tip will not be significantly changed by these secondary pulses.

In the experimental set-up transmitted caustics [132, 139, 165, 226] were used in combination with a Cranz–Schardin high-speed camera. The specimens were subjected to a plane compressive stress pulse created by a projectile shot by an air-gun, which was under an air pressure varying between 1.0 and 2.5 bars. The magnitude of the plane stress pulse was about $\varepsilon_{max} = 2.5 \times 10^3$ μstrain [60]. The magnitude of the reflected pulse from the opposite transverse free boundary of the specimen was 2.5×10^3 μstrain, while the magnitudes of reflected pulses from the notch lips depended on the angles of incidence [208].

For the experimental study of impact phenomena, notched PMMA plates were used. The dimensions of the specimens were $l = 0.300$ m, width $w = 0.040$ m and thickness $d = 0.003$ m (Fig. 125). The angles of the notches φ varied between 25° and 120°, while the angles φ_1 and φ_2 varied between 30° and 175°. The specimens were suspended in a horizontal position by means of thin and flexible filaments, thus forming a modified Hopkinson bar, and were impacted at their transverse firing ends by a steel sphere impinging on a special flat-pulse forming device. Moreover, the firing ends of the strips were sufficiently far away from the region of the notch to assure plane stress waves arriving at this zone. The dynamic properties of PMMA were calculated for the appropriate pulse velocities in simple bar tests [162] and found to be (see Table 4): $E = 4.3 \times 10^9$ N/m², $v = 0.34$ and the stress optical constant for transmitted light rays $c_t = 0.74 \times 10^{-10}$ m²/N.

According to the method of transmitted caustics, a divergent light beam impinges on the specimen in close proximity to the bottom of the notch and the transmitted rays are received on a reference plane, parallel to the plane of the specimen. These rays are deviated in different directions according to the law of refraction and they are concentrated along a strongly illuminated curve, which is called the caustic. From the size and the angular displacement, ϕ, of the axis of symmetry of the caustic relative to the crack, or notch axis, it is possible to calculate the stress intensity factors K_I and K_{II}, provided that the order of singularity at the tip or the apex of the discontinuity is known [139].

The stress intensity factors K_I and K_{II} for the case of a sharp V-notch are given by Rels (512) and (513). Then, the stress intensity factors K_I and K_{II} are calculated according to the theory of dynamic caustics at the propagating crack tips [164, 165], by Rels (601) or (618) and (619).

Figure 128 presents a series of photographs for a PMMA strip, containing a symmetric V-notch of initial length $a_0 = 0.010$ m. The angles of the notch were $\varphi = 120°$, $\varphi_1 = \varphi_2 = 150°$. The air pressure in the gun was 1.0 bar. Figure 129 presents a series of photographs for a PMMA strip, containing a symmetric V-notch of initial

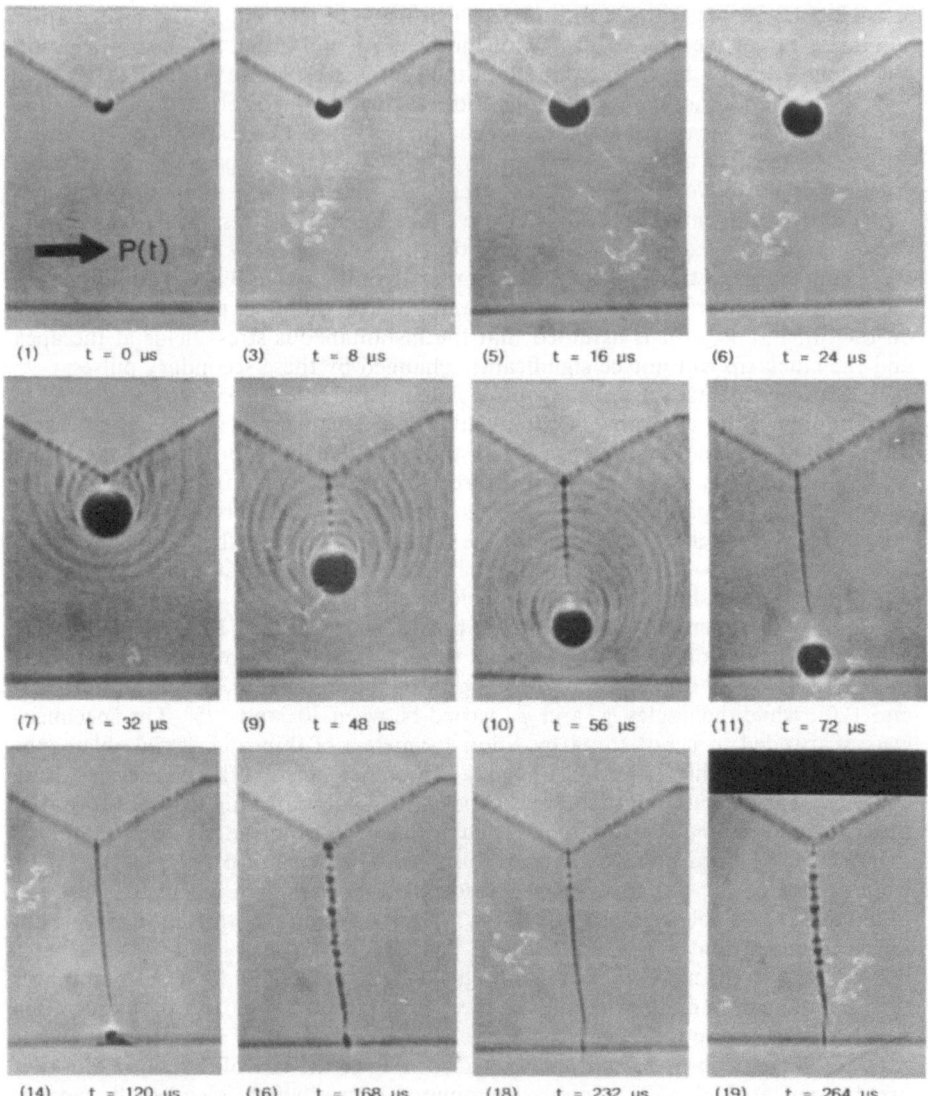

Fig. 128. Series of photographs obtained with a high-speed camera in a V-notched PMMA strip with $a_0 = 0.010$ m, $\varphi = 120°$ and $\varphi_1 = \varphi_2 = 150°$. The projectile shot by the air-gun was under an air pressure of 1.0 bar.

length $a_0 = 0.009$ m. The angles of the notch were in this case $\varphi = 160°$, $\varphi_1 = \varphi_2 = 170°$ and the air pressure in the gun was 2.5 bars.

The series of photographs in Figs 128 and 129 indicate the influence of the arriving tensile stress pulses at the apices of the V-notches. Indeed, at these instants, caustics of the tensile type (almost circular) are formed, which continuously increase in size, since the intensity of the stress pulse is increasing with time. Thus, the values of the respective stress intensity factors at the apices of the notches, K_1, are increasing up

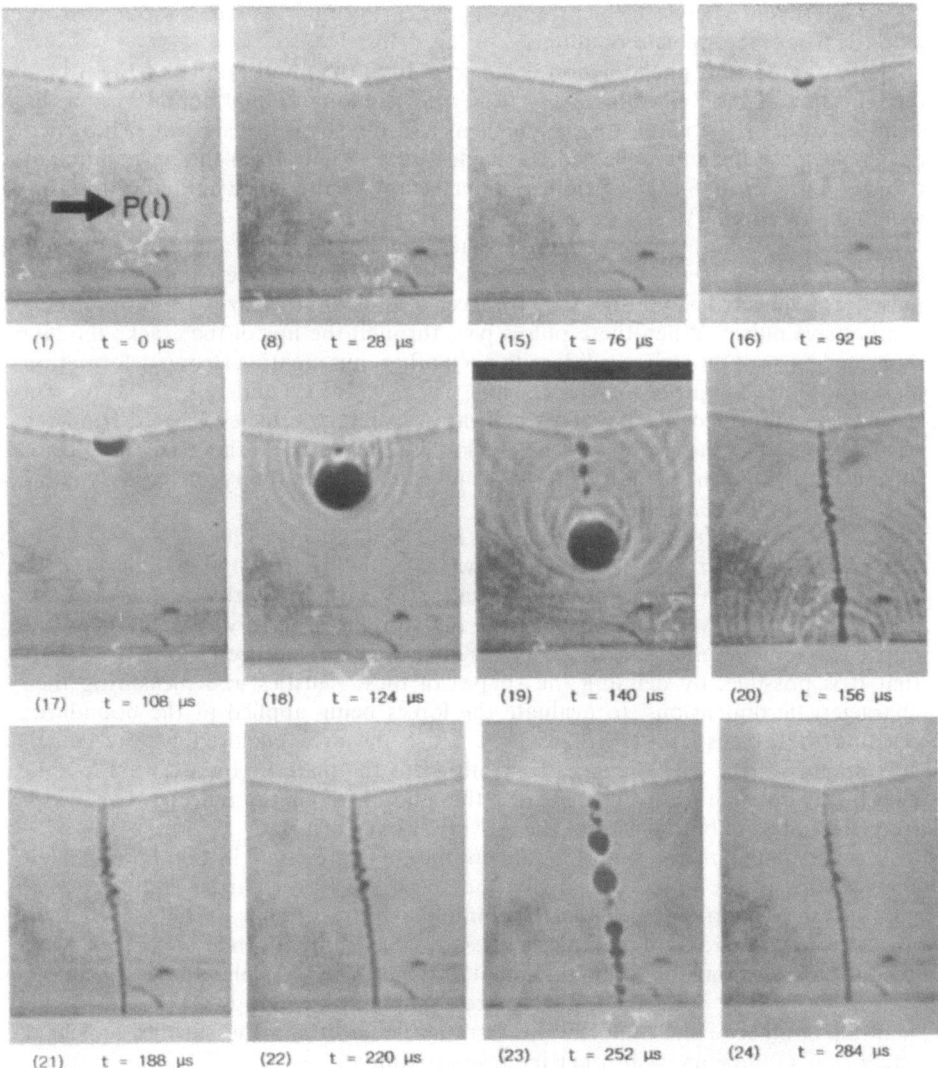

Fig. 129. Series of photographs obtained with high-speed camera in a V-notched PMMA strip with $a_0 = 0.009$ m, $\varphi = 160°$ and $\varphi_1 = \varphi_2 = 170°$. The projectile shot by the air-gun was under an air pressure of 2.5 bars.

to the critical value for initiation of crack propagation at the bottom of the notches (Figs 128(6) and 129(17)).

Subsequently, the cracks propagate with continuously increasing velocities, also radiating stress waves from their tips (Figs 128(7, 9, 10) and 129(18, 19)). These stress waves are of the Rayleigh type as shown in Ref. [225]. When the cracks approach the opposite longitudinal boundaries of the strips, their velocities are progressively and rapidly reduced and the respective values of stress intensity factor are also reduced, as may be determined from the progressive reduction in the sizes of the caustics in these neighbourhoods (Fig. 128(14)). These reductions of c and K_1 are due to interference between the progressing crack tips and the stress waves already reflected from the opposite boundaries of the strips.

Another important phenomenon detected only in the photographs of Fig. 129 is the reflection of the Rayleigh waves at the opposite longitudinal boundaries of the strips as soon as the crack tips reach them and the strips are separated into two pieces. At these instants, reflected Rayleigh waves radiate from the new source of reflected energy, that is the point of intersection of the crack and the opposite longitudinal boundary.

On the other hand, after passage of the extensional stress pulses and the annullment of the crack velocities c and the K_I-stress intensity factors, the extensional stress pulses are reflected from the free transverse ends of the strips and return as compressive pulses. When these pulses pass through the lips of the cracks the two separated parts of the strips collide with each other thus creating strong deformation along the whole length of the crack. Thus, Figs 128(16–19) and 129(20–24) present a series of connected black dots, which are secondary caustics created from the partial contacts of the two lips of the cracks. The variations of their sizes and forms during the evolution of the phenomena indicate the variability of these dynamic after-effects until their total attenuation.

Of importance are two phenomena indicated in Figs 129(20) and (23). In Fig. 129(20) together with the Rayleigh waves radiating from the extremity of the crack there is also a black dot of significant size travelling with the same velocity as the Rayleigh waves, which represents the amount of energy reflected from the opposite boundary and travelling back inside the plate. It has been shown previously that it is possible, by defining the shapes of these caustics and measuring their characteristic dimensions, to evaluate the forces being applied at the boundaries creating these caustics [228, 229, 230]. Since the area enclosed by the caustic corresponds to the elastic energy transmitted to the plate by the external loading [231], it is possible to evaluate readily the amount of energy reflected backwards from the opposite boundaries of the already broken strips.

A similar phenomenon but of a different nature is presented in Fig. 129(23). Here the two lips of the crack that broke the specimen are in partial contact due to zig-zagging of initial paths. Thus, the compressive stress pulses partially close these lips and create areas of contact with strong caustics at their extremities and pseudocaustics along the rest of their contact zones. The same phenomena of contact during the passage of the compressive stress pulse appear in Figs 128(16, 18, 19), but here the contact of the opposite lips is rather smooth and the caustics appear as small beads following each other in close proximity.

The differences in the behaviour of the two strips in Figs 128 and 129 are due to the difference in angle φ of the notches for the two cases ($\varphi = 120°$ and $160°$, respectively). Because of this, the stress pulse in the second case must be much stronger than in the first case, in order to create fracture with such an obtuse notch. Indeed, the stress pulse in the second case is 2.5 times larger than the first one. This explains the stronger deformation phenomena appearing in the strip after its total failure and the large deformations of the lips of the notch as well as the separated lips of the already broken crack.

Figure 130 presents the variations of the crack velocities, c, and the respective stress intensity factors, K_I, versus real time from the instant of the arrival of the pulses in the neighbourhood of the notches for the two cases studied in Figs 128 and 129. First it may be remarked that the loading phenomena of the notches and the crack propagations appear earlier for the sharper notch ($\varphi = 120°$) than for the

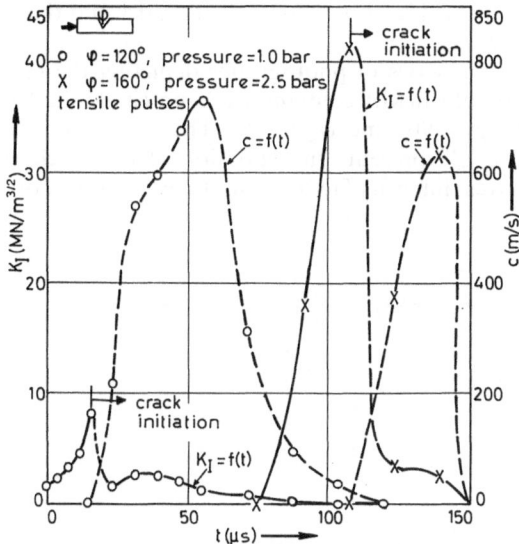

Fig. 130. Variation of the stress intensity factor K_I, and the crack velocity c, versus time for the experiments of Figs 128 and 129.

obtuse notch ($\varphi = 160°$). In both cases there is an abrupt increase in the stress intensity factor up to a maximum when cracks initiate at the bottom of the notches and then an initial abrupt decrease of K_I followed by a smoother and wavy attenuation of K_I presenting a rather shallow secondary maximum. The crack velocities in both cases follow similar bell-shaped curves with smoother slopes for the increasing branches and initially abrupt, afterwards smoothly attenuating, decreasing branches.

The differences in total times for the propagations of the cracks are due to the differences in the intensities of the stress pulses. Indeed, for the 120° notch the total time of crack propagation $t_{120} = 107\,\mu s$, whereas for the 160° notch this time is only $t_{160} = 41\,\mu s$. The ratio t_{120}/t_{160} is approximately 2.5, which is the inverse ratio of the stress pulse intensities to the strips.

From Fig. 130 it may be remarked that, while for the sharper notch, fracture is achieved under small applied pressure and low values of K_I, for the obtuse notch higher applied pressures are necessary to initiate fracture which starts at high values of K_I. Conversely, the average crack velocities are high for the 120° notch and low for the 160° notch. Finally, since the initial notches are symmetric to the transverse direction of the strip and to the front of the applied plane stress pulses, only the K_I stress intensity factor is operative and $K_{II} = 0$ everywhere.

It may also be remarked that while, before the initiation of propagation of the cracks, the values for K_I are in general high, after initiation they are reduced abruptly and to much lower values. This phenomenon may be explained by the fact that the orders of singularities at the bottoms of the notches are much lower than the order of singularity at a crack tip ($\lambda = -1/2$). Thus for $\varphi = 120°$ it may be found from Ref. [132] (Table 6) that $\lambda_{120} = -0.38427$ and for $\varphi = 160°$ that $\lambda_{160} = -0.16530$. The transition of the order of singularity from its value corresponding to the respective notch to the order $\lambda = -1/2$ for a crack tip is made by consuming some internal energy which considerably reduces the caustics and the values of K_I during this transition period.

Figure 131 presents a series of photographs depicting the mode of loading and fracture of a thin strip of PMMA containing a V-notch of initial length $a_0 = 0.005$ m and angles $\varphi = 60°$, $\varphi_1 = 90°$ and $\varphi_2 = 150°$. The applied pressure by the air-gun was 1.3 bars. Figure 132 presents the variation of the crack velocity c, and the components of the stress intensity factor versus time for the test presented in Fig. 131.

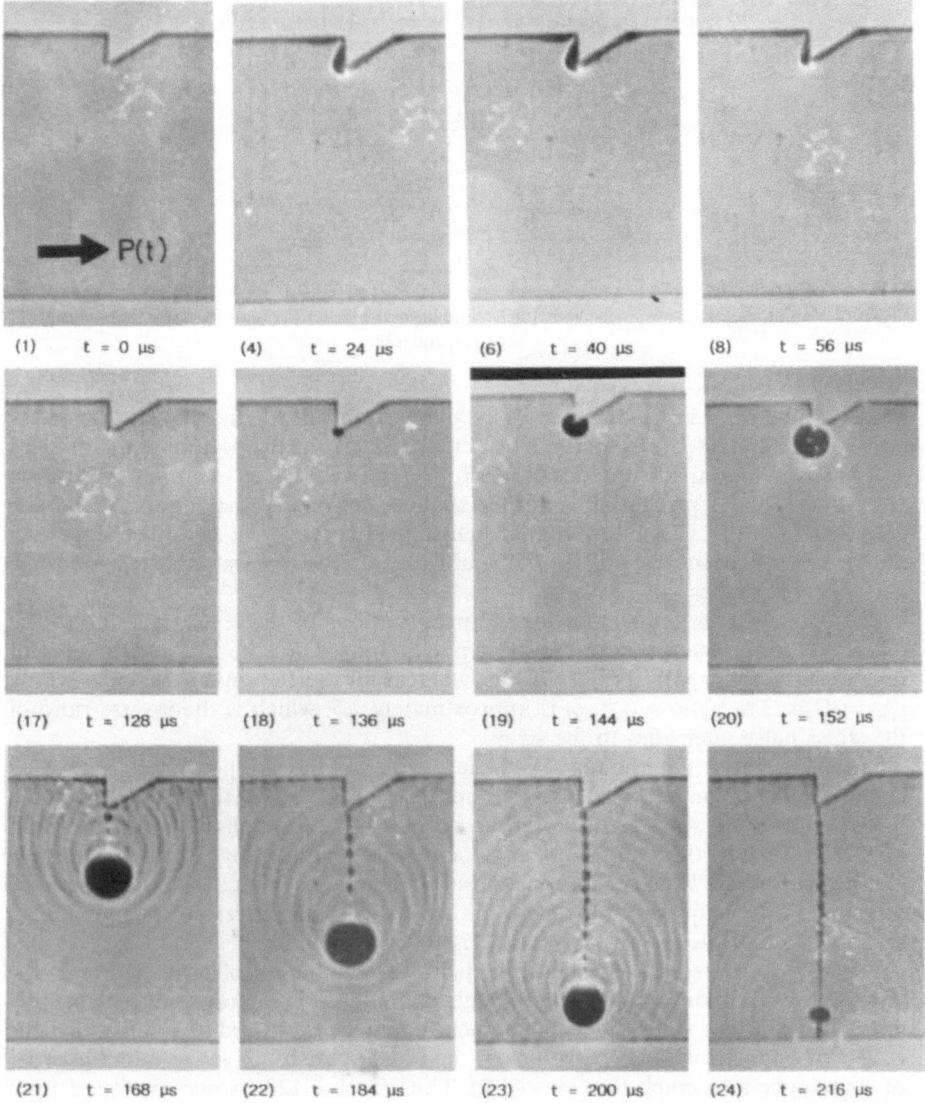

Fig. 131. Series of photographs obtained with a high-speed camera in a V-notched PMMA strip with $a_0 = 0.005$ m, $\varphi = 60°$ and $\varphi_1 = 90°$, $\varphi_2 = 150°$. The projectile shot by the air-gun was under an air pressure of 1.3 bars.

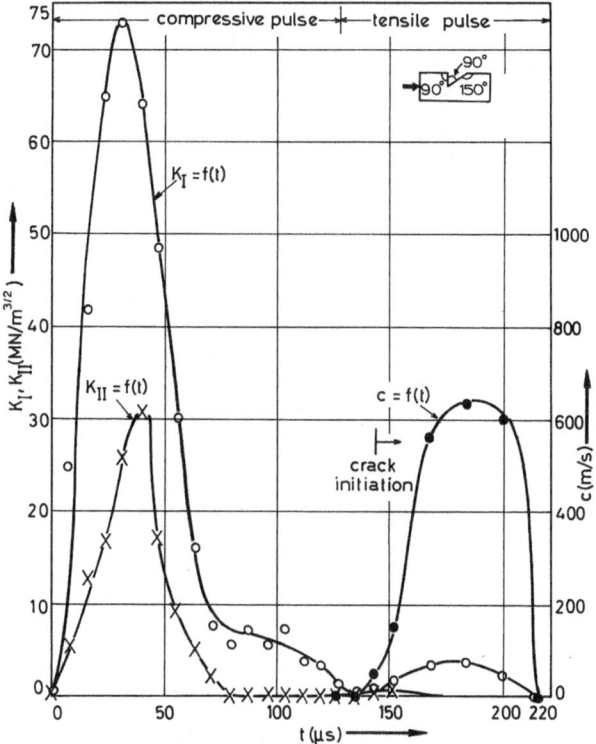

Fig. 132. Variation of the stress intensity factors K_I and K_{II} and the crack velocity c, versus time for the experiment of Fig. 131.

The photographs of Fig. 131 indicate that the orthogonal notch, loaded from the side of its normal lip, is strained according to the model of Fig. 127. The compressive pulse (Figs 131(1–17)) loads the V-notch with a compressive stress and a shear stress component of high value. In this way a compression type caustic is formed along the normal lip of the notch. This caustic also presents an angular displacement due to the shear stress which contributes to the creation of the K_{II} stress intensity factor. Relations (512) and (513) allow the evaluation of the K_I and K_{II} stress intensity factors from the transverse diameter and angular displacement of the caustic.

Subsequently the stress pulse, after a reflection along the opposite transverse boundary of the strip returns to the notch as a tensile pulse and loads accordingly the V-notch. Caustics are formed for the tensile type (almost circular), corresponding to Figs 131(18–20). This loading of the bottom of the notch results in nucleation and propagation of a crack indicated in Figs 131(20–23). Emission of Rayleigh stress waves accompanies the propagation of the crack (Figs 131(21–23)), whereas after the exit of the crack from the opposite longitudinal boundary and the total separation of the strip into two pieces, a backward-travelling caustic in Fig. 131(24) indicates a backward-reflected energy quantity which is proportional to the size of the caustic and which travels with the Rayleigh velocity and does not attenuate during propagation.

The sizes and orientations of the caustics as well as their exact positions suffice to evaluate the crack tip velocities and the instantaneous values of the dynamic K_I and K_{II} stress intensity factors.

Figure 132 presents the variation of the crack tip velocity and the values of the K_I and K_{II} dynamic stress intensity factors, versus time. It is clear from these plots that again K_I and K_{II} attain large values during the passage of the compressive pulse, but these are afterwards reduced considerably in the period of the passage

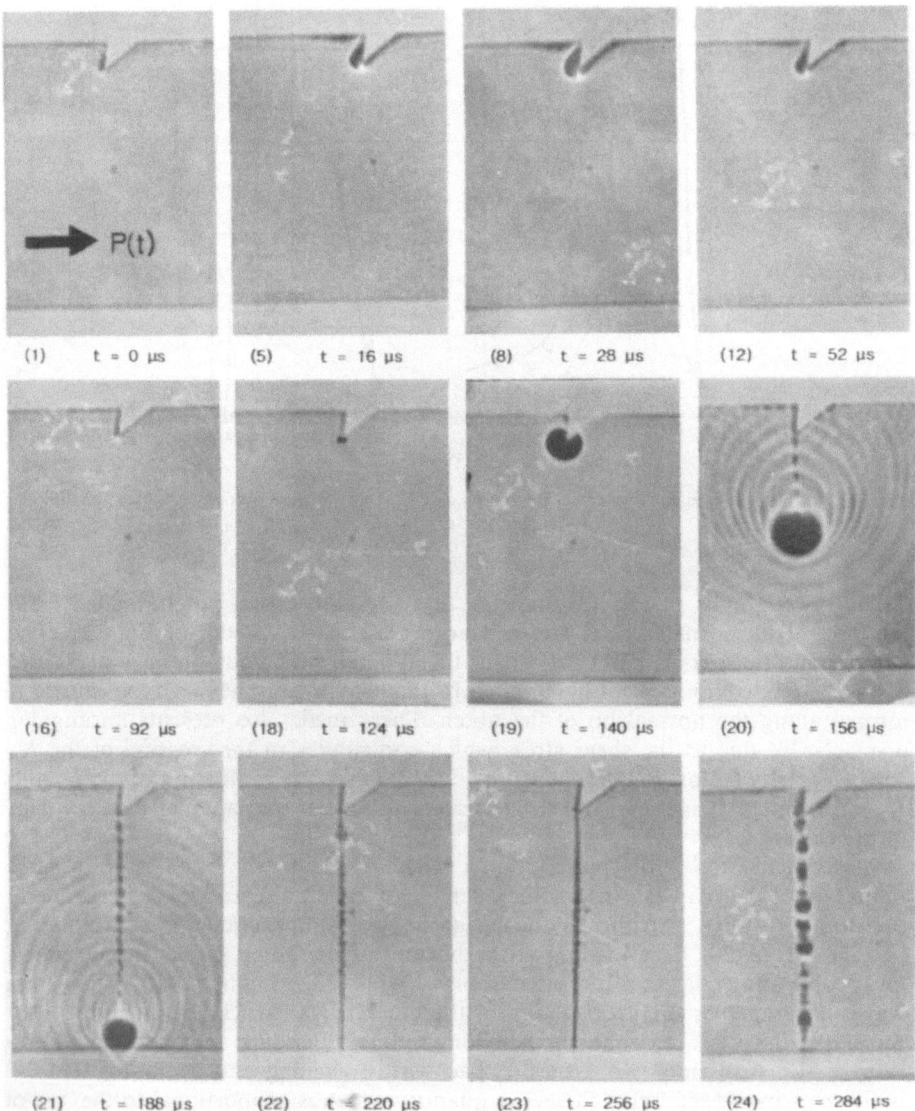

Fig. 133. Series of photographs obtained with a high-speed camera in a V-notched PMMA strip with $a_0 = 0.004\,\mathrm{m}$, $\varphi = 50°$ and $\varphi_1 = 85°$, $\varphi_2 = 140°$. Air pressure was 1.3 bars.

of the tensile stress pulse, whereas the value of the crack propagation velocity remains rather high.

Note an interesting phenomenon appearing in Figs 131(21–24). Secondary caustics are formed at the intersections of the crack path, behind the crack tip with the traces of the Rayleigh waves; these caustics indicate the amount of deformation at the Rayleigh fronts and they may yield the means of evaluating the energies carried with them.

Finally in Figs 133 and 135 asymmetric V-notches cut out in PMMA strips were studied under the influence of initial compressive pulses created by air-gun pressures of 1.3 bars. In Fig. 133 the notch had an initial length $a_0 = 0.004$ m and angles $\varphi = 50°$, $\varphi_1 = 85°$ and $\varphi_2 = 140°$, whereas in Fig. 135 the respective values were $a_0 = 0.0025$ m, $\varphi = 25°$, $\varphi_1 = 30°$ and $\varphi_2 = 175°$.

These two cases are of interest to show the phenomenon of caging-in of the stress pulses inside the protrusions formed by the longitudinal boundaries of the strips and the less oblique lips of the V-notches. As shown in the photographs of Figs 133 and 135, the loading modes of the V-notches follow the patterns of stress distributions indicative of applying first compressive pulses to the notches and afterwards tensile stress pulses. During the period of the influence of the compressive pulse a strong shear loading along the lips of the notches is apparent (see Figs 133(1–16) and 135(3–15)). It is also important to note that during the caging-in phenomena not only the lips of the notches are highly loaded, but also the parts of the longitudinal

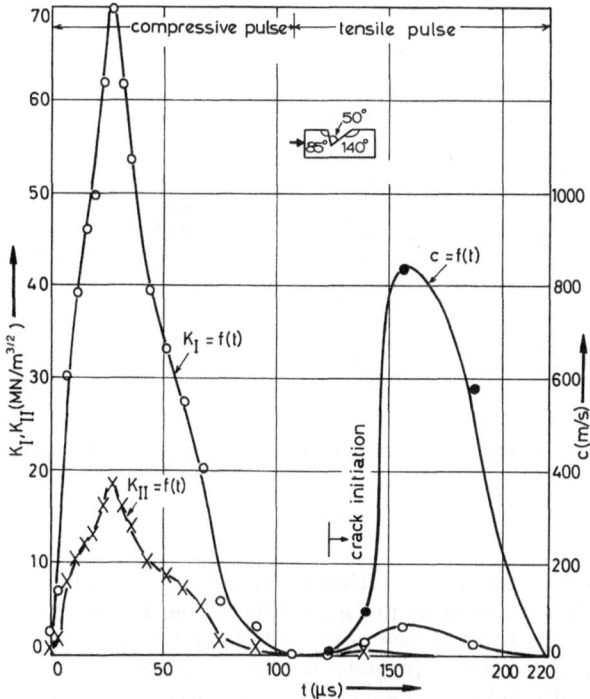

Fig. 134. Variation of the stress intensity factors K_I and K_{II} and the crack velocity c, versus time for the experiment of Fig. 133.

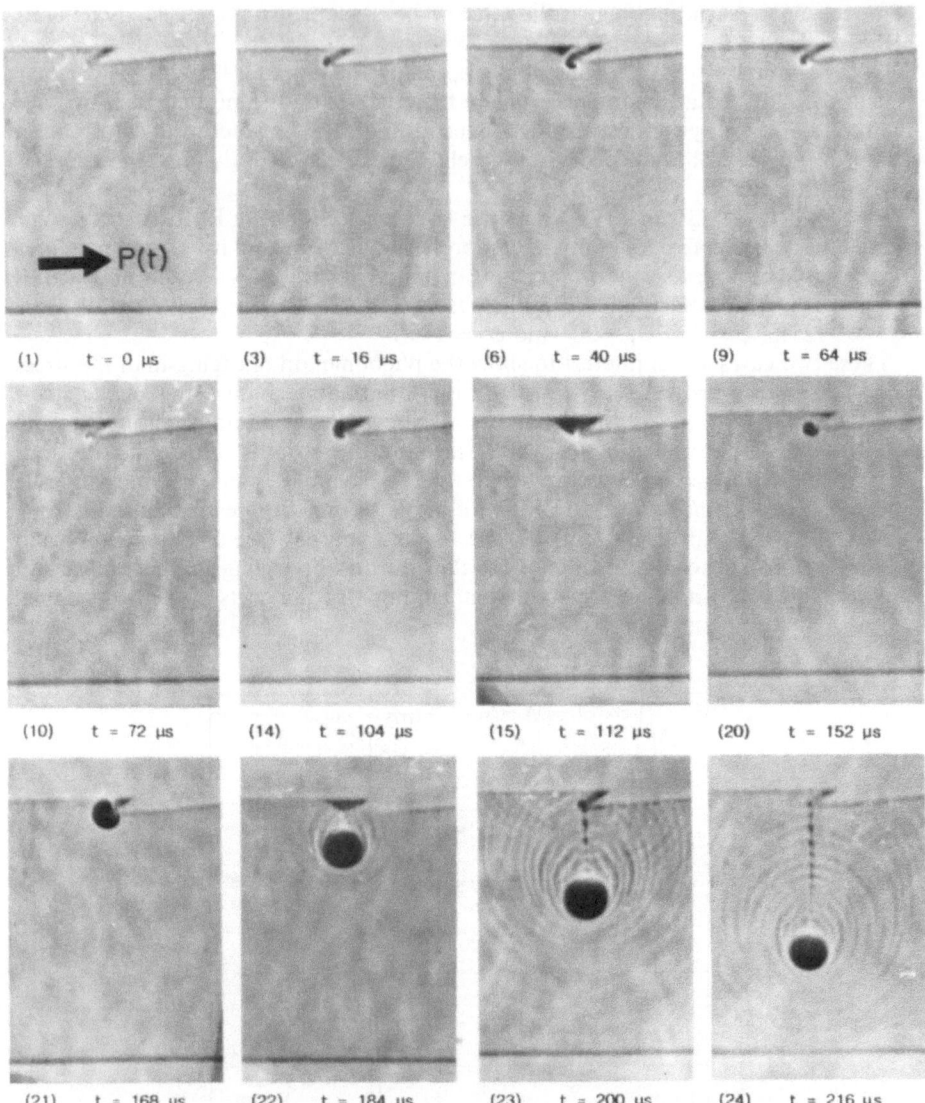

Fig. 135. Series of photographs obtained with a high-speed camera in a V-notched PMMA strip with $a_0 = 0.0025$ m, $\varphi = 25°$ and $\varphi_1 = 30°$, $\varphi_2 = 175°$. Air pressure was 1.3 bars.

boundaries lying on the sides of the incoming pulses, so that intense pseudocaustics appear along these boundaries in Figs 133(5–12) and 135(3–9), the higher loadings of these boundaries being shown in Figs 133(8) and 135(6).

Again during the compressive pulse loadings of the notches the values of K_I and K_{II} increase rapidly and attain high values as shown in Figs 134 and 136, respectively, where the variations of c and of K_I and K_{II} stress intensity factors are plotted versus time.

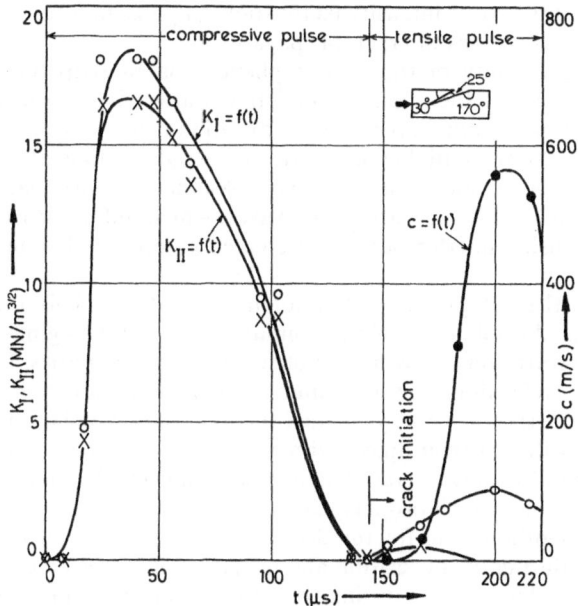

Fig. 136. Variation of the stress intensity factors K_I and K_{II} and the crack velocity c, versus time for the experiment of Fig. 135.

During the phase of loading of the notches with reflected tensile pulses, cracks are initiated at the bottom of the notches, which are propagated straight ahead under lower values of the K_I stress intensity factor for reasons explained already in the first series of tests. The propagating cracks are accompanied along the whole width of the strips by Rayleigh waves and also by reflections of some parts of the energies stored at the cracks which travelled backwards with the Rayleigh wave velocities for PMMA.

An important remark about the test of Fig. 133 is that upon relaxation of the strip after complete fracture (Fig. 133(23)), the returning back-reflected compressive stress pulse compresses the totality of the lips of the broken specimen, creating significant sizes of caustics at points of load concentrations while simultaneously loading both lips of the notch.

This phenomenon of compressive loading of the lips of cracks and faults, which is well described in geomechanics, especially in cases of seismic phenomena, has, for the first time, been repeatably recorded in our experiments. In the tests of Fig. 135 the recording was concentrated in the early phases of the caging-in phenomenon in the acute wedge between the longitudinal boundary and the lip of the notch. This zone is totally under intense deformation, presenting a combined mode-I and mode-II caustic in Fig. 135(14) which is transformed in Fig. 135(15) into a higher order caustic with an oblique cross shape and is incapable of being described by the existing theory of caustics, being a "higher order catastrophe" necessitating a different mathematical treatment.

Finally, it is worthwhile pointing out that in all cases of asymmetric notches the propagating cracks were mainly under mode-I deformation and the values of the

K_{II} components of the stress intensity factor took only insignificant values and then only during the early stages of crack propagation.

The experimental study of the impact phenomena in strips containing sharp V-notches along one of their longitudinal boundaries, based on the method of caustics, revealed important features of the dynamic behaviour of structures.

Thus, reflections of the initial compressive stress pulses along all the boundaries, longitudinal, transverse, and notch lips, gave secondary tensile stress pulses and created longitudinal and transverse stress waves, which diffracted from the bottom of the notches causing interference which created complicated states of loading of the plates.

Depending on the obliqueness of the notch lips and especially that of the front lip first receiving the influence of the coming stress pulse, caging-in phenomena appeared in the protrusions between the notches and the boundaries. These multiplied the phenomena of reflections and sometimes resulted in fracture of these protrusions.

Compressive stress pulses always loaded the lips of the notches and created high value stress intensities at their apices with both K_I and K_{II} components.

Tensile stress pulses always resulted in nucleating cracks emanating from the bottom of the notches, which propagated straight ahead along transverse directions under mode-I conditions and fractured the strips. The values of K_I and K_{II} stress intensity factors during the nucleation and propagation of the cracks were abruptly reduced from their high values, during the influence of the compressive pulses. The K_I stress intensity factor varied smoothly, presenting some flat maxima, whereas the K_{II} stress intensity factor was rapidly attenuated.

The propagation of the cracks was followed by Rayleigh waves creating characteristic traces and developing secondary caustics at their intersections with the lips of the cracks. From these caustics it is possible by an inverse method of caustics to evaluate the elastic energy carried forward by each Rayleigh wave.

Phenomena of reflection of waves from the opposite longitudinal boundaries of the strips are detected and similarly the backward-travelling waves and their caustics along the crack lips. This gives a means of evaluating the reflected elastic energy from the boundaries of the strips.

Finally, an important phenomenon well known in geomechanics, that of the closure of the separated lips of cracks or faults by backward-travelling compressive stress pulses, was readily detected and an analysis of the caustics formed by the partial contacts of opposite lips of the cracks may give a powerful means of studying such phenomena.

Part II
The Det.-Criterion of Fracture

Chapter 4
The Elastic Strain Energy Density

4.1 General Aspects

Study of the distribution of elastic strain energy density at the crack tip is of special interest in understanding the mechanism of fracture.

An extensive theoretical study of the concept of the elastic strain energy density was made by Sih in a series of papers [72–75]. Also, Riedmüller [80] and Theocaris and Papadopoulos [83] have studied the distribution of the elastic strain energy density at the crack tip in isotropic elastic media deformed under modes I and II.

When a body is loaded by tractions and body forces, the movement of the applied loads is converted to work on the body, which is stored in the form of strain energy or elastic strain energy. It is possible to express the energy stored per unit volume of material W, in terms of stress components, strain components or stress and strain components combined.

4.2 The Elastic Strain Energy

The elastic strain energy dW stored in a parallelepiped of volume dV dominating at the strained plate is expressed in terms of the stress components by [72, 90, 232]:

$$\frac{dW}{dV} = \frac{1}{2E}\sigma_{ii}^2 - \frac{v}{E}(\sigma_{xx}\sigma_{yy} + \sigma_{yy}\sigma_{zz} + \sigma_{zz}\sigma_{xx}) + \frac{1}{2G}(\tau_{xy}^2 + \tau_{yz}^2 + \tau_{zx}^2) \qquad (625)$$

where:

$$\sigma_{ii}^2 = \sigma_{xx}^2 + \sigma_{yy}^2 + \sigma_{zz}^2 \qquad (626)$$

For the case of plane-stress conditions, where $\tau_{xz} = \tau_{yz} = \sigma_{zz} = 0$ is valid, Rel. (625) becomes:

$$\frac{dW}{dV} = \frac{1}{2E}(\sigma_{xx}^2 + \sigma_{yy}^2) - \frac{v}{E}\sigma_{xx}\sigma_{yy} + \frac{1}{2G}\tau_{xy}^2 \qquad (627)$$

and by setting:

$$E = 2G(1 + v) \qquad (628)$$

we obtain:

$$\frac{dW}{dV} = \frac{1}{8G}\left[\frac{1-v}{1+v}(\sigma_{xx} + \sigma_{yy})^2 + (\sigma_{xx} - \sigma_{yy})^2 + 4\tau_{xy}^2\right] \tag{629}$$

where G is the shear modulus and v is Poisson's ratio of the elastic and isotropic material of the strained plate.

For the case of plane-strain conditions, where $\gamma_{yz} = \gamma_{xz} = 0$, $\varepsilon_{zz} = 0$ and $\sigma_{zz} = v(\sigma_{xx} + \sigma_{yy})$ are valid, Rel. (625) becomes:

$$\frac{dW}{dV} = \frac{1}{2E}[\sigma_{xx}^2 + \sigma_{yy}^2 + v^2(\sigma_{xx} + \sigma_{yy})^2] - \frac{v}{E}[\sigma_{xx}\sigma_{yy} + v(\sigma_{xx} + \sigma_{yy})^2] + \frac{1}{2G}\tau_{xy}^2 \tag{630}$$

or after some algebra:

$$\frac{dW}{dV} = \frac{1}{8G}[(1-2v)(\sigma_{xx} + \sigma_{yy})^2 + (\sigma_{xx} - \sigma_{yy})^2 + 4\tau_{xy}^2] \tag{631}$$

Introducing the well known relations of elasticity:

$$\sigma_{xx} + \sigma_{yy} = \sigma_1 + \sigma_2$$
$$(\sigma_{xx} - \sigma_{yy})^2 + 4\tau_{xy}^2 = (\sigma_1 - \sigma_2)^2 \tag{632}$$

relations (629) and (631) become:

$$\frac{dW}{dV} = \frac{1}{8G}[\varkappa_{1,2}(\sigma_1 + \sigma_2)^2 + (\sigma_1 - \sigma_2)^2]$$

$$= \frac{1+v}{4E}[\varkappa_{1,2}(\sigma_{xx} + \sigma_{yy})^2 + (\sigma_{xx} - \sigma_{yy})^2 + 4\tau_{xy}^2] \tag{633}$$

where σ_1 and σ_2 are the principal stresses corresponding to the stress tensor σ_{ij}, and $\varkappa_{1,2}$ take the values:

$$\varkappa_1 = \frac{1-v}{1+v}, \qquad \text{for plane stress}$$

$$\varkappa_2 = (1-2v), \quad \text{for plane strain} \tag{634}$$

For a thin elastic and isotropic plate, containing a crack inclined to the axes of symmetry of the plate, and subjected to a biaxial tension at infinity (Fig. 14), the complex stress functions of Muskhelishvili [89] are given by Rels (329) and (330). The stress components at the crack tip are given by Rels (365), (366) and (367). Introducing Rels (365), (366) and (367) into Rels (632), we obtain:

$$\sigma_1 + \sigma_2 = \frac{2}{\sqrt{2\pi r}}\left(K_I\cos\frac{\theta}{2} - K_{II}\sin\frac{\theta}{2}\right) - \lambda \tag{635}$$

$$\sigma_1 - \sigma_2 = \pm\left(\frac{1}{2\pi r}[K_I^2\sin^2\theta + 2K_IK_{II}\sin 2\theta + K_{II}^2(4 - 3\sin^2\theta)] + \lambda^2\right.$$

$$\left. + \frac{\lambda}{\sqrt{2\pi r}}\sin\theta\left[K_I\sin\frac{3\theta}{2} + 4K_{II} + 2K_{II}\cos\frac{3\theta}{2}\right]\right)^{\frac{1}{2}} \tag{636}$$

where K_I and K_{II} are given by Rels (348) and (349) and λ is given by:

$$\lambda = \sigma(1 - k)\cos 2\omega \tag{637}$$

4.2.1 Evaluation of the Elastic Strain Energy Density from the Diameters of the Caustic

The parametric equations of the caustic, for the case of biaxial loading, are given by Rels (383) and (384) and the initial curve is given by Rel. (379). The stress intensity factors K_I and K_{II} are given by Rels (274) and (396).

Using Rels (379), Rels (635) and (636), which give the sum and the difference of principal stresses, may be written as follows [83]:

$$\sigma_1 + \sigma_2 = \frac{2f}{(K_I^2 + K_{II}^2)^{\frac{1}{16}}}\left(K_I \cos\frac{\theta}{2} - K_{II}\sin\frac{\theta}{2}\right) - \lambda \tag{638}$$

and:

$$\sigma_1 - \sigma_2 = \pm\left(\frac{f^2}{(K_I^2 + K_{II}^2)^{\frac{1}{8}}}[K_I^2 \sin^2\theta + 2K_I K_{II}\sin 2\theta + K_{II}^2(4 - 3\sin^2\theta)] \right.$$

$$\left. + \lambda^2 + \frac{f\lambda}{(K_I^2 + K_{II}^2)^{\frac{1}{16}}}\sin\theta\left(K_I\sin\frac{3\theta}{2} + 4K_{II} + 2K_{II}\cos\frac{3\theta}{2}\right)\right)^{\frac{1}{2}} \tag{639}$$

where:

$$f = (2\pi)^{-\frac{1}{2}}(\tfrac{3}{2}C'_{r,t,f})^{-\frac{1}{3}} \tag{640}$$

Introducing Rels (638) and (639) into Eq. (633) we obtain:

$$8G\frac{dW}{dV} = \frac{f^2}{(K_I^2 + K_{II}^2)^{\frac{1}{8}}}\left(4\varkappa_{1,2}\left(K_I \cos\frac{\theta}{2} - K_{II}\sin\frac{\theta}{2}\right)^2 + K_I^2 \sin^2\theta + 2K_I K_{II}\sin 2\theta \right.$$

$$\left. + K_{II}^2(4 - 3\sin^2\theta)\right) - \frac{f\lambda}{(K_I^2 + K_{II}^2)^{\frac{1}{16}}}\left(4\varkappa_{1,2}\left(K_I\cos\frac{\theta}{2} - K_{II}\sin\frac{\theta}{2}\right)\right.$$

$$\left. - \sin\theta\left(K_I\sin\frac{3\theta}{2} + 4K_{II} + 2K_{II}\cos\frac{3\theta}{2}\right)\right) + \lambda^2(1 + \varkappa_{1,2}) \tag{641}$$

Relation (641) gives the distribution of the elastic strain energy density along a small circle representing the singular core, coinciding with the initial curve of the caustic and surrounding the crack tip, when the cracked plate is subjected to a biaxial state of loading at infinity and both K_I and K_{II} are operative [81, 82]. For isotropic, elastic and optically-inert materials the initial circle has a radius r_0 given by Rel. (379).

It has been shown from the previously developed theory, that the size of this circle may be chosen as small as desirable, by decreasing the overall optical constants $C'_{r,t,f}$ and increasing the magnification ratio λ_m. This possibility allows the evaluation of the elastic strain energy distribution around the crack tip along a well defined curve, i.e. the initial curve, which lies as close as possible to the crack tip.

Relation (641) was used to trace on a digital computer and plotter the elastic strain energy density distribution around a crack tip in a plexiglas plate with $v = 0.34$, $d = 0.003$ m, $c_r = -1.70 \times 10^{-10}$ m^2/N, $2a = 0.02$ m, $\sigma = 1$ N/m^2, $\varepsilon = 2$, $z_0 = 2$ m, $z_i = 0.5$ m and $\lambda_m = 5$. Figure 137 presents various distributions of the elastic strain

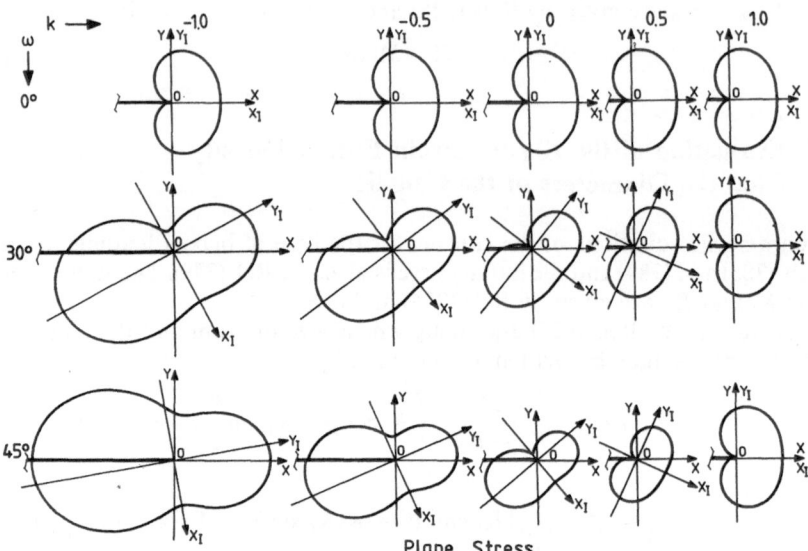

Fig. 137. Plots of the distribution of the elastic strain energy density at the crack tip for plane stress conditions and for k between -1 and 1 and $\omega = 0°$, $30°$ and $45°$.

energy density for plate-stress conditions of the cracked plate when the loading ratio k varies between -1 and 1 for $\omega = 0°$, $30°$, $45°$.

4.3 S_2-Criterion of Fracture

The distribution of strain energy density being stored in the loaded plate and corresponding to the work of the external load, gives information on the expected angle of crack propagation. The evaluation of the expected angle of propagation is more accurate when the evaluation of S is performed closer to the crack tip. The curve closest to the crack tip is the initial curve of the caustic, namely the boundary between the singular region and the rest of the elastic plate. The distribution of S_2 $(S_2 = \mathrm{d}W/r\,\mathrm{d}V)$ along the initial curve possesses extrema.

According to the minimum strain energy density criterion introduced by Sih [72, 75, 76], a crack propagates in the direction of the local minimum of the strain energy density. Thus, by evaluating this minimum along the initial curve of the caustic, the exact direction of crack initiation is determined. This direction is defined by the angle ϑ_{min}^l between the direction of minimum strain energy density and initial crack direction.

The influence of the loading ratio k, as well as the angle ω on the distribution of the elastic strain energy density is clear from Fig. 137. It is worthwhile indicating that the elastic strain energy density distribution always presents a local minimum in front of the crack tip, whose position and magnitude varies in terms of k and ω. Figure 137 shows that, while the magnitude of this minimum depends mainly on k, its position depends exclusively on ω.

The position of this minimum ϑ^l_{min}, may be defined by annulling the partial derivative $(\partial/\partial\theta)(dW/dV)$. By differentiating Rel. (641), one obtains the relation [83]:

$$8G\frac{\partial}{\partial\theta}\left(\frac{dW}{dV}\right) = \frac{f^2}{(K_I^2 + K_{II}^2)^\frac{3}{5}}(-2\varkappa_{1,2}[(K_I^2 - K_{II}^2)\sin\theta + 2K_I K_{II}\cos\theta]$$

$$+ (K_I^2 - 3K_{II}^2)\sin 2\theta + 4K_I K_{II}\cos 2\theta)$$

$$+ \frac{\lambda f}{(K_I^2 + K_{II}^2)^\frac{1}{10}}\left(2\varkappa_{1,2}\left(K_I\sin\frac{\theta}{2} + K_{II}\cos\frac{\theta}{2}\right)\right.$$

$$+ \cos\theta\left(K_I\sin\frac{3\theta}{2} + 2K_{II}\cos\frac{3\theta}{2} + 4K_{II}\right)$$

$$\left.+ \frac{3}{2}\sin\theta\left(K_I\cos\frac{3\theta}{2} - 2K_{II}\sin\frac{3\theta}{2}\right)\right) = 0 \qquad (642)$$

Figures 138 and 139 present the variation of ϑ^l_{min} versus ω for parametric values of the loading ratio k for plane stress (Fig. 138) and plane strain (Fig. 139). In both cases the values of ω vary between 0° and 90°.

The experimental study of Sih's criterion for the propagation of a crack and the direction of the minimum elastic strain energy density, in a plate under plane stress conditions subjected to a general biaxial type of loading at infinity, indicated that the strain energy density distribution depends on the angle of inclination of the crack ω, as well as on the loading ratio k (Fig. 137). The position of the minimum elastic strain energy density, defined by angle ϑ^l_{min}, and therefore the direction of propagation of the crack depends on ω and k, as may be concluded from Figs 138 and 139. For $k \geq 0$ and for ω varying between 0° and 90°, $|\vartheta^l_{min}|$ increases and, having passed through a maximum, it decreases and tends to 0°, for $k = 1$, $\vartheta^l_{min} = 0°$. This means that whatever the angle ω, the crack will spread straight ahead because

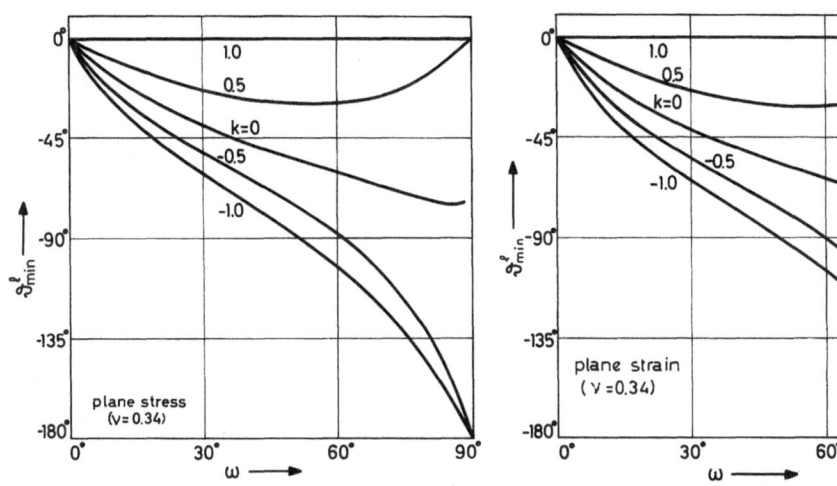

Fig. 138. Variation of ϑ^l_{min} versus ω for various parametric values of k for plane stress conditions.

Fig. 139. Variation of ϑ^l_{min} versus ω for various parametric values of k for plane strain conditions.

$K_{\text{II}} = 0$. For $k < 0$ the angle $|\vartheta^l_{\min}|$ increases and for values of $\omega \geqq 45°$ there exist values of k yielding $K_1 = 0$ and $K_{\text{II}} \neq 0$ (pure shear). At these positions of pure shear a change of sign of $\vartheta^l_{\min}$ appears and then a decrease of its value tending to $0°$ for ω tending to $90°$ is detected.

The corresponding S-criterion for dynamic crack problems has been developed in Ref. [174].

Chapter 5
Det.-Criterion of Fracture

5.1 General Aspects

Various criteria have been introduced for the determination of the conditions of initiation and the prediction of the propagation direction of a crack subjected to combined mode-I and mode-II in-plane deformations. First of all, Griffith [3] introduced a criterion based on physical consideration to determine the conditions to initiate the propagation of a crack. Erdogan and Sih [69] developed the maximum tangential stress criterion, according to which the crack will propagate along the direction of the maximum tangential stress. The maximum energy release rate criterion, the G-criterion [233, 234], represents a generalization of Griffith's original energy release rate concept [3]. Sih [74] developed the minimum elastic strain energy density criterion (S-criterion) which is based directly on the total strain energy density, that is, the sum of its distortional and dilatational components. Theocaris and Papadopoulos [83] proposed the two-term approximation S_2-criterion (Sect. 4.3) which is a modification of the usual S-criterion with a closer approximation, which considers two terms in the series expansion of the stress function [89], instead of only the singular term anticipated by Sih. Modifications to the S-criterion have been made by Theocaris and Andrianopoulos [235] and Wang [236]. Recently, Theocaris and Andrianopoulos [84, 237] proposed a new criterion referred to as the T-criterion. According to the T-criterion, the crack propagates along a direction defined by a maximum of the total energy density, which is also a maximum for the dilatational strain energy density when this distribution is evaluated along a locus of constant distortional strain energy density, as it is on the initial Mises elastic–plastic boundary. Also, the fracture will initiate when the dilatational strain energy density on the elastic–plastic boundary reaches a critical maximum value which is a material property. Yehia [238] proposed as a modification of the T-criterion that the fracture will initiate when the distance from the crack tip to the elastic–plastic boundary in the direction of crack propagation reaches a maximum value.

Recently, Papadopoulos [85, 86, 244–249] proposed a new and simple criterion of fracture, the Det.-criterion. The Det.-criterion of fracture is based on the second stress invariant and for plane-conditions, on the determinant of the stress tensor, Det.(σ_{ij}) [85, 245]. According to the Det.-criterion, the crack propagates in the direction of the maximum value of the determinant of the stress tensor, Det.(σ_{ij}), and the fracture will initiate when the determinant of the stress tensor, Det.(σ_{ij}), on the boundary of the core-region reaches a critical maximum value which is a material property. These conditions of initiation of the crack are proposed as the *Det.-criterion of fracture.*

5.2 Theoretical Consideration

5.2.1 Three-dimensional Crack Problems

The elastic strain energy dW stored in a parallelepiped of volume dV dominating at the strained plate of linear elastic material is expressed by Rel. (625).

Energy has been used for the description of failure of the materials element by yielding. Two theories have been developed: (i) Beltrami–Haigh theory (total energy), and (ii) Hubert von Mises–Huncky theory (distortional energy). According to these theories, failure in a material by yielding occurs when the total or the distortional strain energy density absorbed by the material equals the energy density stored in the material loaded in uniaxial tension at yield. This quantity corresponds to the limiting energy and is regarded as a material constant.

The total elastic strain energy density (Rel. (625)) is divided into two components, the dilatational strain energy density T_V and the distortional strain energy density T_D [90, 232, 239, 240]:

$$\frac{dW}{dV} = T_V + T_D \tag{643}$$

T_V and T_D are given by:

$$T_V = \frac{1 - 2v}{6E}(\sigma_{xx} + \sigma_{yy} + \sigma_{zz})^2 \tag{644}$$

$$T_D = \frac{1 + v}{3E}[(\sigma_{xx} + \sigma_{yy} + \sigma_{zz})^2$$
$$- 3(\sigma_{xx}\sigma_{yy} + \sigma_{xx}\sigma_{zz} + \sigma_{yy}\sigma_{zz} - \tau_{xy}^2 - \tau_{xz}^2 - \tau_{yz}^2)] \tag{645}$$

The stress invariants of the stress tensor σ_{ij} are given by [90, 232]:

$$I_1 = \sigma_{xx} + \sigma_{yy} + \sigma_{zz} \tag{646}$$

$$I_2 = \begin{vmatrix} \sigma_{xx} & \tau_{xy} \\ \tau_{xy} & \sigma_{yy} \end{vmatrix} + \begin{vmatrix} \sigma_{xx} & \tau_{xz} \\ \tau_{xz} & \sigma_{zz} \end{vmatrix} + \begin{vmatrix} \sigma_{yy} & \tau_{yz} \\ \tau_{yz} & \sigma_{zz} \end{vmatrix}$$
$$= \sigma_{xx}\sigma_{yy} + \sigma_{xx}\sigma_{zz} + \sigma_{yy}\sigma_{zz} - \tau_{xy}^2 - \tau_{xz}^2 - \tau_{yz}^2 \tag{647}$$

$$I_3 = \begin{vmatrix} \sigma_{xx} & \tau_{xy} & \tau_{xz} \\ \tau_{xy} & \sigma_{yy} & \tau_{yz} \\ \tau_{xz} & \tau_{yz} & \sigma_{zz} \end{vmatrix} \tag{648}$$

and Rels (644) and (645) are written:

$$T_V = \frac{1 - 2v}{6E}I_1^2 \tag{649}$$

$$T_D = \frac{1+v}{3E}(I_1^2 - 3I_2) \tag{650}$$

Substituting Rel. (649) into Rel. (650), we obtain:

$$T_D = \frac{1+v}{3E}\left(\frac{6E}{1-2v}T_V - 3I_2\right) \tag{651}$$

or:

$$\frac{2E}{1-2v}T_V - \frac{E}{1+v}T_D = I_2 \tag{652}$$

Relation (652), being independent of I_3, is an even function [232].
From Rel. (645), for pure tension, we obtain:

$$T_D = T_{D,0} = \frac{1+v}{3E}\sigma_0^2 = \frac{1+v}{E}k^2 \tag{653}$$

where $T_{D,0}$ is the maximum constant value of T_D, σ_0 is the yield stress of the material for pure tension and k is the yield stress of the material for pure shear.
Also, from Rel. (644), for pure tension and for a crack normal to the direction of loading, we obtain:

$$T_V = T_{V,0} = \frac{1-2v}{6E}(\sigma_\infty^{90°})^2 \tag{654}$$

where $\sigma_\infty^{90°}$ is the fracture stress in pure tension and $T_{V,0}$ is the maximum constant value of T_V which corresponds to fracture.
Substituting Rels (653) and (654) into Rel. (652) we obtain:

$$\tfrac{1}{3}[(\sigma_\infty^{90°})^2 - \sigma_0^2] = \tfrac{1}{3}[(\sigma_\infty^{90°})^2 - 3k^2] = I_2 \tag{655}$$

From Rel. (652) it is concluded that for $T_D = T_{D,0}$ (maximum value), the position and maximum value of T_V depend on the second stress invariant I_2. Therefore, I_2 arranges the two components T_V and T_D of the total elastic strain energy density around the crack tip, so that the difference between T_V and T_D reaches a maximum value when I_2 becomes maximum. Or, at the positions where I_2 becomes maximum, a maximum value of the difference of T_V and T_D occurs and subsequently crack propagation will occur.
From Rel. (655) it is concluded that for $I_2 > 0$, $\sigma_\infty^{90°} > \sigma_0$ or $\sigma_\infty^{90°} > k\sqrt{3}$ while for $I_2 < 0$, $\sigma_\infty^{90°} < \sigma_0$ or $\sigma_\infty^{90°} < k\sqrt{3}$. Therefore, at the position where the distribution of I_2 around the crack tip presents a positive maximum ($T_V > T_D$), crack initiation will occur, while at the position where the distribution of I_2 presents a negative maximum ($T_D > T_V$), yielding of the material will occur.
In the spherical coordinate system (r, θ, ϕ) centred at the crack tip (Fig. 140(a)), the local stresses on an element near the border of an elliptical crack can be expressed by [252]:

$$\sigma_{xx} = \frac{K_I}{\sqrt{2\pi r}}\sqrt{\frac{\kappa+1}{2\lambda\cos\theta}}\left(\frac{2-\kappa+\kappa^2}{2\kappa^3}\right) - \frac{K_{II}}{\sqrt{2\pi r}}\sqrt{\frac{\kappa-1}{2\lambda\cos\theta}}\left(\frac{2+\kappa+3\kappa^2}{2\kappa^3}\right) \tag{656}$$

$$\sigma_{yy} = -\frac{K_I}{\sqrt{2\pi r}}\sqrt{\frac{\kappa+1}{2\lambda\cos\theta}}\left(\frac{2-\kappa-3\kappa^2}{2\kappa^3}\right) + \frac{K_{II}}{\sqrt{2\pi r}}\sqrt{\frac{\kappa-1}{2\lambda\cos\theta}}\left(\frac{2+\kappa-\kappa^2}{2\kappa^3}\right) \tag{657}$$

$$\sigma_{zz} = \frac{K_{\mathrm{I}}}{\sqrt{2\pi r}} \sqrt{\frac{\kappa+1}{2\lambda\cos\theta}} \frac{2v}{\kappa} - \frac{K_{\mathrm{II}}}{\sqrt{2\pi r}} \sqrt{\frac{\kappa-1}{2\lambda\cos\theta}} \frac{2v}{\kappa} \tag{658}$$

$$\tau_{xy} = \frac{K_{\mathrm{I}}}{\sqrt{2\pi r}} \sqrt{\frac{\kappa-1}{2\lambda\cos\theta}} \left(\frac{2+\kappa-\kappa^2}{2\kappa^3}\right) + \frac{K_{\mathrm{II}}}{\sqrt{2\pi r}} \sqrt{\frac{\kappa+1}{2\lambda\cos\theta}} \left(\frac{2-\kappa+\kappa^2}{2\kappa^3}\right) \tag{659}$$

$$\tau_{xz} = -\frac{K_{\mathrm{III}}}{\sqrt{2\pi r}} \sqrt{\frac{\kappa-1}{2\lambda\cos\theta}} \frac{1}{\kappa} \tag{660}$$

$$\tau_{yz} = \frac{K_{\mathrm{III}}}{\sqrt{2\pi r}} \sqrt{\frac{\kappa+1}{2\lambda\cos\theta}} \frac{1}{\kappa} \tag{661}$$

The stress distribution depends on the two spherical angles θ and ϕ (Fig. 140(a)) through the parameters:

$$\lambda = \lambda(\phi, \omega, \text{ crack geometry}), \quad \kappa = \kappa(\theta, \lambda)$$

Substituting Rels (656)–(661) into Rel. (647) we obtain:

$$2\pi r I_2 = K_{\mathrm{I}}^2 D_1 + K_{\mathrm{II}}^2 D_2 + K_{\mathrm{I}} K_{\mathrm{II}} D_{12} + K_{\mathrm{III}}^2 D_3 \tag{662}$$

where:

$$D_1 = -\frac{1}{8\kappa^6 \lambda\cos\theta} [4 + (13 + 8v)\kappa^2 + 3(7 + 4v)\kappa^3 + 11\kappa^4 + (7 - 4v)\kappa^5] \tag{663}$$

$$D_2 = -\frac{1}{4\kappa^3 \lambda\cos\theta} [3 + 2\kappa - \kappa^2 + 8v\kappa(1 - \kappa)] \tag{664}$$

$$D_{12} = \frac{1 - \kappa^2}{2\kappa^3 \lambda^2 \cos^2\theta} [1 + \kappa(1 + 4\kappa)] \tag{665}$$

$$D_3 = -\frac{1}{\kappa\lambda\cos\theta} \tag{666}$$

In the case of $\phi = 0°$, Rels (663)–(666) reduce to the coefficients D_1, D_2, D_{12} for the two-dimensional crack problems which are developed in Refs [85, 244–249] for plane-stress and plane-strain conditions. In these cases the second stress invariant

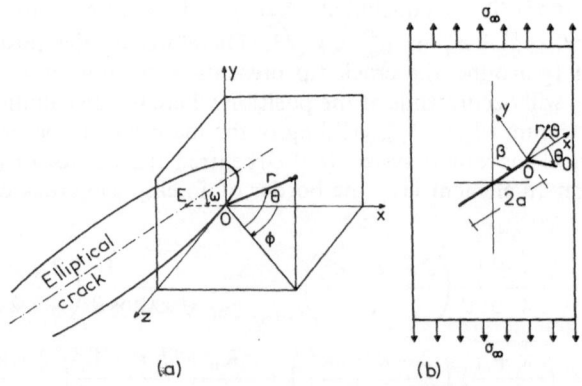

(a)   (b)

Fig. 140. Geometry of a cracked plate and the coordinate system at the crack tip.

becomes $I_2 = \text{Det.}(\sigma_{ij})$ and Rel. (662) becomes:

$$2\pi r I_2 = 2\pi r \, \text{Det.}(\sigma_{ij}) = K_I^2 D_1 + K_{II}^2 D_2 + K_I K_{II} D_{12} \qquad (667)$$

The positions of the maxima of the I_2 distribution around the crack tip can be calculated by [252]:

$$\left.\frac{\partial I_2}{\partial \theta}\right|_{\theta=\theta_0} = 0, \quad \left.\frac{\partial I_2}{\partial \phi}\right|_{\phi=\phi_0} = 0 \qquad (668)$$

and:

$$\left.\frac{\partial^2 I_2}{\partial \theta^2}\right|_{\theta=\theta_0} < 0, \quad \left.\frac{\partial^2 I_2}{\partial \phi^2}\right|_{\phi=\phi_0} < 0, \quad \left(\frac{\partial^2 I_2}{\partial \theta \partial \phi}\right)^2 - \left.\frac{\partial^2 I_2}{\partial \theta^2}\frac{\partial^2 I_2}{\partial \phi^2}\right|_{\substack{\theta=\theta_0 \\ \phi=\phi_0}} > 0 \qquad (669)$$

Relations (668) and (669) describe the Det.-criterion of fracture $(\text{Det.}(\sigma_{ij}) \equiv I_2)$ for three-dimensional crack problems. The Det.-criterion (or I_2-criterion) postulates that the crack propagates along the direction defined by a positive maximum of the second stress invariant, I_2, when this distribution is evaluated around a circle, boundary of the core-region, which is defined by the initial curve of caustics [101] around the crack tip. On the other hand, the T-criterion [237] is based on the distribution of the dilatational strain energy density T_V along a locus of constant distortional strain energy density T_D, the Mises elastic–plastic boundary. So, the direction (θ_0, ϕ_0) for crack initiation is defined by Rels (668) and (669), while the critical stress of crack initiation is calculated by the condition:

$$I_2 = I_{2,cr} \qquad (670)$$

where $I_{2,cr} \equiv \text{Det.}(\sigma_{ij})_{cr}$ is a constant of the material.

5.2.2 Two-dimensional Crack Problems

Plane Stress
For the case under generalized plane-stress conditions, the stress components in the vicinity of the crack tip for a Cartesian coordinate system (Fig. 140(b)) are:

$$\sigma_{xx} \neq 0, \quad \sigma_{yy} \neq 0, \quad \tau_{xy} \neq 0, \quad \sigma_{zz} = \tau_{xz} = \tau_{yz} = 0 \qquad (671)$$

Then, Rels (644) and (645) reduce to the relations:

$$T_V = \frac{1-2v}{6E}(\sigma_{xx} + \sigma_{yy})^2 \qquad (672)$$

$$T_D = \frac{1+v}{3E}[(\sigma_{xx} + \sigma_{yy})^2 - 3(\sigma_{xx}\sigma_{yy} - \tau_{xy}^2)] \qquad (673)$$

and the stress invariants (646)–(648) reduce to the relations:

$$I_1 = \sigma_{xx} + \sigma_{yy} \qquad (674)$$

$$I_2 = \sigma_{xx}\sigma_{yy} - \tau_{xy}^2 = \text{Det.}(\sigma_{ij}) \qquad (675)$$

and Rels (649) and (650) reduce to the relations:

$$T_V = \frac{1-2v}{6E}I_1^2 \qquad (676)$$

$$T_D = \frac{1+v}{3E}[I_1^2 - 3\,\mathrm{Det.}(\sigma_{ij})] \tag{677}$$

Then, Rels (652) and (655) are written [85, 86, 245]:

$$\frac{2E}{1-2v}T_V - \frac{E}{1+v}T_D = \mathrm{Det.}(\sigma_{ij}) \tag{678}$$

$$\tfrac{1}{3}[(\sigma_\infty^{90°})^2 - \sigma_0^2] = \tfrac{1}{3}[(\sigma_\infty^{90°})^2 - 3k^2] = \mathrm{Det.}(\sigma_{ij}) \tag{679}$$

In this case, the second stress invariant is equal to the determinant of the stress tensor $\mathrm{Det.}(\sigma_{ij})$ and the Det.-criterion is developed in Refs [85, 86, 245].

Plane Strain
For the case of plane-strain conditions, the stress components in the vicinity of the crack tip are:

$$\sigma_{xx} \neq 0, \quad \sigma_{yy} \neq 0, \quad \sigma_{zz} = v(\sigma_{xx} + \sigma_{yy}), \quad \tau_{xy} \neq 0, \quad \tau_{xz} = \tau_{yz} = 0 \tag{680}$$

Then, Rels (644) and (645) reduce to the relations:

$$T_V = \frac{(1-2v)(1+v)^2}{6E}(\sigma_{xx} + \sigma_{yy})^2 \tag{681}$$

$$T_D = \frac{1+v}{3E}[(v^2 - v + 1)(\sigma_{xx} + \sigma_{yy})^2 - 3(\sigma_{xx}\sigma_{yy} - \tau_{xy}^2)] \tag{682}$$

and the stress invariants (646)–(648) are written:

$$I_1 = (1+v)(\sigma_{xx} + \sigma_{yy}) \tag{683}$$

$$I_2 = v(\sigma_{xx} + \sigma_{yy})^2 + \sigma_{xx}\sigma_{yy} - \tau_{xy}^2 = v(\sigma_{xx} + \sigma_{yy})^2 + \mathrm{Det.}(\sigma_{ij}) \tag{684}$$

$$I_3 = v(\sigma_{xx} + \sigma_{yy})(\sigma_{xx}\sigma_{yy} - \tau_{xy}^2) = v(\sigma_{xx} + \sigma_{yy})\,\mathrm{Det.}(\sigma_{ij}) \tag{685}$$

where:

$$\mathrm{Det.}(\sigma_{ij}) = \sigma_{xx}\sigma_{yy} - \tau_{xy}^2 \tag{686}$$

Relations (649) and (650) are written:

$$T_V = \frac{1-v}{6E}I_1^2 \tag{687}$$

$$T_D = \frac{1+v}{3E}\left(\frac{(v^2 - v + 1)}{(1+v)^2}I_1^2 - 3\,\mathrm{Det.}(\sigma_{ij})\right) \tag{688}$$

Then, Rels (652)–(655) are written [85, 86, 245]:

$$\frac{2E(v^2 - v + 1)}{(1-2v)(1+v)^2}T_V - \frac{E}{1+v}T_D = \mathrm{Det.}(\sigma_{ij}) \tag{689}$$

$$T_D = T_{D,0} = \frac{(1+v)(v^2 - v + 1)}{3E}\sigma_0^2 = \frac{(1+v)(v^2 - v + 1)}{E}k^2 \tag{690}$$

$$T_V = T_{V,0} = \frac{(1-2v)(1+v)^2}{6E}(\sigma_\infty^{90°})^2 \tag{691}$$

and:

$$\frac{(v^2 - v + 1)}{3}[(\sigma_\infty^{90°})^2 - \sigma_0^2] = \frac{(v^2 - v + 1)}{3}[(\sigma_\infty^{90°})^2 - 3k^2] = \mathrm{Det.}(\sigma_{ij}) \tag{692}$$

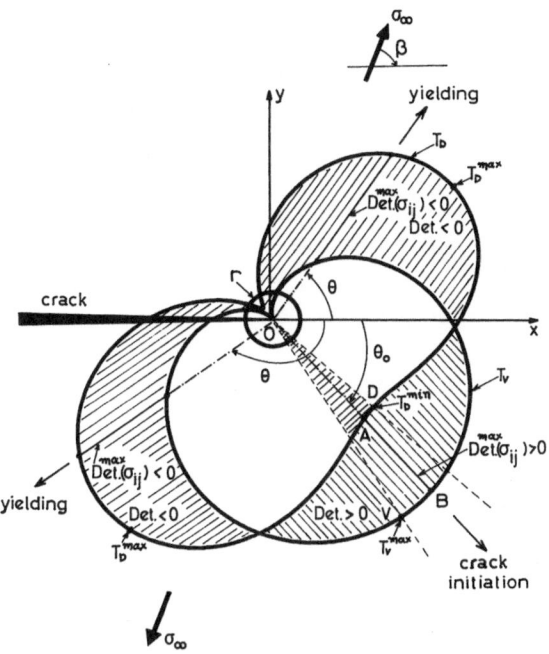

Fig. 141. Distribution of the dilatational T_V and the distortional T_D strain energy density around the crack tip, along the core-region r.

From Rels (655), (672) and (692) it is concluded that the difference of the square of the stresses $\sigma_\infty^{90°}$ and σ_0 depends on I_2, in the three-dimensional case and on Det.(σ_{ij}), in the two-dimensional case, for uncracked specimens.

Figure 141 shows the distribution of the dilatational strain energy density T_V and the distortional strain energy density T_D around the crack tip, along a circle of radius r (core-region). In this figure a narrow region is observed (DOV) in front of the crack tip between the direction of the minimum of T_D and the direction of the maximum of T_V where it is valid that $T_V > T_D$. This means that there is a great possibility that the crack initiates from this region. The exact direction of crack initiation within this region, is defined by the maximum of the difference of the energy densities T_V and T_D. This maximum $((AB)$, Fig. 141), corresponds to a positive maximum of the Det.(σ_{ij}) (max Det.$(\sigma_{ij}) = (AB)$) (Rels (678) and (689)). This direction subtends an angle $(-\theta_0)$ (Fig. 141) with the Ox-axis of the crack. Therefore, the direction of crack initiation is defined by the angle $(-\theta_0)$ which is predicted by the Det.-criterion.

As shown in Fig. 141, there are three regions (shaded areas) where the Det.(σ_{ij}) takes positive values as well as negative values. In the areas where the Det.(σ_{ij}) takes negative values we have yielding of the material. The directions of the yielding of the material, in these regions, are defined by the negative maximum of Det.(σ_{ij}).

Chapter 6
Application of the Det.-Criterion in Plane Crack Problems

6.1 Cracked Plates Under Uniaxial Tension

For a thin elastic and isotropic plate, under conditions of generalized plane stress, containing a slant internal crack of length $2a$ and obliqueness β (Fig. 140(a)), which is subjected at infinity to a tension σ_∞, the stress field in the vicinity of the crack tip is given by [42, 45, 89]:

$$\sigma_{xx} = \frac{K_I}{\sqrt{2\pi r}} \cos\frac{\theta}{2}\left(1 - \sin\frac{\theta}{2}\sin\frac{3\theta}{2}\right) - \frac{K_{II}}{\sqrt{2\pi r}}\sin\frac{\theta}{2}\left(2 + \cos\frac{\theta}{2}\cos\frac{3\theta}{2}\right) + \sigma_\infty \cos 2\beta$$

$$(693)$$

$$\sigma_{yy} = \frac{K_I}{\sqrt{2\pi r}} \cos\frac{\theta}{2}\left(1 + \sin\frac{\theta}{2}\sin\frac{3\theta}{2}\right) + \frac{K_{II}}{\sqrt{2\pi r}}\sin\frac{\theta}{2}\cos\frac{\theta}{2}\cos\frac{3\theta}{2} \qquad (694)$$

$$\tau_{xy} = \frac{K_I}{\sqrt{2\pi r}} \sin\frac{\theta}{2}\cos\frac{\theta}{2}\cos\frac{3\theta}{2} + \frac{K_{II}}{\sqrt{2\pi r}}\cos\frac{\theta}{2}\left(1 - \sin\frac{\theta}{2}\sin\frac{3\theta}{2}\right) \qquad (695)$$

where the stress intensity factors K_I and K_{II} are given by:

$$K_I = \sigma_\infty\sqrt{\pi a}\sin^2\beta, \quad K_{II} = \sigma_\infty\sqrt{\pi a}\sin\beta\cos\beta \qquad (696)$$

6.1.1 For Singular Solution

By substituting Rels (693)–(696) (without the constant term $\sigma_\infty \cos 2\beta$) into Rel. (686), we obtain [85]:

$$\mathrm{Det.}(\sigma_{ij}) = D/2\pi r \qquad (697)$$

where:

$$D = K_I^2 D_1 + K_{II}^2 D_2 + K_I K_{II} D_{12} \qquad (698)$$

with:

$$D_1 = \cos^4\frac{\theta}{2} \qquad (699)$$

$$D_2 = \cos^2 \frac{\theta}{2} \left(3 \sin^2 \frac{\theta}{2} - 1 \right) \tag{700}$$

$$D_{12} = -4 \sin \frac{\theta}{2} \cos^3 \frac{\theta}{2} \tag{701}$$

By substituting the stress intensity factors K_I and K_{II} from Rel. (696) into Rel. (698), we obtain:

$$\frac{D}{\pi a \sigma_\infty^2} = D_1 \sin^4 \beta + D_2 \sin^2 \beta \cos^2 \beta + D_{12} \sin^3 \beta \cos \beta \tag{702}$$

Relation (702) gives the variation of the determinant of the stress tensor around the crack tip. Figures 142(a), (b) and (c) present the variation of the determinant of the stress tensor around the tip for the typical cases of internal crack obliqueness $\beta = 90°$, $30°$ and $5°$ respectively. In Fig. 142(a) the distribution of the determinant of the stress tensor presents a positive maximum at the position $\theta = 0°$, while Fig. 142(b) shows two positive maxima at positions $\theta = -83°$ and $+130°$, and a negative maximum at position $\theta = +31°$. Therefore, the crack will propagate to directions $\theta_0 = 0°$ and $\theta_0 = -83°$, respectively.

The position of the first positive maximum (θ_0) may be defined by annulling the partial derivative of Rel. (698):

$$\frac{\partial D}{\partial \theta} = 0 \tag{703}$$

Figure 143 presents the variation of $-\theta_0$ versus β. In the same figure, the variation of $-\theta_0$, according to T [237], G [234], S [74] and S_2 [83] criteria, are plotted.

The critical stress σ_∞^β of crack initiation is calculated by Rels (670) and (697)–(702):

$$\text{Det.}(\sigma_{ij})_{cr} \bigg|_{\theta = \theta_0} = \frac{(\sigma_\infty^\beta)^2}{2(r/a)^\beta} (D_1 \sin^4 \beta + D_2 \sin^2 \beta \cos^2 \beta + D_{12} \sin^3 \beta \cos \beta) \tag{704}$$

or:

$$\sigma_\infty^\beta \bigg|_{\theta = \theta_0} = \left(\frac{2(r/a)^\beta \text{Det.}(\sigma_{ij})_{cr}}{(D_1 \sin^4 \beta + D_2 \sin^2 \beta \cos^2 \beta + D_{12} \sin^3 \beta \cos \beta)} \right)^{\frac{1}{2}} \tag{705}$$

For $\beta = 90°$ and $\theta = \theta_0 = 0°$ Rel. (705) becomes:

$$\sigma_\infty^{90°} = (2(r/a)^{90°} \text{Det.}(\sigma_{ij})_{cr})^{\frac{1}{2}} \tag{706}$$

where $(r/a)^\beta \ll 1$.

The ratio $(r/a)^\beta$ depends on the angle β and defines the core-region around the crack tip. In static crack problems the ratio $(r/a)^\beta$ is calculated by the initial curve of the caustics [101]:

$$r^\beta = r^{90°}(1 + \kappa^2)^{\frac{1}{3}} \tag{707}$$

where:

$$\kappa = \frac{K_{II}}{K_I} = \cot \beta \quad \text{and} \quad r^{90°} = (\tfrac{2}{3} C_{r,t,f})^{\frac{2}{3}} \tag{708}$$

Then, Rel. (707) becomes:

$$\left(\frac{r}{a} \right)^\beta = \left(\frac{r}{a} \right)^{90°} \sin^{0.4} \beta \tag{709}$$

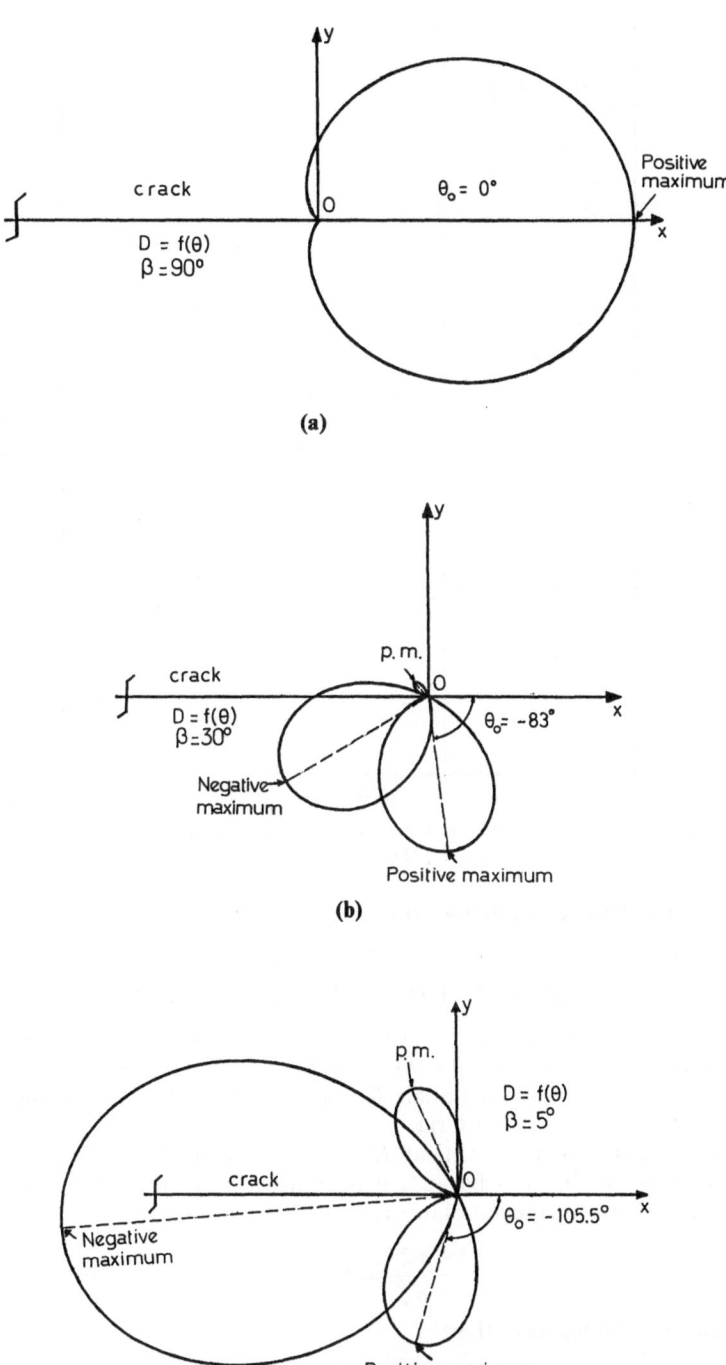

Fig. 142. Distribution of the determinant of the stress tensor around the tip of an internal crack; obliqueness: **(a)** $\beta = 90°$, **(b)** $\beta = 30°$ and **(c)** $\beta = 5°$.

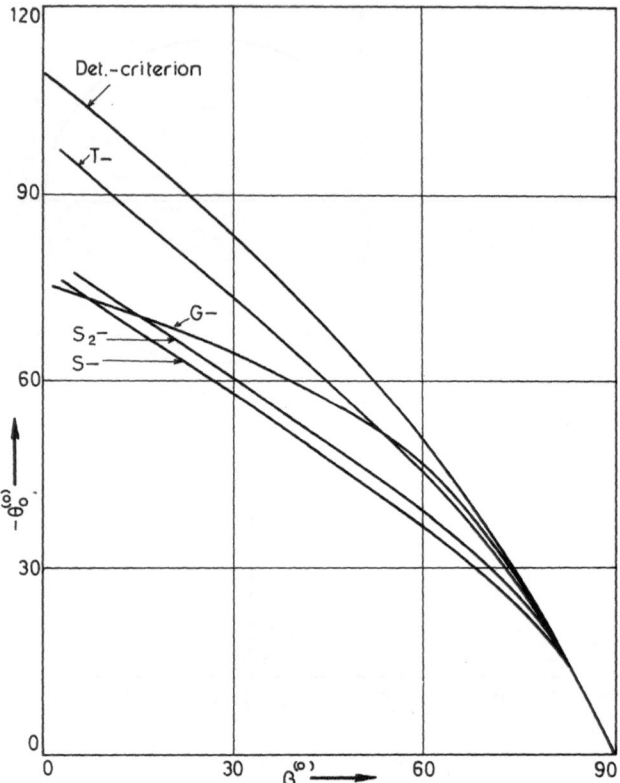

Fig. 143. Variation of $-\theta_0$ versus β for an internal oblique crack, according to Det.-criteria of fracture
G [234], T [237], S [74] and S_2 [83].

From Rels (705), (706) and (709) we obtain:

$$\left.\frac{\sigma_\infty^\beta}{\sigma_\infty^{90°}}\right|_{\theta=\theta_0} = \frac{\sin^{0.2}\beta}{[D_1\sin^4\beta + D_2\sin^2\beta\cos^2\beta + D_{12}\sin^3\beta\cos\beta]^{\frac{1}{4}}} \qquad (710)$$

Relation (710) gives the critical stress for crack initiation σ_∞^β normalized to its characteristic value $\sigma_\infty^{90°}$ for $\beta = 90°$. Figure 144 presents the variation of the critical fracture stress versus the angle β, according to Det.-criterion and σ_θ [69], G [234], S [74] and T [237] criteria.

It should be noted that the critical fracture stress could not take the very high values appearing in Fig. 144 because it is subject to an upper limit, the critical fracture stress of the uncracked plate, σ_∞^{uncr}. The relation between $\sigma_\infty^{90°}$ and σ_∞^{uncr} is:

$$\frac{\sigma_\infty^{uncr}}{\sigma_\infty^{90°}} = \delta > 1 \qquad (711)$$

Then, from Fig. 144 we have that:

$$\sigma_\infty^{90°} \leqq \sigma_\infty^\beta < \sigma_\infty^{uncr} \qquad (712)$$

or:

$$1 \leqq \frac{\sigma_\infty^\beta}{\sigma_\infty^{90°}} < \delta, \quad \text{for } 0 < \beta \leqq 90° \qquad (713)$$

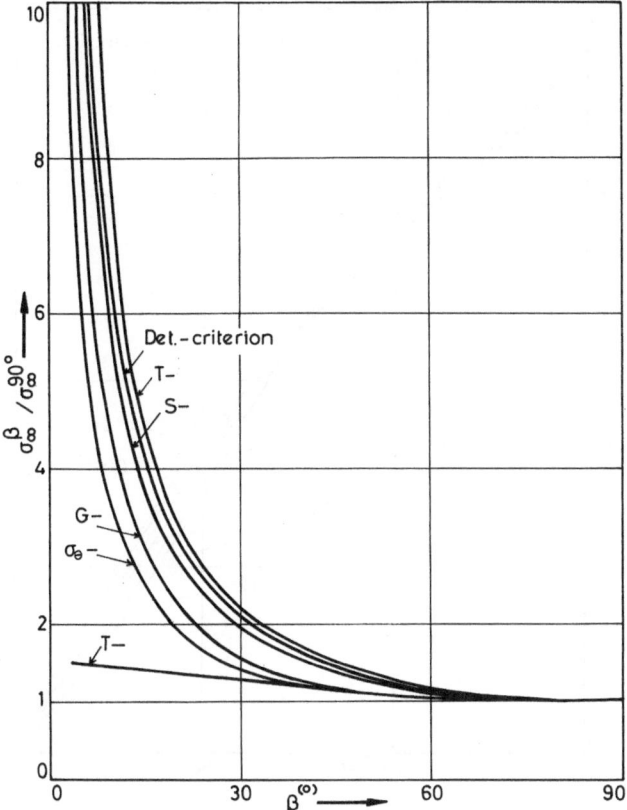

Fig. 144. Fracture stress σ_∞^β versus β, reduced to its values $\sigma_\infty^{90°}$ for $\beta = 90°$, according to Det.-, σ_θ [69], G [234], S [74] and T [237] criteria of fracture.

From Rel. (713), we can conclude that Rel. (710) cannot take values higher than the ratio δ, because above this value the cracked plate may fracture other than at the crack tip. Therefore, the curves of Fig. 144 are limited by the ratio δ.

6.1.2 For Solution with Constant Term

Comparable relations to Rels (702) and (710) can be obtained for the solution with a constant term of the stress σ_{xx} ($\sigma_\infty \cos 2\beta$) [85]:

$$D_c = \frac{D}{4\pi a \sigma_\infty^2} = D_1 \sin^4 \beta + D_2 \sin^2 \beta \cos^2 \beta + D_{12} \sin^3 \beta \cos \beta$$

$$+ \sqrt{2(r/a)^{90°}} \cos 2\beta \sin^{1.2}\beta \left(\sin \beta \cos\frac{\theta}{2}\left(1 + \sin\frac{\theta}{2}\sin\frac{3\theta}{2}\right) \right.$$

$$\left. + \cos \beta \sin\frac{\theta}{2}\cos\frac{\theta}{2}\cos\frac{3\theta}{2} \right) \tag{714}$$

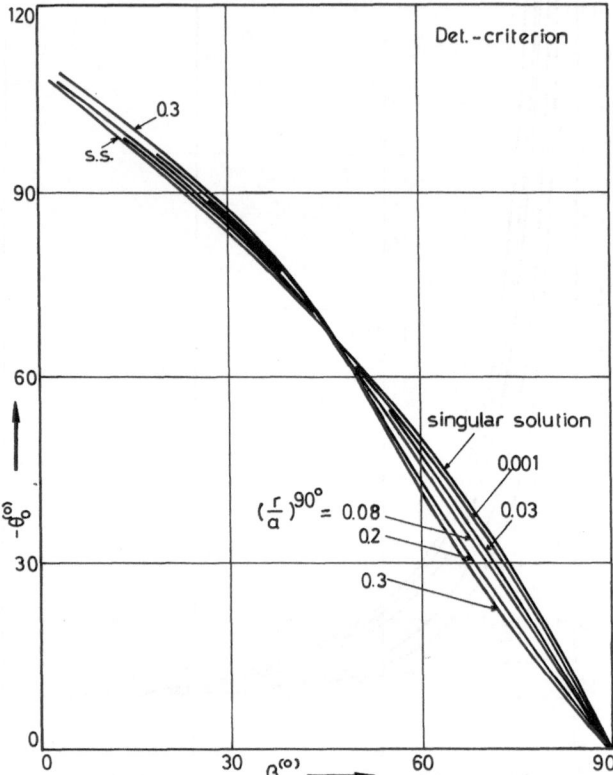

Fig. 145. Variation of $-\theta_0$ versus β for an internal oblique crack, according to Det.-criterion for singular solution and for $(r/a)^{90°} = 0.001, 0.03, 0.08, 0.2, 0.3$.

and:

$$\left.\frac{\sigma_\infty^\beta}{\sigma_\infty^{90°}}\right|_{\theta=\theta_0} = \frac{(1 - \sqrt{2(r/a)^{90°}})^{\frac{1}{4}} \sin^{0.2}\beta}{\sqrt{D_c}} \tag{715}$$

Figure 145 presents the variation of $-\theta_0$ versus β for various values of the ratio $(r/a)^{90°}$, while Fig. 146 presents the variation of the critical fracture stress versus β for various values of the ratio $(r/a)^{90°}$, according to the Det.-criterion.

It is worth noting that the constant term of the stress field does not significantly influence the direction $-\theta_0$ of initiation of crack propagation (Fig. 145), while it strongly influences the critical fracture stress (Fig. 146). Therefore, for various values of the ratio $(r/a)^{90°}$, i.e. for various types of materials, brittle or ductile, the critical fracture stress varies considerably.

Figure 146 includes experimentally measured fracture stresses for various materials [56, 241–243].

From this study it is concluded that the second stress invariant is the main factor regulating the mode of fracture.

The distribution of the determinant of the second stress invariant along a circle around the crack tip presents positive and negative maxima. The crack will propagate along the direction of the positive maxima and mainly along the direction of the

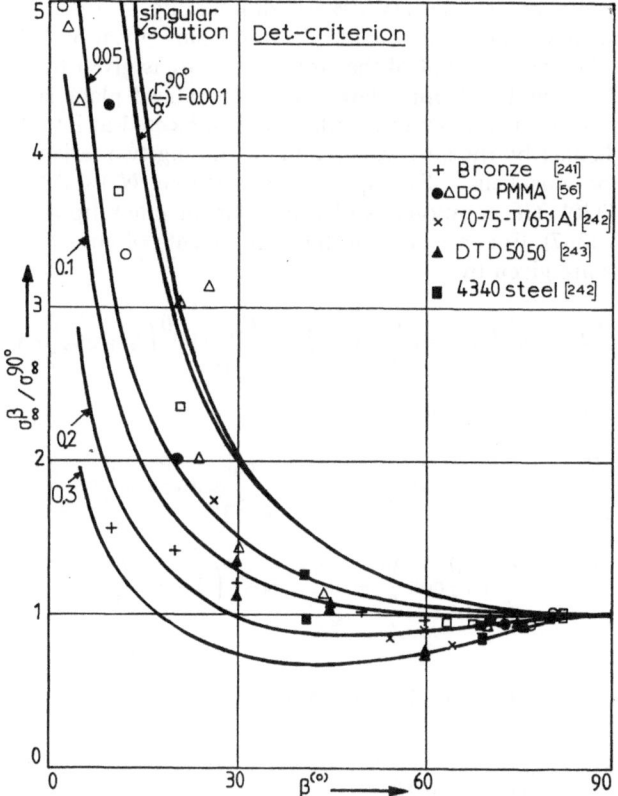

Fig. 146. Fracture stress σ_∞^β reduced to its value $\sigma_\infty^{90°}$ for $\beta = 90°$, versus β, according to Det.-criterion for singular solution and for $(r/a)^{90°} = 0.001, 0.05, 0.1, 0.2, 0.3$.

local maximum $(-\theta_0)$ because in this direction the initiation of the crack takes place with the minimum of critical stress.

The critical stress of initiation depends strongly on the ratio $(r/a)^{90°}$. Therefore, when the ratio $(r/a)^{90°}$ increases the critical stress decreases considerably. This means that if the core-region at the crack tip is large (ductile materials) the critical stress will be very low.

6.2 Cracked Plates Under Biaxial Loading

In this section crack initiation under biaxial loading is studied. The crack extension angle and the critical stress of fracture are determined for various values of the loading angle β and for various values of the biaxial factor k. The Det.-criterion developed by the author was used [85, 86, 245]. According to the Det.-criterion, the crack propagates in the direction of the maximum value of the determinant of the stress tensor. The Det.-criterion is simpler and more powerful than other fracture criteria used up to now.

The elastic strain energy density dW/dV dominating at the strained plate is expressed by Rel. (633). The components T_V and T_D are given by Rels (676), (677), (687) and (688). The determinant of the stress tensor σ_{ij} is given by Rel. (686).

For a thin elastic and isotropic plate under generalized plane-stress conditions containing a slant internal crack of length $2a$ and subjected at infinity to a biaxial state of stress defined by the stresses σ_∞ and $k\sigma_\infty$ along two adjacent sides of the plate (Fig. 14), the Muskhelishvili complex stress functions $\Phi(z)$ and $\Omega(z)$ are given by Rels (329) and (330). The components of stresses at the crack tip are given by Rels (365), (366) and (367). For $\omega = 90° - \beta$, the components of stresses and the stress intensity factors are given by:

$$\sigma_{xx} \simeq \frac{K_\text{I}}{\sqrt{2\pi r}} \cos\frac{\theta}{2}\left(1 - \sin\frac{\theta}{2}\sin\frac{3\theta}{2}\right) - \frac{K_\text{II}}{\sqrt{2\pi r}}\sin\frac{\theta}{2}\left(2 + \cos\frac{\theta}{2}\cos\frac{3\theta}{2}\right)$$
$$+ \sigma_\infty(1-k)\cos 2\beta \tag{716}$$

$$\sigma_{yy} \simeq \frac{K_\text{I}}{\sqrt{2\pi r}}\cos\frac{\theta}{2}\left(1 + \sin\frac{\theta}{2}\sin\frac{3\theta}{2}\right) + \frac{K_\text{II}}{\sqrt{2\pi r}}\sin\frac{\theta}{2}\cos\frac{\theta}{2}\cos\frac{3\theta}{2} \tag{717}$$

$$\tau_{xy} \simeq \frac{K_\text{I}}{\sqrt{2\pi r}}\sin\frac{\theta}{2}\cos\frac{\theta}{2}\cos\frac{3\theta}{2} + \frac{K_\text{II}}{\sqrt{2\pi r}}\cos\frac{\theta}{2}\left(1 - \sin\frac{\theta}{2}\sin\frac{3\theta}{2}\right) \tag{718}$$

$$K_\text{I} = \frac{\sigma_\infty\sqrt{\pi a}}{2}[(1+k) - (1-k)\cos 2\beta] \tag{719}$$

$$K_\text{II} = \frac{\sigma_\infty\sqrt{\pi a}}{2}(1-k)\sin 2\beta \tag{720}$$

By substituting Rels (716)–(720) into Rel. (686), we obtain [245]:

$$\text{Det.}(\sigma_{ij}) = \sigma_{xx}\sigma_{yy} - \tau_{xy}^2 = \frac{(\sigma_\infty^{\beta,k})^2}{8(r/a)^\beta}D \tag{721}$$

with:

$$D = [(1+k) - (1-k)\cos 2\beta]^2 D_1 + (1-k)^2\sin^2 2\beta\, D_2 + [(1+k) - (1-k)\cos 2\beta]$$
$$\times (1-k)\sin 2\beta\, D_{12} + 2\sqrt{2(r/a)^{\beta,k}}(1-k)\cos 2\beta$$
$$\times \left([(1+k) - (1-k)\cos 2\beta]\cos\frac{\theta}{2}\left(1 + \sin\frac{\theta}{2}\sin\frac{3\theta}{2}\right)\right.$$
$$\left. + (1-k)\sin 2\beta\sin\frac{\theta}{2}\cos\frac{\theta}{2}\cos\frac{3\theta}{2}\right) \tag{722}$$

where D_1, D_2, D_{12} are given by Rels (699), (700) and (701).

The ratio $(r/a)^{\beta,k}$ depends on the angle β and on the biaxiality factor k and defines the core-region around the crack tip. This ratio is calculated by the law of variation of the initial curve of the caustics [101]:

$$(r/a)^{\beta,k} = (r/a)^{90°}[k^2 + (1-k^2)\sin^2\beta]^{0.2} \tag{723}$$

Therefore, Rel. (722) becomes:

$$D = [(1 + k) - (1 - k)\cos 2\beta]^2 D_1 + (1 - k)^2 \sin^2 2\beta \, D_2 + [(1 + k) - (1 - k)\cos 2\beta]$$
$$\times (1 - k)\sin 2\beta \, D_{12} + 2(2(r/a)^{90°})^{\frac{1}{2}}(1 - k)[k^2 + (1 - k^2)\sin^2 \beta]^{0.1} \cos 2\beta$$
$$\times \left([(1 + k) - (1 - k)\cos 2\beta]\cos \frac{\theta}{2}\left(1 + \sin \frac{\theta}{2}\sin \frac{3\theta}{2}\right)\right.$$
$$\left. + (1 - k)\sin 2\beta \sin \frac{\theta}{2}\cos \frac{\theta}{2}\cos \frac{3\theta}{2}\right) \tag{724}$$

Relation (724) gives the variation of the determinant of the stress tensor around the crack tip. Figures 147(a), (b) and (c) present the variation of the determinant of the stress tensor around the crack tip for typical cases of internal crack obliqueness $\beta = 90°$, $45°$ and $10°$, respectively, and for $k = 0$ and $(r/a)^{90°} = 0.1$.

In Fig. 147(a) the distribution of Det.$(\sigma_{ij}) = f(\theta)$ presents a positive maximum at the position $\theta_0 = 0°$ and two negative maxima at positions $\theta = 96°$ and $-96°$. According to the Det.-criterion, crack initiation will occur at position $\theta_0 = 0°$ and yielding will occur at the positions $\theta = 96°$ and $-96°$. Likewise, in the case of Fig. 147(b) crack initiation will occur at the position $\theta_0 = -68°$, while in the case

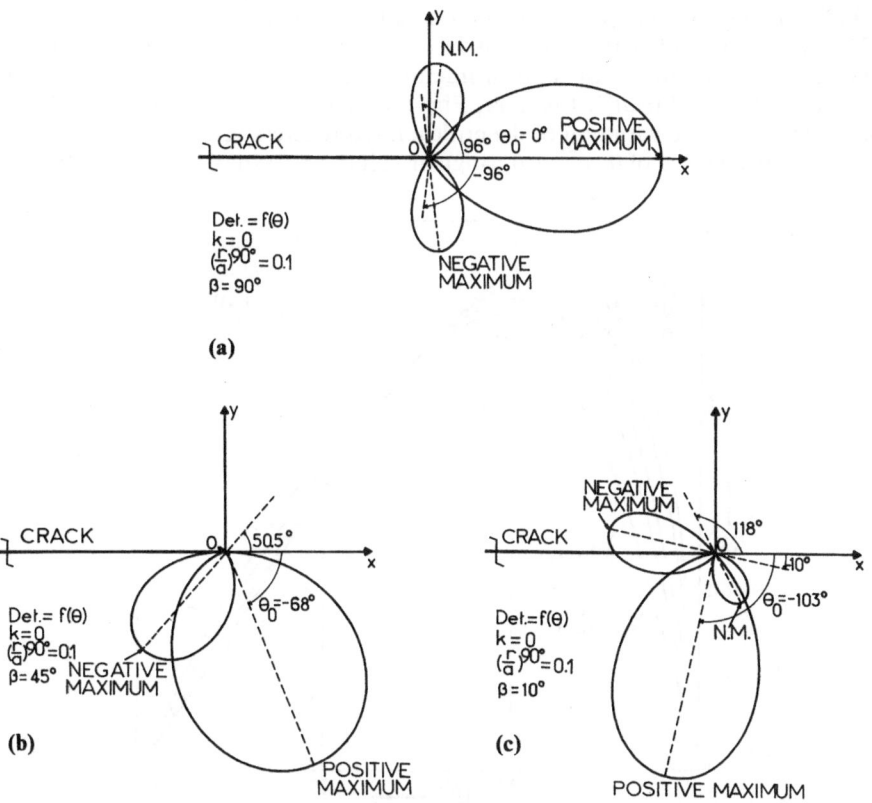

Fig. 147. Distribution of the determinant of the stress tensor around the crack tip for $k = 0$, $(r/a)^{90°} = 0.1$ and for **(a)** $\beta = 90°$, **(b)** $\beta = 45°$, **(c)** $\beta = 10°$.

of Fig. 147(c) crack initiation will occur at the position $\theta_0 = -103°$. The positions of the maxima of the distribution $\text{Det.}(\sigma_{ij}) = f(\theta)$ may be defined by Rel. (703).

The critical stress $\sigma_\infty^{\beta,k}$ of crack initiation is calculated by Rels (670) and (721):

$$\text{Det.}(\sigma_{ij})_{\text{cr}}\bigg|_{\theta=\theta_0} = \frac{(\sigma_\infty^{\beta,k})^2}{8(r/a)^{\beta,k}} D \tag{725}$$

or:

$$\sigma_\infty^{\beta,k}\bigg|_{\theta=\theta_0} = \left(\frac{8(r/a)^{\beta,k}\,\text{Det.}(\sigma_{ij})_{\text{cr}}}{D}\right)^{\frac{1}{2}} \tag{726}$$

For $\beta = 90°$, $\theta = \theta_0 = 0°$ and $k = 0$, Rel. (726) becomes:

$$\sigma_\infty^{90°,0} = \left(\frac{2(r/a)^{90°}\,\text{Det.}(\sigma_{ij})_{\text{cr}}}{1-(2(r/a)^{90°})^{\frac{1}{2}}}\right)^{\frac{1}{2}} \tag{727}$$

From Rels (726), (727) and (723) we obtain:

$$\frac{\sigma_\infty^{\beta,k}}{\sigma_\infty^{90°,0}}\bigg|_{\theta=\theta_0} = \frac{2(1-\sqrt{2(r/a)^{90°}})^{\frac{1}{2}}[k^2+(1-k^2)\sin^2\beta]^{0.1}}{\sqrt{D}} \tag{728}$$

Relation (728) gives the critical stress $\sigma_\infty^{\beta,k}$ of crack initiation normalized to its characteristic value $\sigma_\infty^{90°,0}$ for $\beta = 90°$ and $k = 0$. For $k = 0$ and $(r/a)^{90°} = 0$, Rels (724) and (728) reduce to those of the singular solution [85] (Section 6.1.1).

Figure 145 presented the variation of $-\theta_0$ versus β for $k = 0$, while Fig. 146 presented the variation of the critical fracture stress versus β for $k = 0$ for various parametric values of $(r/a)^{90°}$. Likewise, Fig. 148 presents the variation of $-\theta_0$ and Fig. 149 presents the variation of the critical fracture stress versus β for $k = 0.5$ and for parametric values of $(r/a)^{90°} = 0$ (singular solution), 0.001, 0.05, 0.1, 0.2, 0.3.

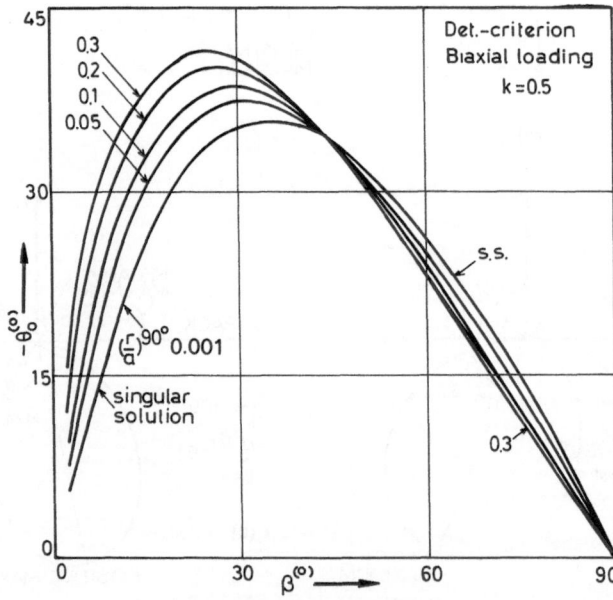

Fig. 148. Variation of $-\theta_0$ versus β according to Det.-criterion for $k = 0.5$, for singular solution and for $(r/a)^{90°} = 0.001, 0.05, 0.1, 0.2, 0.3$.

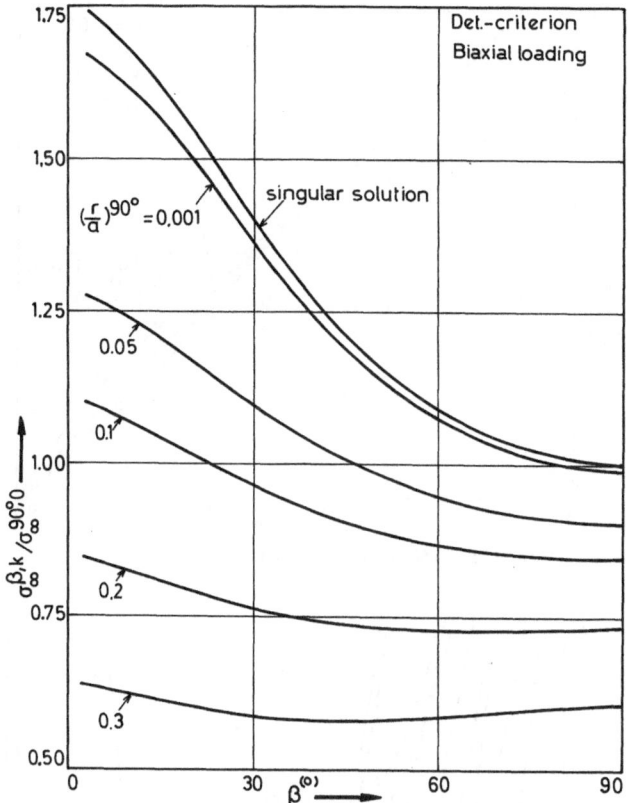

Fig. 149. Variation of fracture stress $\sigma_\infty^{\beta,k}$ reduced to its values $\sigma_\infty^{90°,0}$ versus β according to Det.-criterion for $k = 0.5$, for singular solution and for $(r/a)^{90°} = 0.001$, 0.05, 0.1, 0.2, 0.3.

For the case where $k < 0$ we can suppose that the stress intensity factor K_I is: (i) $K_I < 0$ (for artificial slits) or (ii) $K_I = 0$ (for natural cracks), for definite values of β. In case (i) Rels (724) and (728) are validated; but in case (ii) Rels (724) and (728) become:

$$D = (1 - k)^2 \sin 2\beta \left[D_2 \sin 2\beta + 2(2(r/a)^{90°})^{\frac{1}{2}} \left(\frac{1-k}{2} \right)^{0.2} \right.$$

$$\left. \times \sin^{0.2} 2\beta \cos 2\beta \sin \frac{\theta}{2} \cos \frac{\theta}{2} \cos \frac{3\theta}{2} \right] \tag{729}$$

and:

$$\left. \frac{\sigma_\infty^{\beta,k}}{\sigma_\infty^{90°,0}} \right|_{\theta = \theta_0} = \frac{2^{0.8}(1 - \sqrt{2(r/a)^{90°}})^{\frac{1}{2}}(1-k)^{0.2} \sin^{0.2} 2\beta}{\sqrt{D}} \tag{730}$$

In the case of $k = -0.5$, Rels (729) and (730) are valid for $\beta \leq 35.26°$ and in case of $k = -1.0$ these relations are valid for $\beta \leq 45°$.

Figure 150 presents the variation of $-\theta_0$ versus β for $k = -0.5$ and for the cases where the stress intensity factor K_I is positive, negative or equal to zero. Figure 151

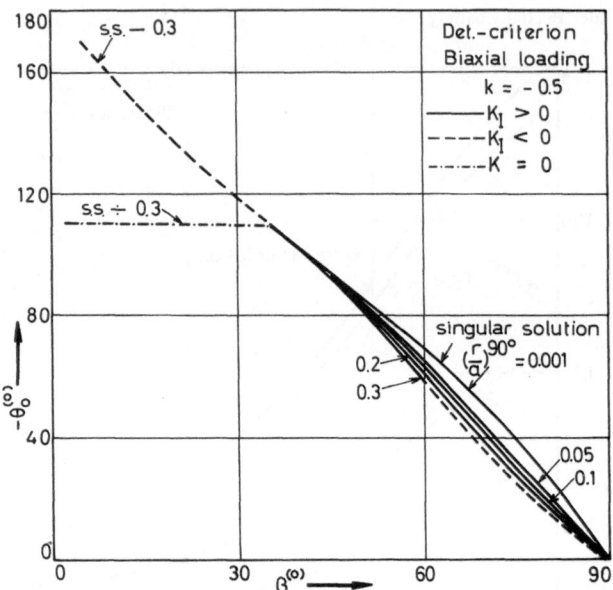

Fig. 150. As in Fig. 148, for $k = -0.5$.

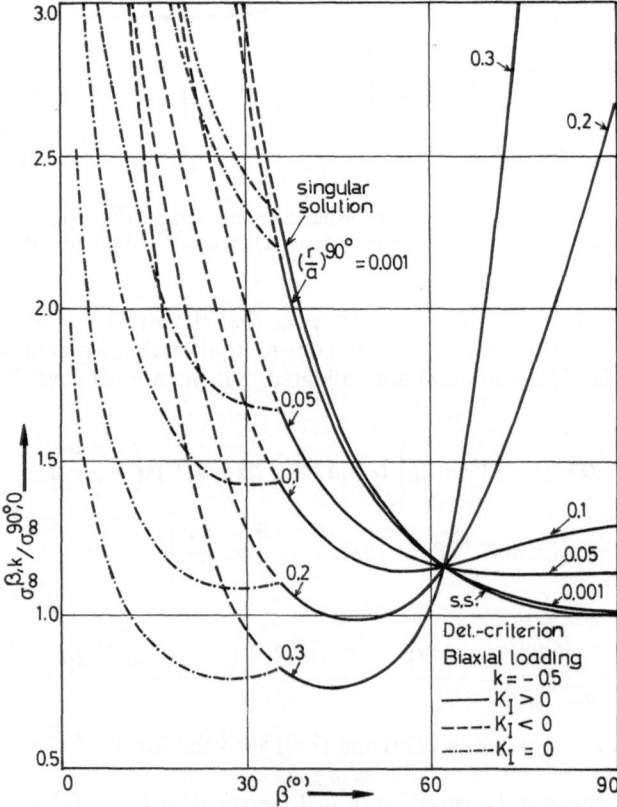

Fig. 151. As in Fig. 149, for $k = -0.5$.

presents the variation of the critical fracture stress versus β for $k = -0.5$ and for the same case.

From this study it is concluded that the prediction of crack initiation is readily accomplished by consideration of the determinant of the stress tensor around the crack tip. The determinant of the stress tensor, as a fracture criterion, is made more flexible and applicable by the parameter $(r/a)^{90°}$ which depends on the material and takes higher values for ductile than for brittle materials. $(r/a)^{90°}$ has a stronger influence upon the value of the critical fracture stress than upon the angle of crack initiation and therefore explains the dispersion of the experimental values of critical fracture stress which appear in various materials [85]. As shown by the figures, when the ratio $(r/a)^{90°}$ increases, the critical fracture stress decreases considerably.

6.3 Cracked Plates Under Biaxial Loading With Higher-Order Approximation Solution

In this section the crack initiation under biaxial loading is examined by giving the expressions for stresses in a series form, which are equally valid near and far from the crack tip. Based on these expressions, the crack extension angle and the critical stress of fracture are determined by use of the Det.-criterion [249].

For a thin elastic and isotropic plate under plane-stress conditions, containing a slant internal crack of length $2a$ and subjected to a biaxial state of stress at infinity, the elastic-stress components in the vicinity of the crack tip are given by the so-called singular solution [43]. Also, it has been shown [42, 43, 45] that the constant second term in the series representation of stresses is significant. The constant term has been used as a second parameter by Irwin [46], Bradley and Kobayashi [253], C.W. Smith [254], and Theocaris and Papadopoulos [83, 101, 241]. In order to improve further the accuracy by using data from the far field of the crack tip, three-parameter methods were introduced [255, 256]. Six and more parameter methods for the evaluation of stress intensity factors were introduced by Rossmanith *et al.* [257] and Theocaris *et al.* [258].

The Muskhelishvili complex stress functions $\Phi(z)$ and $\Omega(z)$ are given by Rels (329) and (330). The components of stresses at the crack tip may be derived by Rels (333), (334) and (335). The stress intensity factors K_I and K_{II} for biaxial loading are given by Rels (719) and (720).

After some algebra the stress components σ_{ij} are given in an increasing-order power series, as follows [258]:

$$\sigma_{xx} = \sigma_\infty(1-k)\cos 2\beta + \frac{K_I}{2\sqrt{\pi a}}\left((r/2a)^{-\frac{1}{2}}\cos\frac{\theta}{2}\left(1 - \sin\frac{\theta}{2}\sin\frac{3\theta}{2}\right)\right.$$

$$+ \frac{3}{2}(r/2a)^{\frac{1}{2}}\cos\frac{\theta}{2}\left(1 + \sin^2\frac{\theta}{2}\right) + \sum_{n=1}^{\infty}(r/2a)^{n+\frac{1}{2}}C_n\left[\cos\left(n+\frac{1}{2}\right)\theta\right.$$

$$\left.\left. - \left(n+\frac{1}{2}\right)\sin\theta\sin\left(n-\frac{1}{2}\right)\theta\right]\right) + \frac{K_{II}}{2\sqrt{\pi a}}\left(-(r/2a)^{-\frac{1}{2}}\sin\frac{\theta}{2}\left(2 + \cos\frac{\theta}{2}\cos\frac{3\theta}{2}\right)\right.$$

$$+ \frac{3}{2}(r/2a)^{\frac{1}{2}} \sin\frac{\theta}{2}\left(2 + \cos^2\frac{\theta}{2}\right) + \sum_{n=1}^{\infty}(r/2a)^{n+\frac{1}{2}}C_n\left[2\sin\left(n+\frac{1}{2}\right)\theta\right.$$

$$\left.+ \left(n+\frac{1}{2}\right)\sin\theta\cos\left(n-\frac{1}{2}\right)\theta\right]\right) \tag{731}$$

$$\sigma_{yy} = \frac{K_I}{2\sqrt{\pi a}}\left((r/2a)^{-\frac{1}{2}}\cos\frac{\theta}{2}\left(1 + \sin\frac{\theta}{2}\sin\frac{3\theta}{2}\right) + \frac{3}{2}(r/2a)^{\frac{1}{2}}\cos^3\frac{\theta}{2}\right.$$

$$\left.+ \sum_{n=1}^{\infty}(r/2a)^{n+\frac{1}{2}}C_n\left[\cos\left(n+\frac{1}{2}\right)\theta + \left(n+\frac{1}{2}\right)\sin\theta\sin\left(n-\frac{1}{2}\right)\theta\right]\right)$$

$$+ \frac{K_{II}}{2\sqrt{\pi a}}\left((r/2a)^{-\frac{1}{2}}\sin\frac{\theta}{2}\cos\frac{\theta}{2}\cos\frac{3\theta}{2} - \frac{3}{2}(r/2a)^{\frac{1}{2}}\sin\frac{\theta}{2}\cos^2\frac{\theta}{2}\right.$$

$$\left.- \sum_{n=1}^{\infty}(r/2a)^{n+\frac{1}{2}}C_n\left(n+\frac{1}{2}\right)\sin\theta\cos\left(n-\frac{1}{2}\right)\theta\right) \tag{732}$$

$$\tau_{xy} = \frac{K_I}{2\sqrt{\pi a}}\left((r/2a)^{-\frac{1}{2}}\sin\frac{\theta}{2}\cos\frac{\theta}{2}\cos\frac{3\theta}{2} - \frac{3}{2}(r/2a)^{\frac{1}{2}}\sin\frac{\theta}{2}\cos^2\frac{\theta}{2}\right.$$

$$\left.- \sum_{n=1}^{\infty}(r/2a)^{n+\frac{1}{2}}C_n\left(n+\frac{1}{2}\right)\sin\theta\cos\left(n-\frac{1}{2}\right)\theta\right)$$

$$+ \frac{K_{II}}{2\sqrt{\pi a}}\left((r/2a)^{-\frac{1}{2}}\cos\frac{\theta}{2}\left(1 - \sin\frac{\theta}{2}\sin\frac{3\theta}{2}\right) + \frac{3}{2}(r/2a)^{\frac{1}{2}}\cos\frac{\theta}{2}\left(1 + \sin^2\frac{\theta}{2}\right)\right.$$

$$\left.+ \sum_{n=1}^{\infty}(r/2a)^{n+\frac{1}{2}}C_n\left[\cos\left(n+\frac{1}{2}\right)\theta - \left(n+\frac{1}{2}\right)\sin\theta\sin\left(n-\frac{1}{2}\right)\theta\right]\right) \tag{733}$$

where:

$$C_n = (-1)^n \frac{2n+3}{2n+2}\frac{1\times3\times\cdots\times(2n-1)}{2\times4\times\cdots\times(2n)} \tag{734}$$

By substituting Rels (719), (720) and (731)–(734) into Rel. (686), we obtain [245, 249]:

$$\text{Det.}(\sigma_{ij}) = \sigma_{xx}\sigma_{yy} - \tau_{xy}^2 = (\sigma_\infty^{\beta,k})^2\, f(k,(r/a)^{\beta,k}, n, \beta, \theta) \tag{735}$$

where:

$$(r/a)^{\beta,k} = (r/a)^{90°}[k^2 + (1-k^2)\sin^2\beta]^{0.2} \tag{736}$$

The ratio $(r/a)^{\beta,k}$ depends on the angle β and the biaxiality factor k and defines the core-region around the crack tip. This ratio was calculated from the initial curve of the caustics [101]. The ratio $(r/a)^{90°}$ is the core-region for the case of $\beta = 0°$. This value is independent of the biaxiality factor k.

The crack extension angles θ_0 for each value of k, β, $(r/a)^{90°}$ and the necessary number of stress terms n, are defined by $\partial\,\text{Det.}(\sigma_{ij})/\partial\theta = 0$, where the necessary number of stress terms n, is taken as the minimum number of terms of the stress power series for convergence for each θ and $(r/a)^{90°}$.

The critical stress $\sigma_\infty^{\beta,k}$ of crack initiation is calculated by Rels (670), (735) and (736):

$$\sigma_\infty^{\beta,k}\bigg|_{\theta=\theta_0} = \left(\frac{\text{Det.}(\sigma_{ij})_{cr}}{f(k,(r/a)^{\beta,k}, n, \beta, \theta)}\right)^{\frac{1}{2}} \tag{737}$$

For $\beta = 90°$, $\theta = \theta_0 = 0°$ and $k = 0$, rel. (737) becomes:

$$\sigma_\infty^{90°,0}\bigg|_{\theta=\theta_0} = \left(\frac{\mathrm{Det}.(\sigma_{ij})_{\mathrm{cr}}}{f(k=0,(r/a)^{90°},n,\beta=90°,\theta=0°)}\right)^{\frac{1}{2}} \qquad (738)$$

From Rels (737) and (738), we obtain:

$$\frac{\sigma_\infty^{\beta,k}}{\sigma_\infty^{90°,0}}\bigg|_{\theta=\theta_0} = \left(\frac{f(k=0,(r/a)^{90°},n,\beta=90°,\theta=0°)}{f(k,(r/a)^{\beta,k},n,\beta,\theta)}\right)^{\frac{1}{2}} \qquad (739)$$

Relation (739) gives the critical stress $\sigma_\infty^{\beta,k}$ of crack initiation normalized to its characteristic value $\sigma_\infty^{90°,0}$ for $\beta = 90°$ and for $k = 0$. For $k = 0$, $(r/a)^{90°} = 0$ and $n = 0$, Rels (735) and (739) are reduced to those of the singular solution [85].

In order to examine the influence of more terms in the stress power series upon the crack extension angle and the critical fracture stress, Eqs (731)–(733) were solved by computer for various values of $(r/a)^{90°}$ and k. The number of terms necessary for the stress power series to converge for each value of β and θ is calculated. Subsequently, Eqs (735) and (739) were solved by the computer for the respective stresses which are calculated from Eqs (731)–(733).

Figure 152 presents the necessary number of terms to be used for the σ_{ij} stresses versus the angle of loading β for $k = 0$ and $(r/a)^{90°} = 0.05, 0.1, 0.2$ and 0.3. In this figure we observe that the necessary number of stress terms increases as the angle of loading β and the ratio $(r/a)^{90°}$ increase. According to the necessary number of stress terms from Fig. 152 for each angle of loading β, Figs 153 and 154 present the variation of crack extension angle $-\theta_0$ versus the angle of loading β for $k = 0$ (uniaxial tension) and $(r/a)^{90°} = 0.05, 0.2$ and $0.1, 0.3$, respectively. In the same figures the curve of the singular solution and the curves of the two-term approximations are drawn for comparison. In these figures we observe that for higher-order approximations (dotted lines) the angles $-\theta_0$ are slightly reduced in ranges of β $0°–30°$ and $60°–90°$ and are considerably reduced in the range $30°–60°$.

Figures 155 and 156 present the variation of fracture stress $\sigma_\infty^{\beta,k}$ reduced to its values $\sigma_\infty^{90°,0}$ for $\beta = 90°$ and $k = 0$, versus the angle of loading β for $k = 0$ (uniaxial tension) and $(r/a)^{90°} = 0.05, 0.2$ and $0.1, 0.3$, respectively. In the same figures the curve of the singular solution and the curves of two-term approximation are drawn for

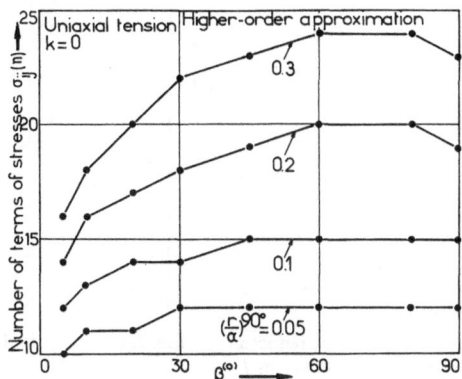

Fig. 152. The necessary number of terms to be used for the σ_{ij} stresses versus the angle of loading β for $k = 0$ and $(r/a)^{90°} = 0.05, 0.1, 0.2, 0.3$.

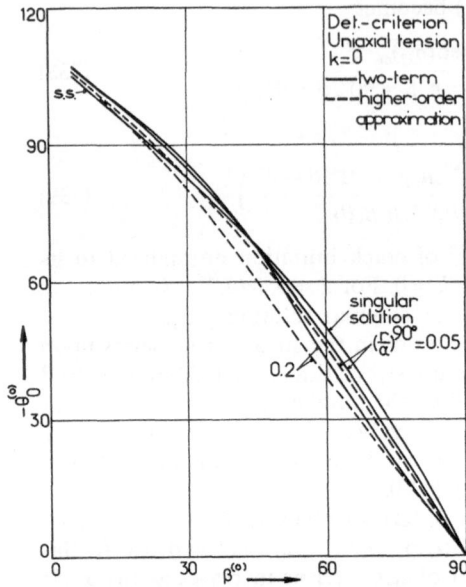

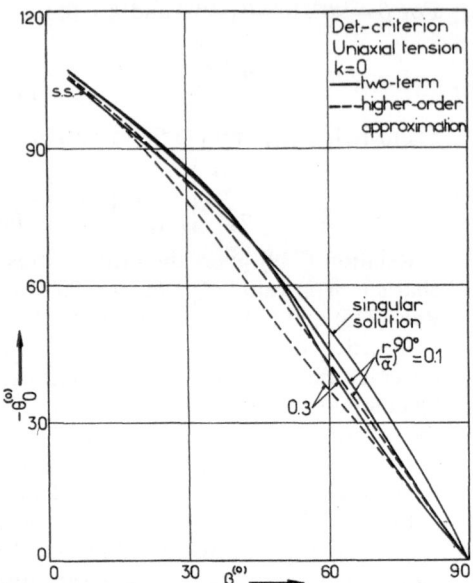

Fig. 153. Variation of $-\theta_0$ versus β according to Det.-criterion for $k=0$ and for singular solution, two-term approximation, higher-order approximation and for $(r/a)^{90°} = 0.05$, 0.2.

Fig. 154. As in Fig. 153, for $(r/a)^{90°} = 0.1$, 0.3.

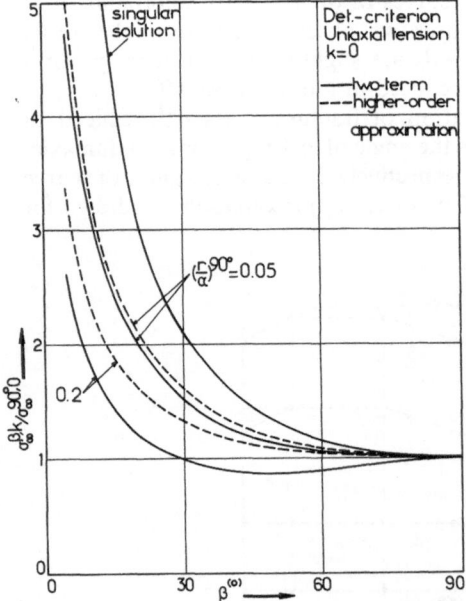

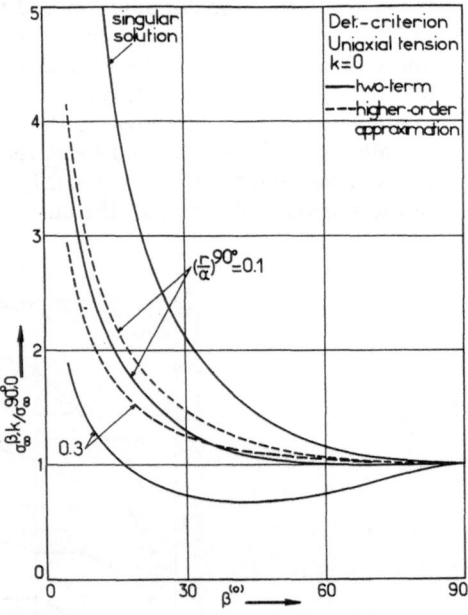

Fig. 155. Variation of fracture stress $\sigma_\infty^{\beta,k}$ reduced to its values $\sigma_\infty^{90°,0}$ for $\beta = 90°$ and $k = 0$, versus β according to Det.-criterion for $k = 0$ and for singular solution, two-term approximation, higher-order approximation and for $(r/a)^{90°} = 0.05$, 0.2.

Fig. 156. As in Fig. 155, for $(r/a)^{90°} = 0.1$, 0.3.

comparison. In these figures we observe that for higher-order approximation (dotted lines) the critical fracture stresses $\sigma_\infty^{\beta,k}/\sigma_\infty^{90°,0}$ remain higher than for the two-term approximation and lower than the singular solution.

Figures 157 and 158 present the variation of the stresses σ_{ij} reduced to value at infinity σ_∞ for $\theta = \theta_0$ versus the angle of loading β for $k = 0$ (uniaxial tension) and $(r/a)^{90°} = 0.05, 0.1$ and $0.2, 0.3$, respectively. In Fig. 157, the curves of stresses σ_{ij} of

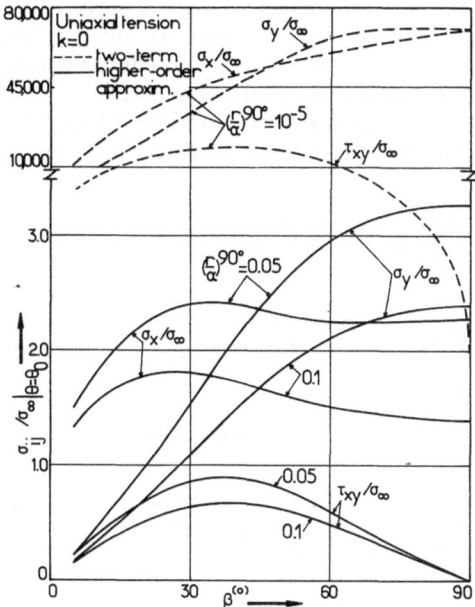

Fig. 157. Variation of the stresses σ_{ij} reduced to value at infinity σ_∞ for $\theta = \theta_0$ versus β for $k = 0$ and for two-term approximation (dotted lines), higher-order approximation and for $(r/a)^{90°} = 0.05, 0.1$.

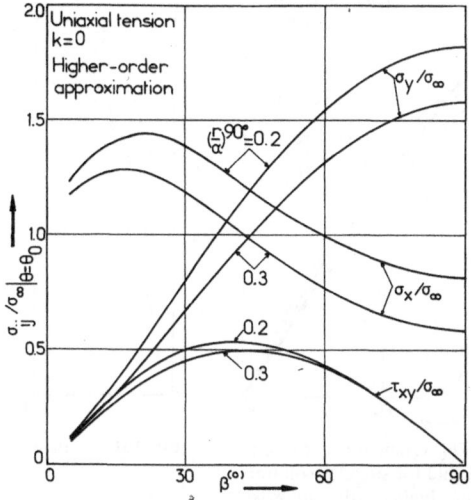

Fig. 158. As in Fig. 157, for $(r/a)^{90°} = 0.2, 0.3$.

two-term approximation (dotted line) for $(r/a)^{90°} = 10^{-5}$ (close to singular solution) are drawn. In these figures we observe that for the higher-order approximation, the stresses have been reduced considerably and mainly for large values of ratio $(r/a)^{90°}$ (far from the crack tip). Also, the same variation can be observed for biaxial loading. Figure 159 presents the necessary number of terms to be used for the σ_{ij} stresses versus β for $k = 0.5$ (biaxial loading) and $(r/a)^{90°} = 0.05, 0.1, 0.2$ and 0.3. Figures 160 and 161 present the variation of $-\theta_0$ versus β for $k = 0.5$ and $(r/a)^{90°} = 0.05, 0.2$ and $0.1, 0.3$, respectively. In the same figures the curves of the two-term approximation are drawn for comparison. In these figures we observe that for higher-order approximation (dotted lines) the angles $-\theta_0$ are reduced considerably for β 0°–60°. Figure 162 presents the variation of fracture stress $\sigma_\infty^{\beta,k}$ reduced to its values $\sigma_\infty^{90°,0}$ for $\beta = 90°$ and $k = 0$, versus β for $k = 0.5$ and for singular solution, two-term approximation, higher-order approximation (dotted lines) and for $(r/a)^{90°} = 0.05, 0.1$,

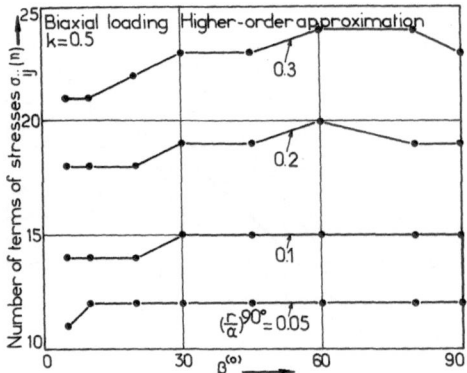

Fig. 159. The necessary number of terms to be used for the σ_{ij} stresses versus the angle of loading β for $k = 0.5$ and $(r/a)^{90°} = 0.05, 0.1, 0.2, 0.3$.

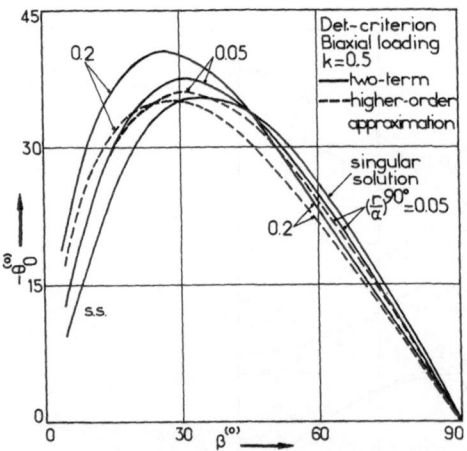

Fig. 160. Variation of $-\theta_0$ versus β according to Det.-criterion for $k = 0.5$ and for singular solution, two-term approximation, higher-order approximation and for $(r/a)^{90°} = 0.05, 0.2$.

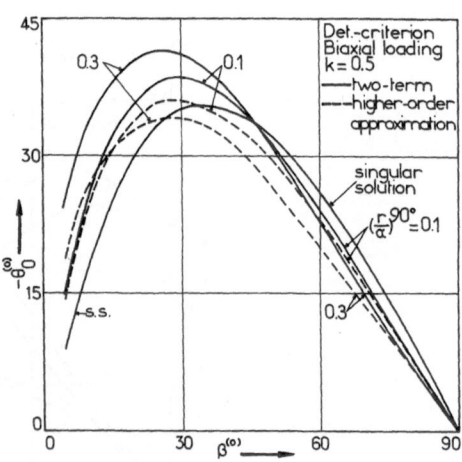

Fig. 161. As in Fig. 160, for $(r/a)^{90°} = 0.1, 0.3$.

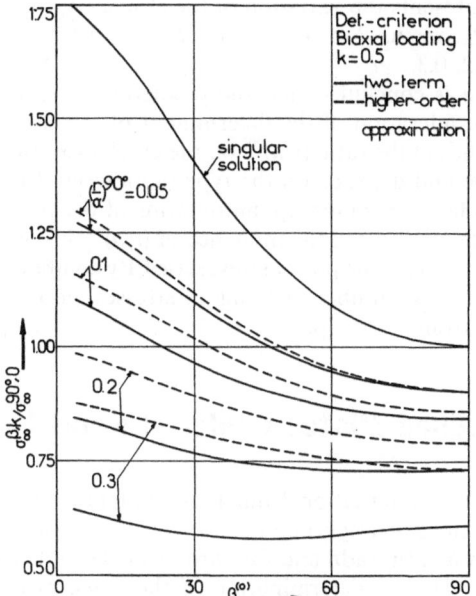

Fig. 162. Variation of fracture stress $\sigma_\infty^{\beta,k}$ reduced to its values $\sigma_\infty^{90°,0}$ for $\beta = 90°$ and $k = 0$, versus β according to Det.-criterion for $k = 0.5$ and for singular solution, two-term approximation, higher-order approximation and for $(r/a)^{90°} = 0.05, 0.1, 0.2, 0.3$.

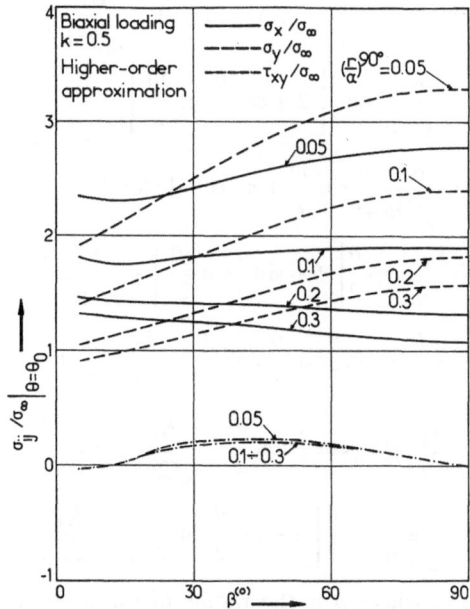

Fig. 163. Variation of the stresses σ_{ij} reduced to value at infinity σ_∞ for $\theta = \theta_0$ versus β for $k = 0.5$, for higher-order approximation and for $(r/a)^{90°} = 0.05, 0.1, 0.2, 0.3$.

0.2, 0.3. Figure 163 presents the variation of the stresses σ_{ij} reduced to the value at infinity σ_∞ for $\theta = \theta_0$, versus β for $k = 0.5$, for higher-order approximation and for $(r/a)^{90°} = 0.05, 0.1, 0.2, 0.3$.

From this study it is concluded that the prediction of crack initiation is easily accomplished by consideration of the determinant of the stress tensor around the crack tip. The influence of the ratio $(r/a)^{90°}$ on the crack extension angle and fracture stress is considerable and depends on the type of material. But, for large values of $(r/a)^{90°}$, i.e. positions far from crack tip, more terms of the power stress series must be taken for accurate results. So, the influence of $(r/a)^{90°}$ is counterbalanced by the necessary number of terms of the power stress series. By combining the polar distance $(r/a)^{90°}$ with the necessary number of terms of stresses, good and accurate results can be obtained far from crack tip.

6.4 Blunt-Notched Plates Under Biaxial Loading

In this section the crack initiation from blunt notches under biaxial loading is studied. The crack extension angle and the critical fracture stress are determined for notches with finite notch tip radii and for various angles of loading and for various biaxiality factors [247]. The determination of the crack extension angle and the critical fracture stress is made by using the Det.-criterion. Relevant studies have been made by Francis [259, 260] and Sih [261]. Results regarding the determination of critical fracture stress and crack extension angle are given by Theocaris [241].

For a cracked plate, with finite notch tip radius ρ, under biaxial loading (Fig. 164) the stress field in the vicinity of the notch tip is given by [44, 45, 262, 263]:

$$\sigma_{xx} = \frac{K_{\mathrm{I}}}{\sqrt{2\pi r}} \cos\frac{\theta}{2}\left[1 - \sin\frac{\theta}{2}\sin\frac{3\theta}{2}\right] - \frac{K_{\mathrm{I}}}{\sqrt{2\pi r}}\frac{\rho}{2r}\cos\frac{3\theta}{2}$$

$$- \frac{K_{\mathrm{II}}}{\sqrt{2\pi r}}\sin\frac{\theta}{2}\left[2 + \cos\frac{\theta}{2}\cos\frac{3\theta}{2}\right]$$

$$+ \frac{K_{\mathrm{II}}}{\sqrt{2\pi r}}\frac{\rho}{2r}\sin\frac{3\theta}{2} + \sigma_\infty(1 - k)\cos 2\beta \tag{740}$$

$$\sigma_{yy} = \frac{K_{\mathrm{I}}}{\sqrt{2\pi r}}\cos\frac{\theta}{2}\left[1 + \sin\frac{\theta}{2}\sin\frac{3\theta}{2}\right] + \frac{K_{\mathrm{I}}}{\sqrt{2\pi r}}\frac{\rho}{2r}\cos\frac{3\theta}{2}$$

$$+ \frac{K_{\mathrm{II}}}{\sqrt{2\pi r}}\sin\frac{\theta}{2}\cos\frac{\theta}{2}\cos\frac{3\theta}{2} - \frac{K_{\mathrm{II}}}{\sqrt{2\pi r}}\frac{\rho}{2r}\sin\frac{3\theta}{2} \tag{741}$$

$$\tau_{xy} = \frac{K_{\mathrm{I}}}{\sqrt{2\pi r}}\sin\frac{\theta}{2}\cos\frac{\theta}{2}\cos\frac{3\theta}{2} - \frac{K_{\mathrm{I}}}{\sqrt{2\pi r}}\frac{\rho}{2r}\sin\frac{3\theta}{2}$$

$$+ \frac{K_{\mathrm{II}}}{\sqrt{2\pi r}}\cos\frac{\theta}{2}\left[1 - \sin\frac{\theta}{2}\sin\frac{3\theta}{2}\right] - \frac{K_{\mathrm{II}}}{\sqrt{2\pi r}}\frac{\rho}{2r}\cos\frac{3\theta}{2} \tag{742}$$

for $0 < \rho/a \ll r/a \ll 1.0$, where ρ is the radius of the notch tip (Fig. 164) and the components of the stress intensity factor $K = K_{\mathrm{I}} - iK_{\mathrm{II}}$ are given by Rels (719) and (720).

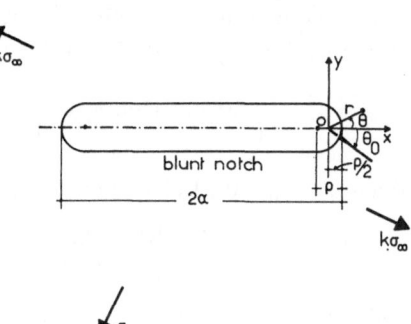

Fig. 164. Blunt notch under biaxial loading.

By substituting Rels (740)–(742) and (719), (720) into Rel. (686), we obtain [247]:

$$\frac{8(r/a)^{\beta,k}}{(\sigma_\infty^{\beta,k,\rho})^2}\,\mathrm{Det.}(\sigma_{ij}) = D + D_{bl} \tag{743}$$

which gives the variation of the determinant of the stress tenser around the notch tip; where $(r/a)^{\beta,k}$ is given by Rel. (723), and:

$$
\begin{aligned}
D = &\left[(1+k)-(1-k)\cos 2\beta\right]^2 D_1 + (1-k)^2 D_2 \sin^2 2\beta + \left[(1+k)-(1-k)\cos 2\beta\right] \\
&\times (1-k)D_{12}\sin 2\beta + 2(2(r/a)^{90°})^{\frac{1}{2}}(1-k)[k^2+(1-k^2)\sin^2\beta]^{0.1}\cos 2\beta \\
&\times \left(\left[(1+k)-(1-k)\cos 2\beta\right]\cos\frac{\theta}{2}\left(1+\sin\frac{\theta}{2}\sin\frac{3\theta}{2}\right) \right. \\
&\left. + (1-k)\sin 2\beta \sin\frac{\theta}{2}\cos\frac{\theta}{2}\cos\frac{3\theta}{2}\right)
\end{aligned}
\tag{744}
$$

$$
\begin{aligned}
D_{bl} = &\left[(1+k)-(1-k)\cos 2\beta\right]^2 D_{1b} + (1-k)^2 D_{2b}\sin^2 2\beta + \left[(1+k)-(1-k)\cos 2\beta\right] \\
&\times (1-k)D_{12b}\sin 2\beta + \frac{2\sqrt{2}(1-k)\cos 2\beta}{((r/a)^{90°})^{\frac{1}{2}}[k^2+(1-k^2)\sin^2\beta]^{0.1}}(\rho/a) \\
&\times \left(\left[(1+k)-(1-k)\cos 2\beta\right]\cos\frac{3\theta}{2} - (1-k)\sin 2\beta\sin\frac{3\theta}{2}\right)
\end{aligned}
\tag{745}
$$

where D_1, D_2, D_{12} are given by Rels (699)–(701), and:

$$D_{1b} = -\left(\frac{(\rho/a)}{2(r/a)^{90°}[k^2+(1-k^2)\sin^2\beta]^{0.2}}\right)^2 \tag{746}$$

$$D_{2b} = -\left(\frac{(\rho/a)}{2(r/a)^{90°}[k^2+(1-k^2)\sin^2\beta]^{0.2}}\right)^2 + \frac{(\rho/a)\cos\theta}{(r/a)^{90°}[k^2+(1-k^2)\sin^2\beta]^{0.2}} \tag{746a}$$

$$D_{12b} = \frac{(\rho/a)\sin\theta}{(r/a)^{90°}[k^2+(1-k^2)\sin^2\beta]^{0.2}} \tag{747}$$

The distribution of $Det.(\sigma_{ij}) = f(\theta)$ presents positive and negative maxima. According to the Det.-criterion [245], crack initiation will occur at the position θ_0 of positive maximum, while yielding will occur at the positions of negative maxima. The position of the positive maximum may be defined by Rel. (703).

The critical fracture stress $\sigma_\infty^{\beta,k,\rho}$ of crack initiation is calculated by Rels (670) and (743):

$$\sigma_\infty^{\beta,k,\rho}\bigg|_{\theta=\theta_0} = \left(\frac{8(r/a)^{\beta,k}\,Det.(\sigma_{ij})_{cr}}{D + D_{bl}}\right)^{\frac{1}{2}} \tag{748}$$

For $\beta = 90°$, $k = 0$, $\rho = 0$ and $\theta = \theta_0 = 0°$, Rel. (748) becomes:

$$\sigma_\infty^{90°,0,0} = \left(\frac{2(r/a)^{90°}\,Det.(\sigma_{ij})_{cr}}{1 - \sqrt{2(r/a)^{90°}}}\right)^{\frac{1}{2}} \tag{749}$$

From Rels (748), (749) and (723), we obtain:

$$\frac{\sigma_\infty^{\beta,k,\rho}}{\sigma_\infty^{90°,0,0}}\bigg|_{\theta=\theta_0} = \frac{2(1 - \sqrt{2(r/a)^{90°}})^{\frac{1}{2}}[k^2 + (1 - k^2)\sin^2\beta]^{0.1}}{\sqrt{D + D_{bl}}} \tag{750}$$

or for $\beta = 90°$, $k = 0$, $\theta = \theta_0 = 0°$ and for each ρ, Rel. (748) becomes:

$$\sigma_\infty^{90°,0,\rho} = \left(\frac{2(r/a)^{90°}\,Det.(\sigma_{ij})_{cr}}{1 - (2(r/a)^{90°})^{\frac{1}{2}} - \left(\frac{(\rho/a)}{4(r/a)^{90°}}\right)^2 - \frac{\sqrt{2}(\rho/a)}{((r/a)^{90°})^{\frac{1}{2}}}}\right)^{\frac{1}{2}} \tag{751}$$

and from Rels (748) and (723):

$$\frac{\sigma_\infty^{\beta,k,\rho}}{\sigma_\infty^{90°,0,\rho}}\bigg|_{\theta=\theta_0} = \frac{2\left[1 - (2(r/a)^{90°})^{\frac{1}{2}} - \left(\frac{(\rho/a)}{4(r/a)^{90°}}\right)^2 - \frac{\sqrt{2}(\rho/a)}{((r/a)^{90°})^{\frac{1}{2}}}\right]^{\frac{1}{2}}[k^2 + (1 - k^2)\sin^2\beta]^{0.1}}{\sqrt{D + D_{bl}}} \tag{752}$$

Relation (750) gives the critical fracture stress $\sigma_\infty^{\beta,k,\rho}$ of crack initiation normalized to its characteristic value $\sigma_\infty^{90°,0,0}$ for $\beta = 90°$, $k = 0$ and $\rho = 0$. Likewise, Rel. (752) gives $\sigma_\infty^{\beta,k,\rho}$ normalized to its characteristic value $\sigma_\infty^{90°,0,\rho}$ for $\beta = 90°$, $k = 0$ and for each ρ. For $\rho = 0$, $(r/a)^{90°} = 0$ and $k = 0$, Rels (743) and (750) are changed into those of the singular solution [85, 86].

In order to examine the influence of the radius of the notch tip on the crack extension angle and the critical fracture stress, Eqs (743), (750) and (752) were solved for various values of radius ρ/a of notch tip for angles of loading β between 0° and 90°.

Figure 165 presents the variation of the crack extension angle $-\theta_0$ versus the angle of loading β for $k = 0$ (uniaxial tension), for $(r/a)^{90°} = 0.05$ and for radii of notch tip $\rho/a = 0$ (sharp crack), 0.002, 0.01, 0.02, 0.03, 0.04, 0.05. In the same figure the curve of the singular solution is drawn. From this figure we observe that as the radius of the notch tip increases the crack extension angle decreases, mainly in the range of small angles of loading β. Experimental measurements of crack extension angle for various materials are included in this figure. The experimental points were taken from Ref. [241]. The dispersion of the experimental points can be explained by the influence of the radius of notch tip, because after loading the sharp notch becomes blunted.

Figure 166 presents the variation of the fracture stress $\sigma_\infty^{\beta,k,\rho}$ reduced to its values $\sigma_\infty^{90°,0,0}$ for $\beta = 90°$, $k = 0$ and $\rho = 0$ (sharp crack) versus the angle of loading

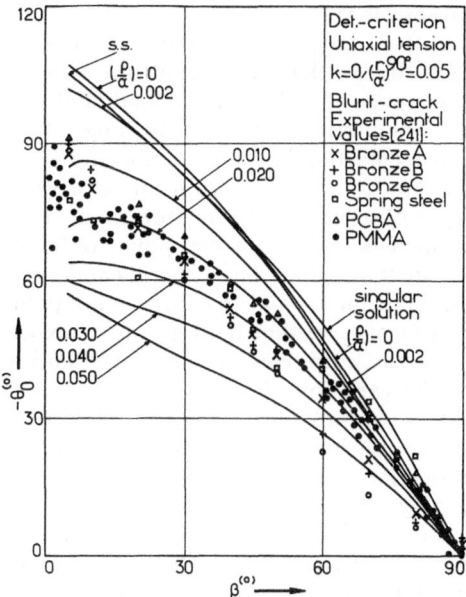

Fig. 165. Variation of $-\theta_0$ versus β according to the Det.-criterion for $k = 0$, $(r/a)^{90°} = 0.05$, for singular solution and for $\rho/a = 0$ (sharp crack), 0.002, 0.010, 0.020, 0.030, 0.040, 0.050.

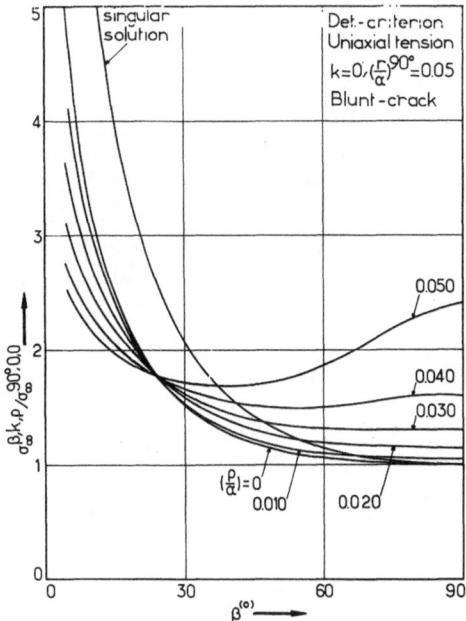

Fig. 166. Variation of fracture stress $\sigma_\infty^{\beta,k,\rho}$ reduced to its values $\sigma_\infty^{90°,0,0}$ versus β according to the Det.-criterion for $k = 0$, $(r/a)^{90°} = 0.05$, for singular solution and for $\rho/a = 0$ (sharp crack), 0.010, 0.020, 0.030, 0.040, 0.050.

β for $k = 0$ (uniaxial tension), for ratio $(r/a)^{90°} = 0.05$ and for radii of notch tip $\rho/a = 0$ (sharp crack), 0.01, 0.02, 0.03, 0.04, 0.05. The curve of the singular solution is also drawn. In this figure we observe that the fracture stress increases as the radius of notch tip increases in the range of high angles of loading β, while the fracture stress decreases as the radius of notch tip increases in the range of small angles of loading β. Likewise, Fig. 167 presents the variation of the fracture stress $\sigma_\infty^{\beta,k,\rho}$ reduced to its values $\sigma_\infty^{90°,0,\rho}$ for $\beta = 90°$, $k = 0$, versus β for each value of radius of notch tip ρ. From this figure we observe that the fracture stress decreases considerably as the radius of notch tip increases. Experimental points of fracture stress for various materials are included in this figure. The experimental points were taken from Ref. [241] and most of these correspond to experimental points of crack extension angles in Fig. 165. We can observe that both groups of experimental points lie in the same zone of radii of notch tip between $\rho/a = 0.020$ and $\rho/a = 0.050$.

Figure 168 presents the variation of the crack extension angle $-\theta_0$ versus the angle β for $k = 0$ (uniaxial tension), for ratio $(r/a)^{90°} = 0.1$, for singular solution and for radii of notch tip $\rho/a = 0$ (sharp crack), 0.002, 0.010, 0.020, 0.030, 0.050, 0.080, 0.1. In this figure we observe that as the ratio ρ/a increases the crack extension angle reduces considerably. The same experimental points of crack extension angle of Ref. [241] were placed in this figure, in which we can observe that experimental points lie mainly in the zone between $\rho/a = 0.030$ and $\rho/a = 0.080$.

Figure 169 presents the variation of the fracture stress $\sigma_\infty^{\beta,k,\rho}$ reduced to its values $\sigma_\infty^{90°,0,\rho}$ for $\beta = 90°$, $k = 0$ and for each value of radius of notch tip ρ. We can observe that the fracture stress decreases since the curve of $\rho/a = 0$ lies lower than

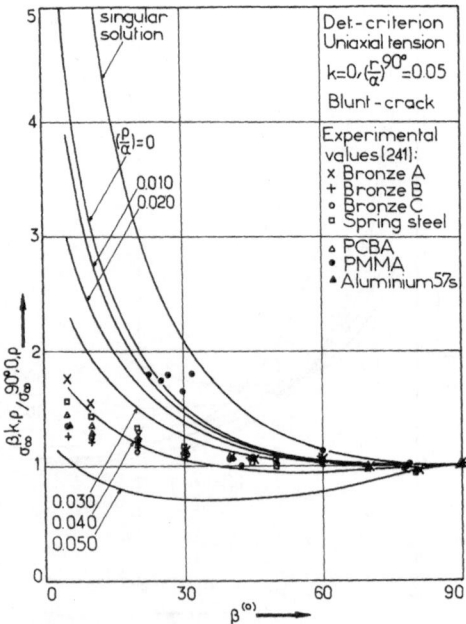

Fig. 167. Variation of fracture stress $\sigma_\infty^{\beta,k,\rho}$ reduced to its values $\sigma_\infty^{90°,0,\rho}$ versus β according to the Det.-criterion for $k = 0$, $(r/a)^{90°} = 0.05$, for singular solution and for $\rho/a = 0$ (sharp crack), 0.010, 0.020, 0.030, 0.040, 0.050.

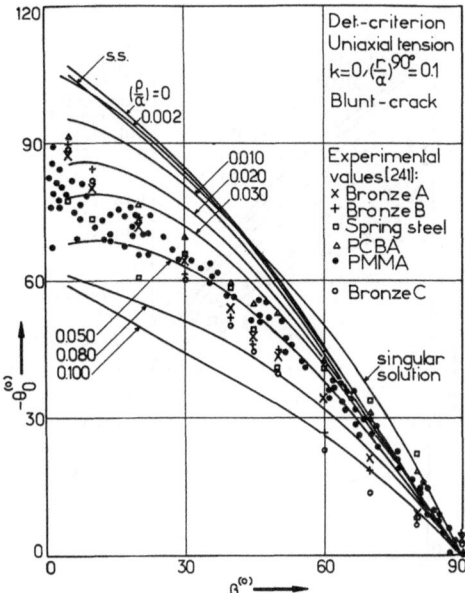

Fig. 168. Variation of $-\theta_0$ versus β according to the Det.-criterion for $k = 0$, $(r/a)^{90°} = 0.1$, for singular solution and for $\rho/a = 0$ (sharp crack), 0.002, 0.010, 0.020, 0.030, 0.050, 0.080, 0.1.

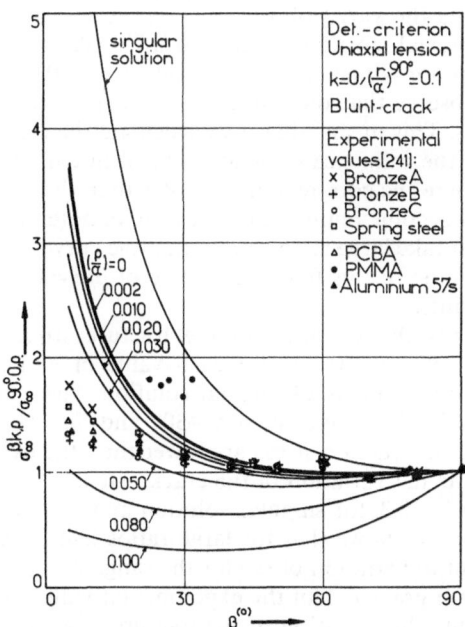

Fig. 169. Variation of fracture stress $\sigma_\infty^{\beta,k,\rho}$ reduced to its values $\sigma_\infty^{90°,0,\rho}$ versus β according to the Det.-criterion for $k = 0$, $(r/a)^{90°} = 0.1$, for singular solution and for $\rho/a = 0$ (sharp crack), 0.002, 0.010, 0.020, 0.030, 0.050, 0.080, 0.1.

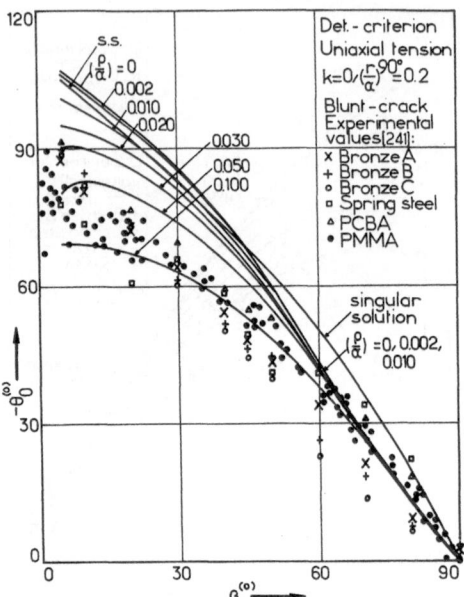

Fig. 170. As in Fig. 168, for $(r/a)^{90°} = 0.2$.

in Fig. 167 [84]. Also, we can observe that experimental points of the fracture stress lie mainly in the zone between $\rho/a = 0.030$ and $\rho/a = 0.080$.

Figure 170 presents the variation of the crack extension angle $-\theta_0$ versus the angle β for $k = 0$ (uniaxial tension), the ratio $(r/a)^{90°} = 0.2$, for singular solution and for radii of notch tip $\rho/a = 0$ (sharp crack), 0.002, 0.010, 0.020, 0.030, 0.050, 0.1. In this figure we can observe that the influence of radius of notch tip is weaker than in the previous cases. This means that if we calculate the crack extension angle far from the notch tip, the influence of small radii of notch tip will be weaker. For ductile materials, where the core-region is greater than that of brittle materials, the initial sharp notches become blunt with large radii of notch tip. Therefore, for large ratios $(r/a)^{90°}$ we must take large ratios ρ/a so that the curves of $\theta_0 = f(\beta)$ go through the experimental points. In this figure the curve of $\rho/a = 0.1$ goes mainly through the experimental points.

Figure 171 presents the variation of the fracture stress $\sigma_\infty^{\beta,k,\rho}$ reduced to its values $\sigma_\infty^{90°,0,\rho}$ for $\beta = 90°$, $k = 0$ and for each value of ρ. The curves $\sigma_\infty^{\beta,k,\rho} = f(\beta)$ lie very low relative to the curve of singular solution. These curves go through the experimental points in the ranges of β 5°–30° and 70°–90°. For β 30°–70° the theoretical values of the fracture stress are lower than the experimental values.

Figure 172 presents the variation of the crack extension angle $-\theta_0$ versus β for $k = 0$, for ratio $(r/a)^{90°} = 0.3$, for singular solution and for $\rho/a = 0.002$, 0.010, 0.020, 0.030, 0.050. This figure shows that for large ratios $(r/a)^{90°}$ the values of $-\theta_0$ are virtually independent of variation of ρ/a for the range of β 50°–90°. The theoretical values of $-\theta_0$ remain greater than the experimental values for β 5°–60°.

Figure 173 presents the variation of the fracture stress $\sigma_\infty^{\beta,k,\rho}$ versus β for $k = 0$, for ratio $(r/a)^{90°} = 0.3$, for singular solution and for $\rho/a = 0$, 0.010, 0.020, 0.030, 0.050. For $(r/a)^{90°} = 0.3$ the theoretical values of the fracture stress are lower than the experimental values for all angles of loading β outside the range 5°–10°.

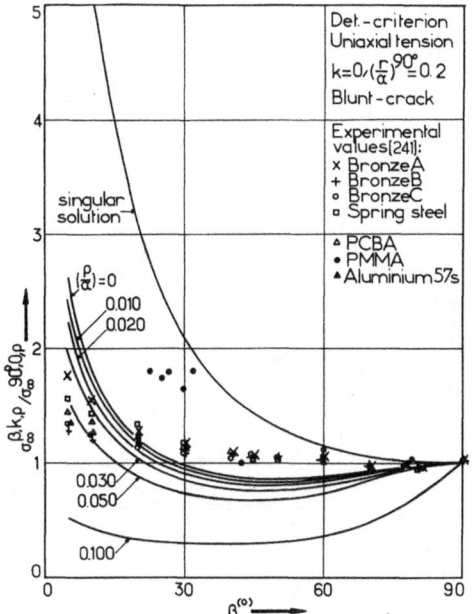

Fig. 171. As in Fig. 169, for $(r/a)^{90°} = 0.2$.

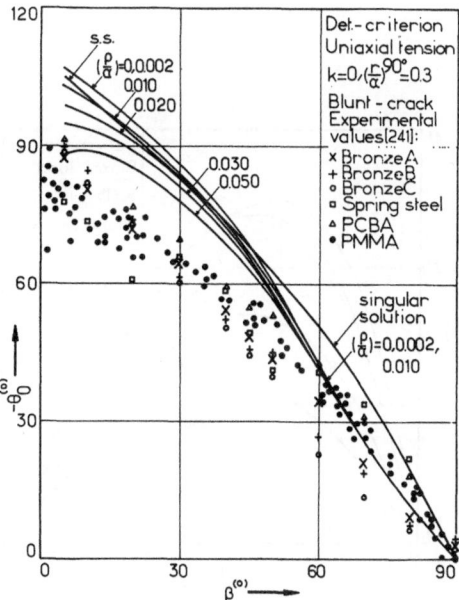

Fig. 172. As in Fig. 165, for $(r/a)^{90°} = 0.3$.

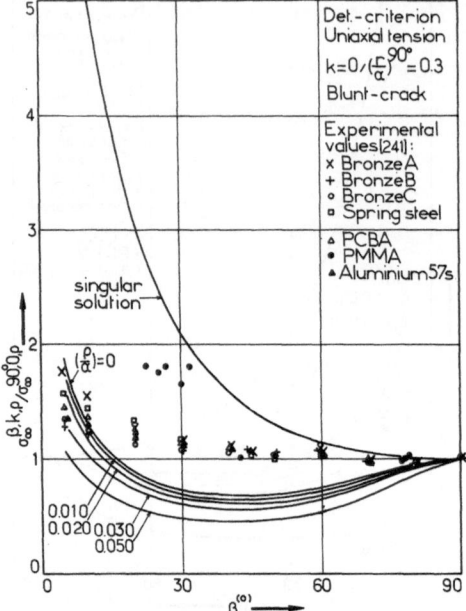

Fig. 173. As in Fig. 167, for $(r/a)^{90°} = 0.3$.

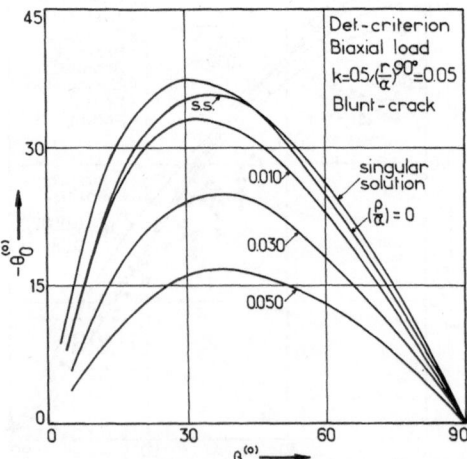

Fig. 174. Variation of $-\theta_0$ versus β according to the Det.-criterion for $k = 0.5$, $(r/a)^{90°} = 0.05$, for the singular solution and for $\rho/a = 0$ (sharp crack), 0.010, 0.030, 0.050.

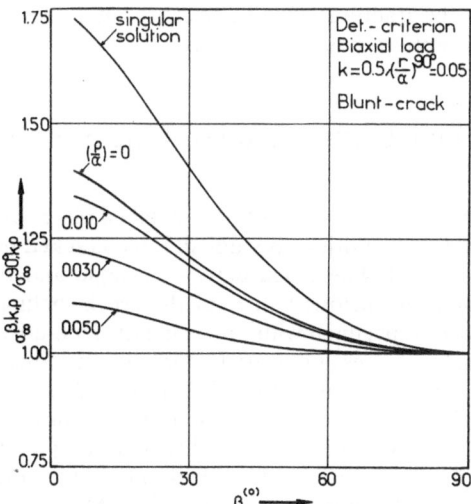

Fig. 175. Variation of fracture stress $\sigma_\infty^{\beta,k,\rho}$ reduced to its values $\sigma_\infty^{90°,k,\rho}$ versus β according to the Det.-criterion for $k = 0.5$, $(r/a)^{90°} = 0.05$, for singular solution and for $\rho/a = 0$ (sharp crack), 0.010, 0.030, 0.050.

Figure 174 presents the variation of the crack extension angle $-\theta_0$ versus β for $k = 0.5$ (biaxial loading), for ratio $(r/a)^{90°} = 0.05$, for the singular solution and for $\rho/a = 0.010, 0.030, 0.050$. In this figure we can observe that the crack extension angle decreases as the radius of notch tip increases. The curves of $-\theta_0 = f(\beta)$ present a maximum in the range β 30°–40°.

Figure 175 presents the variation of the fracture stress $\sigma_\infty^{\beta,k,\rho}$ reduced to its values $\sigma_\infty^{90°,k,\rho}$ for $\beta = 90°$ and for each value of k and ρ. As illustrated, the fracture stress decreases considerably as the radius of notch tip increases and the biaxiality factor k increases. Likewise, the angle of extension $-\theta_0$ and the fracture stress $\sigma_\infty^{\beta,k,\rho}$ decrease, as the radius of notch tip increases and the biaxiality factor k takes negative values.

The following conclusions can be drawn from the results presented: (i) The influence of the ratio $(r/a)^{90°}$ on the crack extension angle and fracture stress is considerable and depends on the type of material. For brittle materials with small core-regions, we can take low values of the ratio $(r/a)^{90°}$, while for ductile materials with large core-regions, we must take large values of the ratio $(r/a)^{90°}$ in order to have coincidence of the theoretical results with the experimental. (ii) The influence of the radius of notch tip is considerable and depends strongly on the ratio $(r/a)^{90°}$. The influence is strong for very low values of the ratio $(r/a)^{90°}$ and weak for larger values. In both cases, the values of $-\theta_0$ and $\sigma_\infty^{\beta,k,\rho}$ decrease considerably. (iii) Better coincidence of the theoretical and experimental results is observed in Figs 168 and 170 for $k = 0$ (uniaxial tension), for ratio $(r/a)^{90°} = 0.1$ and for the zone of radii of notch tip between $\rho/a = 0.030$ and $\rho/a = 0.050$. (iv) For large values of the ratio $(r/a)^{90°}$, the variation of $-\theta_0$ and $\sigma_\infty^{\beta,k,\rho}$ is very weak because the measurements are made far from the notch tip where the stress field decreases.

From this study it may be concluded that we have to estimate the three parameters: core-region, ratio $(r/a)^{90°}$ and radius of notch tip, ρ/a, in order to calculate the crack extension angles and the fracture stresses, because after loading the sharp crack becomes blunt with a finite radius of notch tip.

6.5 Edge-Cracked Plates Under Uniaxial Tension

In this section the influence of geometry of edge-cracked plates on crack propagation under mixed mode loading conditions is studied. Edge-cracked plates under uniaxial tension are examined [244]. The crack extension angle and the critical stress of fracture are determined for various values of the crack inclination angle and the ratio of the crack length to specimen width by using the Det.-criterion in conjunction with experimental results of stress intensity factors in finite edge-cracked plates with oblique cracks relative to the applied load [121] (Sect. 2.7).

Results regarding the determination of fracture toughness and crack extension angle in finite width specimens are given by Ewing et al. [264]. They determined, both theoretically and experimentally, the direction of initial crack propagation and the load to failure in a finite width edge-cracked PMMA plate under tension and bending. The experimental determination of the fracture angle was made by using a projection microscope, while the theoretical results were obtained by employing the minimum potential energy theorem in conjunction with the eigen-series solution of the crack problem. Also, a theoretical study of the influence of specimen geometry on crack extension angle based on the S-criterion is given by Gdoutos [265].

For an edge-cracked plate under mixed mode loading, the stress field in the vicinity of the crack tip is given by Rels (693)–(695), and the stress intensity factors K_I and K_{II} are given by:

$$K_I = f_I \sigma_\infty \sqrt{\pi a} \sin^2 \beta, \quad K_{II} = f_{II} \sigma_\infty \sqrt{\pi a} \sin \beta \cos \beta \tag{753}$$

where f_I and f_{II} are the correction factors of K_I and K_{II} respectively, and are given by the nomograms in Ref. [121]. The correction factors f_I and f_{II} depend on the ratio (h/w) of the free length to width of the specimen, the ratio (a/w) of the crack length to specimen width and the angle of loading β.

6.5.1 For Singular Solution

As in Section 6.1.1, by substituting Rels (693)–(695) (without the constant term $\sigma_\infty \cos 2\beta$) and Rels (753) into Rel. (686), we obtain [244]:

$$\text{Det.}(\sigma_{ij}) = \frac{D}{2\pi r} \tag{754}$$

with:

$$\frac{D}{\pi a \sigma_\infty^2} = f_I^2 D_1 \sin^4 \beta + f_{II}^2 D_2 \sin^2 \beta \cos^2 \beta + f_I f_{II} D_{12} \sin^3 \beta \cos \beta \tag{755}$$

where D_1, D_2, D_{12} are given by Rels (699)–(701). The critical stress σ_∞^β of crack initiation is calculated by Rels (670), (754) and (755) [85]:

$$\left. \frac{\sigma_\infty^\beta}{\sigma_\infty^{90°}} \right|_{\theta = \theta_0} = \left(\frac{(r/a)^\beta (f_I^{90°})^2}{(r/a)^{90°} [f_I^2 D_1 \sin^4 \beta + f_{II}^2 D_2 \sin^2 \beta \cos^2 \beta + f_I f_{II} D_{12} \sin^3 \beta \cos \beta]} \right)^{\frac{1}{2}} \tag{756}$$

The ratio $(r/a)^\beta$ depends on the angle β and defines the core-region around the crack tip. In static crack problems the ratio $(r/a)^\beta$ is calculated by the initial curve of the caustic [101]:

$$r^\beta = (\tfrac{2}{3} C'_{r,t,f})^{\frac{2}{3}}(K_I^2 + K_{II}^2)^{\frac{1}{3}} = (\tfrac{2}{3} C'_{r,t,f}\sigma\sqrt{\pi a})^{\frac{2}{3}}\sin^{0.4}\beta(f_I^2\sin^2\beta + f_{II}^2\cos^2\beta)^{0.2} \quad (757)$$

but:

$$r^{90°} = (\tfrac{2}{3} C'_{r,t,f}\sigma_\infty\sqrt{\pi a})^{\frac{2}{3}}(f_I^{90°})^{0.4} \quad (758)$$

then, Rel. (757) becomes:

$$(r/a)^\beta = (r/a)^{90°}(f_I^{90°})^{-0.4}\sin^{0.4}\beta(f_I^2\sin^2\beta + f_{II}^2\cos^2\beta)^{0.2} \quad (759)$$

and Rel. (756) becomes:

$$\left.\frac{\sigma_\infty^\beta}{\sigma_\infty^{90°}}\right|_{\theta=\theta_0} = \frac{(f_I^{90°})^{0.8}\sin^{0.2}\beta(f_I^2\sin^2\beta + f_{II}^2\cos^2\beta)^{0.1}}{[f_I^2 D_1\sin^4\beta + f_{II}^2 D_2\sin^2\beta\cos^2\beta + f_I f_{II}D_{12}\sin^3\beta\cos\beta]^{\frac{1}{2}}} \quad (760)$$

Relation (760) gives the critical stress of crack initiation σ_∞^β normalized to its characteristic value $\sigma_\infty^{90°}$ for $\beta = 90°$.

6.5.2 For Solution with Constant Term

Relations corresponding to Rels (755) and (760) can be obtained for the case of solution with the constant term $(\sigma_\infty\cos 2\beta)$ of the stress σ_{xx}:

$$D_c = \frac{D}{4\pi a(\sigma_\infty^\beta)^2} = f_I^2 D_1\sin^4\beta + f_{II}^2 D_2\sin^2\beta\cos^2\beta + f_I f_{II}D_{12}\sin^3\beta\cos\beta$$

$$+ (2(r/a)^{90°})^{\frac{1}{2}}(f_I^2\sin^2\beta + f_{II}^2\cos^2\beta)^{0.1}(f_I^{90°})^{-0.2}\sin^{1.2}\beta\cos 2\beta$$

$$\times \left(f_I\sin\beta\cos\frac{\theta}{2}\left(1 + \sin\frac{\theta}{2}\sin\frac{3\theta}{2}\right) + f_{II}\cos\beta\sin\frac{\theta}{2}\cos\frac{\theta}{2}\cos\frac{3\theta}{2}\right) \quad (761)$$

and:

$$\left.\frac{\sigma_\infty^\beta}{\sigma_\infty^{90°}}\right|_{\theta=\theta_0} = \frac{((f_I^{90°})^2 - (2(r/a)^{90°})^{\frac{1}{2}}f_I^{90°})^{\frac{1}{2}}\sin^{0.2}\beta(f_I^2\sin^2\beta + f_{II}^2\cos^2\beta)^{0.1}}{(f_I^{90°})^{0.2}\sqrt{D_c}} \quad (762)$$

Relation (761) gives the variation of the stress tensor determinant around the crack tip, while Rel. (762) gives the critical stress of crack initiation σ_∞^β normalized to its characteristic value $\sigma_\infty^{90°}$ for $\beta = 90°$. The position of the positive

Table 9. Correction factors f_I, f_{II} of stress intensity factors K_I, K_{II} respectively, for edge-cracked plates, from nomograms of Figs 40 and 41 (Sect. 2.7) [121]. Specimen length to width ratio $= 2$

$\beta^{(°)}$	$a/w = 0.1$		$a/w = 0.2$		$a/w = 0.3$		$a/w = 0.4$		$a/w = 0.5$	
	f_I	f_{II}	f_I	f_{II}	f_I	f_{II}	f_I	f_{II}	f_I	f_{II}
20	2.43	0.843	2.8	0.935	3.05	1.0	3.55	1.08	3.8	1.1
30	1.725	0.765	2.025	0.835	2.15	0.885	2.425	0.93	2.6	0.965
45	1.28	0.69	1.6	0.81	1.775	0.86	2.0	0.945	2.475	1.155
60	1.1	0.635	1.45	0.825	1.655	0.88	2.15	1.13	2.65	1.39
80	1.05	0.71	1.35	0.86	1.65	0.985	2.2	1.33	2.73	1.58
90	1.075	0	1.3	0	1.6	0	2.075	0	2.8	0

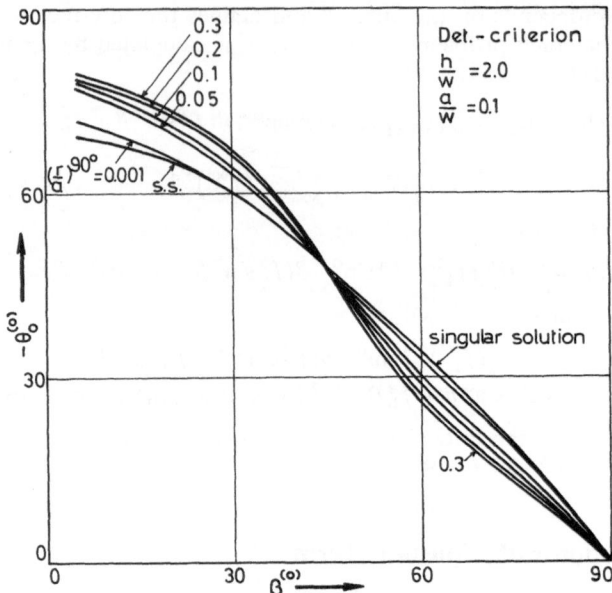

Fig. 176. Variation of $-\theta_0$ versus β for an edge oblique crack, according to the Det.-criterion for singular solution and for $(r/a)^{90°} = 0.001, 0.05, 0.1, 0.2, 0.3$ and for specimens with $h/w = 2.0$, $a/w = 0.1$.

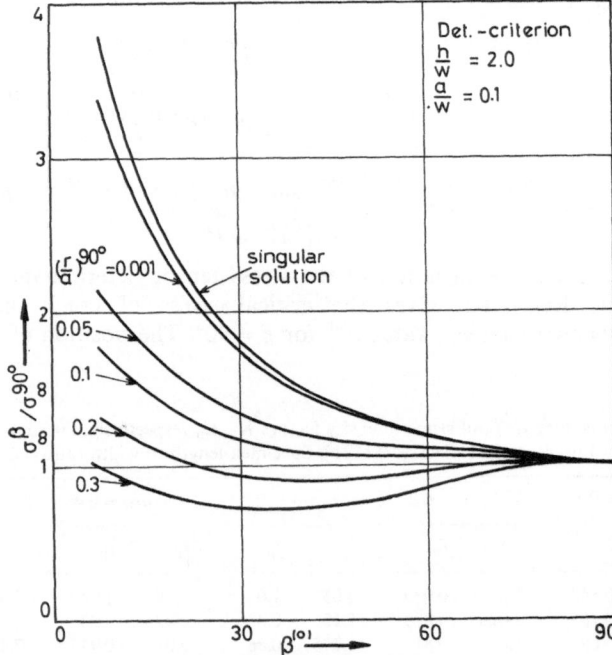

Fig. 177. Variation of fracture stress σ_∞^β versus β, reduced to its values $\sigma_\infty^{90°}$ for $\beta = 90°$, according to the Det.-criterion for singular solution and for $(r/a)^{90°} = 0.001, 0.05, 0.1, 0.2, 0.3$ and for specimens with $h/w = 2.0$ and $a/w = 0.1$.

Table 10. Tension data for angled cracks in PMMA [264]

h/w	a/w	$\beta^{(\circ)}$	$-\theta_0^{(\circ)}$	$\sigma_\infty^\beta/\sigma_\infty^{86^\circ}$
	0.22	86.0	2.7	1.0
	0.22	74.85	21.4	0.96
	0.26	68.3	22.91	0.77
	0.31	55.63	27.63	0.92
2.4	0.26	45.58	46.0	1.14
	0.41	35.6	56.0	1.07
	0.14	25.46	50.8	2.55
	0.46	15.0	66.83	1.68
	0.18	59.36	27.2	1.29
	0.31	19.0	64.7	1.66

maximum (θ_0) may be defined by annulling the partial derivative of Rels (755) and (761).

For the case of finite width edge-cracked plates loaded in tension considered in this paper the correction factors f_{I}, f_{II} were obtained from the experimental results of Ref. [121] (Sect. 2.7). For this study the correction factors f_{I}, f_{II} are given in Table 9. Then, for ratios $h/w = 2.0$, $a/w = 0.1$, 0.2, 0.3, 0.4, 0.5 and $(r/a)^{90^\circ} = 0.001$, 0.05, 0.1, 0.2, 0.3 and for β 90°–5°, Rels (760) and (762) were solved by computer.

Figures 176 and 177 present the variation of the crack extension angle $-\theta_0$ and the fracture stress σ_∞^β versus the angle of loading β, respectively, for ratios $h/w = 2.0$ and $a/w = 0.1$, and for values of $(r/a)^{90^\circ} = 0$ (singular solution), 0.001, 0.05, 0.1, 0.2 and 0.3. From these figures we observe that the crack extension angle $-\theta_0$ decreases as the ratio $(r/a)^{90^\circ}$ increases in the interval $45° < \beta < 90°$ and increases as the ratio $(r/a)^{90^\circ}$ increases in the interval $0° < \beta < 45°$, while the fracture stress σ_∞^β decreases rapidly as the ratio $(r/a)^{90^\circ}$ increases for all intervals of β.

Finally, the experimental results of Ewing et al. [264] are given in Table 10 for comparison.

From the results presented the following conclusions can be drawn: (i) For all cases studied the crack extension angle $-\theta_0$ decreases as the ratio a/w increases. (ii) The crack extension angle $-\theta_0$ increases in the interval $0° < \beta < 45°$ and decreases in the interval $45° < \beta < 90°$, as the ratio $(r/a)^{90^\circ}$ increases. (iii) For all cases studied the fracture stress σ_∞^β decreases as the angle of loading β decreases and the ratio $(r/a)^{90^\circ}$ increases. (iv) For all cases studied the fracture stress σ_∞^β increases as the ratio a/w increases.

6.6 Edge Blunt-Notched Plates Under Uniaxial Tension

In this section a study of the dependence of the crack extension angle under mixed mode loading upon the geometry of the specimen and the finite notch tip radius takes place [248]. Edge-notched plates under tension are studied. Determination of the crack extension angle and the critical fracture stress is made by using the Det.-criterion in conjunction with experimental results of stress intensity factors in finite edge-cracked plates with oblique cracks relative to the applied load [121]

(Sect. 2.7). Results regarding the determination of fracture toughness and crack extension angle in finite width specimens are given by Ewing et al. [264].

For an edge-notched plate, with finite notch tip radius ρ, under mixed mode loading the stress field in the vicinity of the notch tip is given by Rels (740), (741) and (742) [44, 45, 262, 263] and the stress intensity factors K_I and K_{II} are given by Rels (753) [121]. By applying the Det.-criterion of fracture for this case, we obtain [248]:

$$\frac{8(r/a)^\beta}{(\sigma_\infty^{\beta,\rho})^2} \mathrm{Det.}(\sigma_{ij}) = D + D_{bI} \tag{763}$$

where [244]:

$$(r/a)^\beta = (r/a)^{90°}(f_I^{90°})^{-0.4} \sin^{0.4}\beta (f_I^2 \sin^2\beta + f_{II}^2 \cos^2\beta)^{0.2} \tag{764}$$

and:

$$D = f_I^2 D_1 \sin^4\beta + f_{II}^2 D_2 \sin^2\beta \cos^2\beta + f_I f_{II} D_{12} \sin^3\beta \cos\beta$$

$$+ (2(r/a)^{90°})^{\frac{1}{2}} \cos 2\beta \left(f_I \sin^2\beta \cos\frac{\theta}{2} \left(1 + \sin\frac{\theta}{2}\sin\frac{3\theta}{2} \right) \right.$$

$$\left. + f_{II}\sin\beta\cos\beta\sin\frac{\theta}{2}\cos\frac{\theta}{2}\cos\frac{3\theta}{2} \right) \tag{765}$$

$$D_{bI} = f_I^2 D_{1b}\sin^4\beta + f_{II}^2 D_{2b}\sin^2\beta\cos^2\beta + f_I f_{II} D_{12b}\sin^3\beta\cos\beta$$

$$+ \sqrt{2(r/a)^\beta}\,\frac{(\rho/a)\cos 2\beta}{2(r/a)^\beta}\left(f_I\sin^2\beta\cos\frac{3\theta}{2} - f_{II}\sin\beta\cos\beta\sin\frac{3\theta}{2} \right) \tag{766}$$

and:

$$D_1 = \cos^4\frac{\theta}{2} \tag{767}$$

$$D_2 = \cos^2\frac{\theta}{2}\left(3\sin^2\frac{\theta}{2} - 1 \right) \tag{768}$$

$$D_{12} = -4\sin\frac{\theta}{2}\cos^3\frac{\theta}{2} \tag{769}$$

$$D_{1b} = -\left(\frac{(\rho/a)}{2(r/a)^\beta} \right)^2 \tag{770}$$

$$D_{2b} = -\left(\frac{(\rho/a)}{2(r/a)^\beta} \right)^2 + \left(\frac{(\rho/a)\cos\theta}{(r/a)^\beta} \right) \tag{771}$$

$$D_{12b} = \frac{(\rho/a)\sin\theta}{(r/a)^\beta} \tag{772}$$

Relation (763) gives the variation of the determinant of the stress tensor around the notch tip. The distribution $\mathrm{Det.}(\sigma_{ij}) = f(\beta)$ around the notch tip presents positive and negative maxima. According to the Det.-criterion [85, 245], crack initiation will occur at the positions θ_0 of positive maxima, while yielding will occur at the positions of negative maxima. The positions of the positive maxima of the distribution $\mathrm{Det.}(\sigma_{ij}) = f(\theta)$ may be defined by the condition $\partial D/\partial\theta = 0$.

The critical fracture stress $\sigma_\infty^{\beta,\rho}$ of crack initiation is calculated by Rels (670) and

(763):

$$\left.\frac{\sigma_\infty^{\beta;\rho}}{\sigma_\infty^{90°,0}}\right|_{\theta=\theta_0} = \frac{2\sqrt{(r/a)^\beta((f_I^{90°})^2 - (2(r/a)^{90°})^{\frac{1}{2}}f_I^{90°})^{\frac{1}{2}}}}{\sqrt{(r/a)^{90°}}\sqrt{D+D_{bI}}}$$

(773)

or for $\beta = 90°$ and for each value of ρ, Rel. (773) becomes:

$$\left.\frac{\sigma_\infty^{\beta,\rho}}{\sigma_\infty^{90°,\rho}}\right|_{\theta=\theta_0} = \frac{2\sqrt{(r/a)^\beta}\left((f_I^{90°})^2 - (2(r/a)^{90°})^{\frac{1}{2}}f_I^{90°} - f_I^{90°}\left(\frac{(\rho/a)}{2(r/a)^{90°}}\right)^2 - f_I^{90°}\frac{(\rho/a)}{(2(r/a)^{90°})^{\frac{1}{2}}}\right)^{\frac{1}{2}}}{((r/a)^{90°})^{\frac{1}{2}}\sqrt{D+D_{bI}}}$$

(774)

Relation (773) gives the critical fracture stress $\sigma_\infty^{\beta,\rho}$ of crack initiation normalized to its characteristic value $\sigma_\infty^{90°,0}$ for $\beta = 90°$ and $\rho = 0$. Likewise, Rel. (774) gives the critical fracture stress $\sigma_\infty^{\beta,\rho}$ of crack initiation normalized to its characteristic value $\sigma_\infty^{90°,\rho}$ for $\beta = 90°$ and for each value of ρ. For $\rho = 0$ and $(r/a)^{90°} = 0$, Rels (763) and (773) are changed into those of the singular solution [244].

For the case of finite width edge-notched plates, loading in tension was considered. The correction factors f_I and f_{II} were obtained from the experimental results of Ref. [121], and are given in Table 9. In order to examine the influence of the radius of the notch tip on the crack extension angle and the critical fracture stress, Eqs (763) and (773) were solved for various values of radius ρ/a of notch tip and for specimens with ratios $h/w = 2.0$, $a/w = 0.1, 0.2, 0.3, 0.5$ and $(r/a)^{90°} = 0.05, 0.1, 0.2, 0.3$ and for β 90°–5°.

Figure 178 presents the variation of the crack extension angle $-\theta_0$ versus the angle of loading β for specimens with ratios $h/w = 2.0$, $a/w = 0.1$ and $(r/a)^{90°} = 0.05$ and for radii of notch tip $\rho/a = 0$ (sharp crack), 0.010, 0.020, 0.030, 0.040 and 0.050. In the same figure the curve of the singular solution is drawn. In this figure we observe that the crack extension angle decreases as the radius of notch tip increases. Figure 179 presents the variation of the fracture stress $\sigma_\infty^{\beta,\rho}$ reduced to its values $\sigma_\infty^{90°,\rho}$ for $\beta = 90°$ and for each value of radius of notch tip ρ, versus the angle of

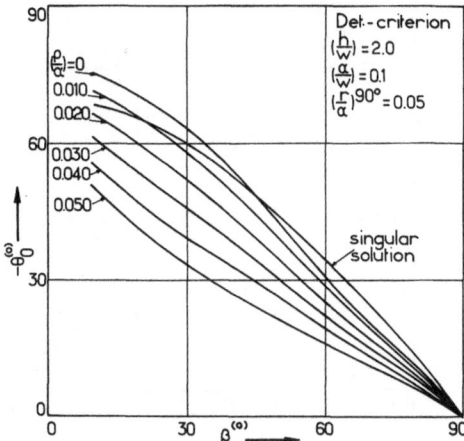

Fig. 178. Variation of $-\theta_0$ versus β according to the Det.-criterion for specimens with $h/w = 2.0$, $a/w = 0.1$ and for $(r/a)^{90°} = 0.05$, for singular solution and for $\rho/a = 0$ (sharp crack), 0.010, 0.020, 0.030, 0.040 and 0.050.

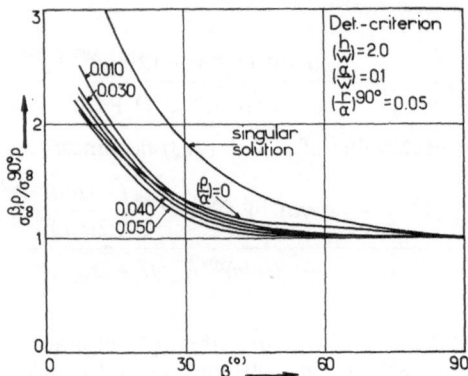

Fig. 179. Variation of fracture stress $\sigma_\infty^{\beta,\rho}$ reduced to its values $\sigma_\infty^{90°,\rho}$ versus β according to the Det.-criterion for specimens with $h/w = 2.0$, $a/w = 0.1$ and for $(r/a)^{90°} = 0.05$, for singular solution and for $\rho/a = 0$ (sharp crack), 0.010, 0.030, 0.040 and 0.050.

loading β for specimens with ratios $h/w = 2.0$, $a/w = 0.1$ and $(r/a)^{90°} = 0.05$ and for radii of notch tip $\rho/a = 0$ (sharp crack), 0.010, 0.030, 0.040 and 0.050. In this figure we also observe that the fracture stress decreases considerably as the radius of notch tip increases. The position of the curves relative to the curve of the singular solution depends strongly on the ratio $(r/a)^{90°}$. Likewise, Fig. 180 presents the variation of the crack extension angle $-\theta_0$ versus β for ratio $(r/a)^{90°} = 0.1$ and Fig. 181 presents the variation of the fracture stress $\sigma_\infty^{\beta,\rho}/\sigma_\infty^{90°,\rho}$ versus β for the ratio $(r/a)^{90°} = 0.1$. We observe that the reduction of the crack extension angles are weaker than in Fig. 178, while the reduction of the fracture stress is stronger than in Fig. 179. The variation of the crack extension angles and the fracture stresses depends on the ratio a/w, of the initial crack length to width of the specimen.

Finally, the experimental results of Ewing *et al.* [264] are given in Table 10, for comparison with the theoretical results.

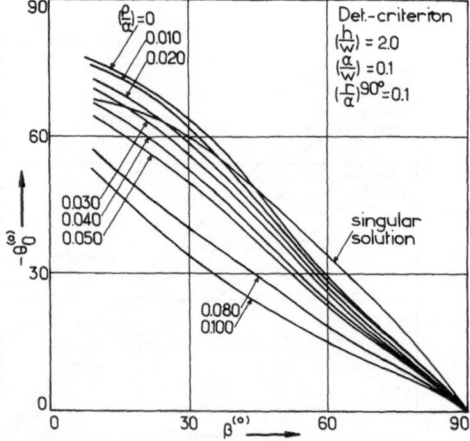

Fig. 180. Variation of $-\theta_0$ versus β according to the Det.-criterion for specimen with $h/w = 2.0$, $a/w = 0.1$ and for $(r/a)^{90°} = 0.1$, for singular solution and for $\rho/a = 0$ (sharp crack), 0.010, 0.020, 0.030, 0.040, 0.050, 0.080 and 0.100.

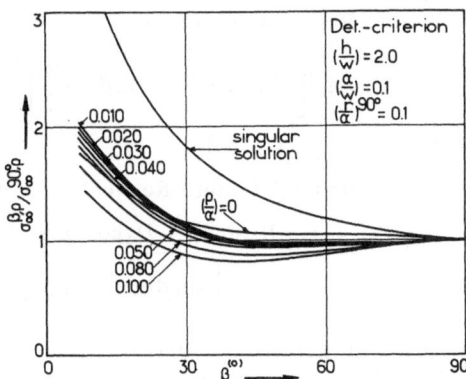

Fig. 181. Variation of fracture stress $\sigma_\infty^{\beta,\rho}$ reduced to its values $\sigma_\infty^{90°,\rho}$ versus β according to the Det.-criterion for specimen with $h/w = 2.0$, $a/w = 0.1$ and for $(r/a)^{90°} = 0.1$, for singular solution and for $\rho/a = 0$ (sharp crack), 0.010, 0.020, 0.030, 0.040, 0.050, 0.080 and 0.100.

A study of the dependence of the crack extension angle on specimen geometry and the radius of the notch tip was made. Conclusions drawn were: (i) The influence of the ratio $(r/a)^{90°}$ on the crack extension angle and the fracture stress is considerable and depends on the type of material. (ii) For each value of $(r/a)^{90°}$ the crack extension angles $-\theta_0$ and fracture stress $\sigma_\infty^{\beta,\rho}/\sigma_\infty^{90°,\rho}$ decrease as the ratios a/w and ρ/a increase. (iii) The influence of the radius of notch tip is strong for very low values of the ratio $(r/a)^{90°}$ and is weak for greater values. The same variation can be observed for the ratio a/w.

From this study it is concluded that we have to estimate the parameters: core-region, $(r/a)^{90°}$, geometry of specimen, h/w and a/w, and radius of notch tip, ρ/a, for the calculation of the crack extension angles and the fracture stresses because after loading the sharp crack becomes blunt with a finite radius of notch tip.

6.7 Cracked Plates Under Stress-Assisted Diffusion

In this section crack initiation under stress-assisted diffusion is studied [144, 145]. Around the crack tip a core-region is developed by stress-assisted diffusion. This core-region is a strongly brittle area influenced by the sum of the principal stresses at the crack tip and by phenomenological coefficients. Based on this assumption, the crack extension angle is determined by using the Det.-criterion. This study is applied to various materials with different phenomenological coefficients.

According to stress-assisted diffusion theory [147–151] (Sect. 2.10), the distribution of the concentration ρ around the crack tip is given by Rel. (514). This concentration depends on the sum of the stresses $\sigma_{xx} + \sigma_{yy}$.

6.7.1 For Singular Solution

The sum of stresses $\sigma_{xx} + \sigma_{yy}$ from Rels (693) and (694) for singular solution is:

$$\sigma_{xx} + \sigma_{yy} = \frac{2\sigma_\infty \sin \beta}{\sqrt{2r/a}} \left[\sin \beta \cos \frac{\theta}{2} - \cos \beta \sin \frac{\theta}{2} \right] \qquad (775)$$

By substituting Rel. (775) into Rel. (514), we obtain:

$$\frac{r}{a} = \left(\frac{2B\sigma_\infty \sin\beta \left[\sin\beta \cos\frac{\theta}{2} - \cos\beta \sin\frac{\theta}{2} \right]}{\sqrt{2}[(\rho/\rho_0)^{1/4} - 1]} \right)^2 \tag{776}$$

By substituting Rels (693)–(695) and (776) into Rel. (686), we obtain:

$$\text{Det.}(\sigma_{ij}) = \frac{[(\rho/\rho_0)^{1/4} - 1]^2}{4B^2} \times \frac{D_1 \sin^2\beta + D_2 \cos^2\beta + D_{12} \sin\beta \cos\beta}{\left[\sin\beta \cos\frac{\theta}{2} - \cos\beta \sin\frac{\theta}{2} \right]^2} \tag{777}$$

where D_1, D_2, D_{12} are given by Rels (699)–(701).

Relation (777) gives the distribution of the determinant of the stress tensor $\text{Det.}(\sigma_{ij})$ on a curve, which is given by Rel. (776), around the crack tip [85].

6.7.2 For Solution with Constant Term

The sum of stresses $\sigma_{xx} + \sigma_{yy}$ from Rels (693) and (694) for the stresses with constant term $(+ \sigma_\infty \cos 2\beta)$ is:

$$\sigma_{xx} + \sigma_{yy} = \frac{2\sigma_\infty \sin\beta}{\sqrt{2r/a}} \left[\sin\beta \cos\frac{\theta}{2} - \cos\beta \sin\frac{\theta}{2} \right] + \sigma_\infty \cos 2\beta \tag{778}$$

By substituting Rel. (778) into Rel. (514), we obtain:

$$\frac{r}{a} = \left(\frac{2B\sigma_\infty \sin\beta \left[\sin\beta \cos\frac{\theta}{2} - \cos\beta \sin\frac{\theta}{2} \right]}{\sqrt{2}[(\rho/\rho_0)^{1/4} - 1 - B\sigma_\infty \cos 2\beta]} \right)^2 \tag{779}$$

By substituting Rels (693)–(695) and (779) into Rel. (686), we obtain:

$$\text{Det.}(\sigma_{ij}) = \frac{[(\rho/\rho_0)^{1/4} - 1 - B\sigma_\infty \cos 2\beta]^2}{4B^2} \times \frac{D_c}{\sin^2\beta \left[\sin\beta \cos\frac{\theta}{2} - \cos\beta \sin\frac{\theta}{2} \right]^2} \tag{780}$$

where:

$$D_c = D_1 \sin^4\beta + D_2 \sin^2\beta \cos^2\beta + D_{12} \sin^3\beta \cos\beta + \sqrt{2r/a} \cos 2\beta \sin\beta$$

$$\times \left[\sin\beta \cos\frac{\theta}{2}\left(1 + \sin\frac{\theta}{2}\sin\frac{3\theta}{2} \right) + \cos\beta \sin\frac{\theta}{2}\cos\frac{\theta}{2}\cos\frac{3\theta}{2} \right] \tag{781}$$

According to Rel. (670) it is possible to calculate the critical stress of fracture.

Applying Rels (777) and (780) for the materials of Table 11 and according to relation $\partial D/\partial\theta = 0$, the angle θ_0 of the initial crack extension is determined.

In order to examine the influence of stress-assisted diffusion on the crack extension angle, Rels (777) and (780) were solved for various materials (Table 11) and for angles of loading β between 0° and 90° (Fig. 140(b)).

Figure 182 presents the variation of the crack extension angle $-\theta_0$ versus the angle of loading β for hydrogen diffusion in iron with properties as in Table 11 and for ratio $\rho/\rho_0 = 1.5$, 5, 10 and 50. In the same figure the curves for the singular

Table 11. Properties of materials [152]

Material	Environment	A	$B(m^2/N)$	$\sigma_u(N/m^2)$
(i) iron	H_2	1.31	8.715×10^{-10}	4.0×10^8
(ii) 4130 steel	H_2	2	1.6×10^{-10}	1.6×10^8
(iii) 36G2S	H_2S	2.5	1.6×10^{-10}	9.9×10^8
(iv) Grade 250 Mar.	H_2S	2.5	1.6×10^{-10}	17.2×10^8
(v) 65G	H_2S	1.4	1.6×10^{-10}	7.4×10^8

solution without diffusion (dotted line) and with diffusion (continuous line) are drawn. From this figure we observed that the crack extension angle decreases as the stress-assisted diffusion rises and influences the phenomenon of crack initiation. As the ratio ρ/ρ_0 increases, the crack extension angle tends towards the limit value of the singular solution. The ratio ρ/ρ_0 depends on the phenomenological coefficients A and B and is a constant of the material.

Figure 183 presents the variation of the crack extension angle $-\theta_0$ versus the angle of loading β for hydrogen diffusion in 4130 steel with properties as in Table 11 and for ratio $\rho/\rho_0 = 1.5, 5, 10$ and 50. The curves of singular solution are also drawn. In this figure we observe the same behaviour of curves as in Fig. 182. For all values of ratio ρ/ρ_0, the crack extension angle for $\beta = 45°$ is $\theta_0 = -60°$.

Figure 184 presents $-\theta_0$ versus β curves for H_2S diffusion in 36G2S with properties as in Table 11 and for ratio $\rho/\rho_0 = 1.5, 5, 10$ and 50. In this figure we can observe the same behaviour of curves as in previous figures.

Figure 185 presents $-\theta_0$ versus β curves for H_2S diffusion in Grade 250 Mar

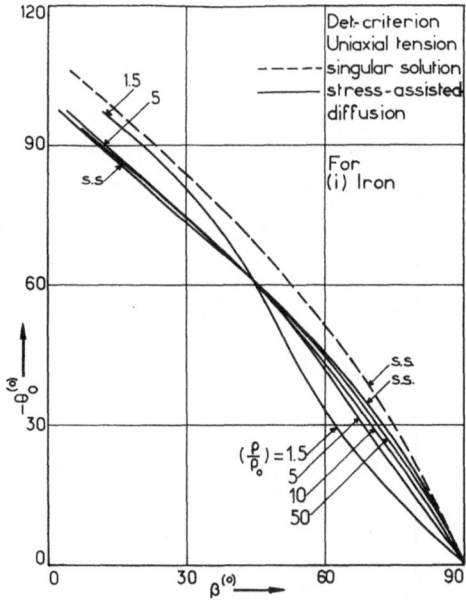

Fig. 182. Variation of $-\theta_0$ versus β according to the Det.-criterion for iron, for singular solution and for ratio $\rho/\rho_0 = 1.5, 5, 10$ and 50.

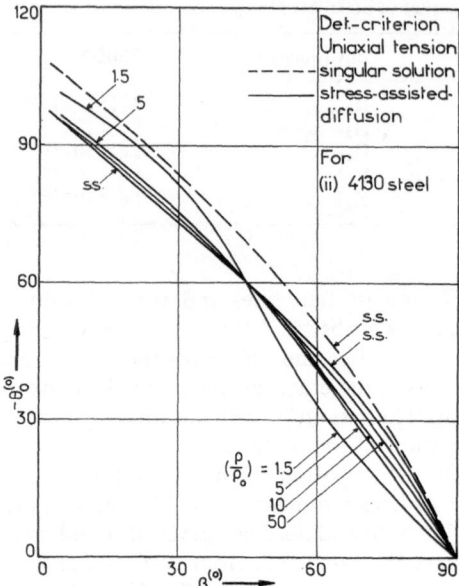

Fig. 183. As in Fig. 182, for 4130 steel.

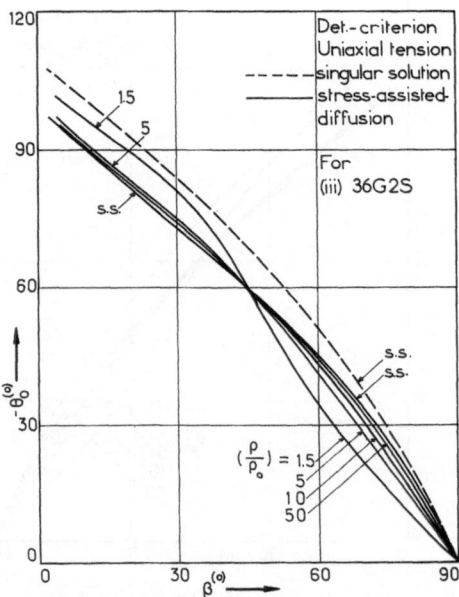

Fig. 184. As in Fig. 182, for 36G2S.

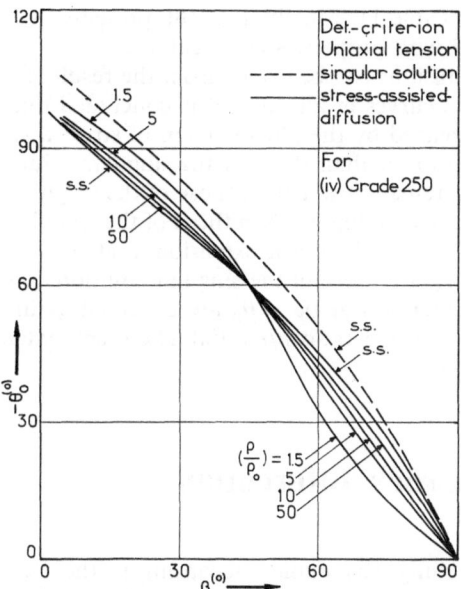

Fig. 185. As in Fig. 182, for Grade 250 Mar.

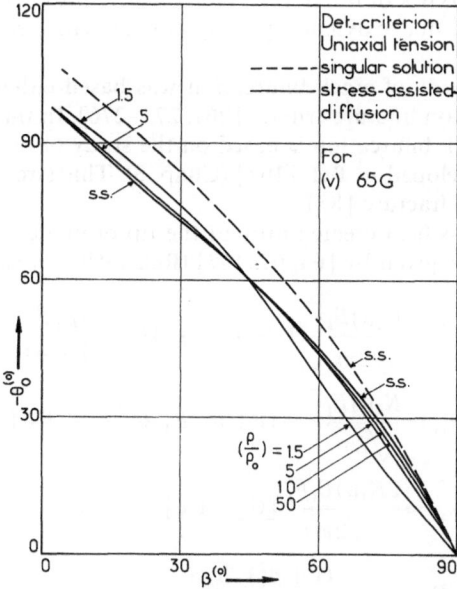

Fig. 186. As in Fig. 182, for 65G.

with properties as in Table 11. Finally, Fig. 186 presents $-\theta_0$ versus β curves for H_2S diffusion in 65 G with properties as in Table 11.

The following conclusions can be drawn from the results presented: (i) From the same behaviour of the curves in all cases, it is concluded that the crack extension angle is strongly influenced by the phenomenon of stress-assisted diffusion. (ii) All values of $-\theta_0$ remain lower than those of the singular solution without diffusion. (iii) All values of $-\theta_0$ remain lower than those of the singular solution for $\beta \geqq 45°$. (iv) All values of $-\theta_0$ remain higher than those of the singular solution for $\beta \leqq 45°$. (v) For all cases, for $\beta = 45°$ the crack extension angle $\theta_0 = 60°$. (vi) For all cases, the limit values of $-\theta_0$ are those of the singular solution. (vii) For all cases, small differences of crack extension angle $-\theta_0$ are observed. (viii) The crack extension angle depends mainly on the ratio ρ/ρ_0. (ix) The crack extension angle does not depend on the material.

6.8 Dynamic Crack Bifurcation

Dynamic crack-branching was studied according to the Det.-criterion of fracture [246]. In this section the stress field for a non-uniformly propagating crack under mode-I deformation is adopted. Predictions concerning both the critical crack velocity and the bifurcation angle are given.

The most interesting phenomenon during the crack propagation is the bifurcation of a moving crack. There are many causes of crack bifurcation. In homogeneous plates one of the significant factors for bifurcation is the crack velocity [266, 267], while in non-homogenous plates, e.g. plates containing inclusions, as occurs with particulates and fibre-composites, an additional main factor for bifurcation is the existence of several types of interfaces [177, 268–271]. Furthermore, the variation of the elastic modulus due to dynamic propagation of a crack causes crack bifurcation [167, 272].

The theoretical study of crack bifurcation was based either on critical stress or strain [266, 171], or on fracture criteria [267, 273–276]. In this section a theoretical investigation of crack bifurcation is based on the study of the elastodynamic stress field which was developed in Ref. [167] (Chap. 3). This stress field was applied to the Det.-criterion of fracture [85].

The singular stress field created around the tip of mode-I crack propagating at a variable velocity is given by [64, 67, 167] (Rels (559)–(561), Chap. 3):

$$\sigma_{xx} = \frac{K_1(t) B_1}{\sqrt{2\pi r}} \left[(1 + 2\beta_1^2 - \beta_2^2) F_1 - \frac{4\beta_1 \beta_2}{1 + \beta_2^2} F_2 \right] \tag{782}$$

$$\sigma_{yy} = \frac{K_1(t) B_1}{\sqrt{2\pi r}} \left[-(1 + \beta_2^2) F_1 + \frac{4\beta_1 \beta_2}{1 + \beta_2^2} F_2 \right] \tag{783}$$

$$\tau_{xy} = \frac{2 K_1(t) B_1 \beta_1}{\sqrt{2\pi r}} [G_2 - G_1] \tag{784}$$

$$B_1 = \frac{(1 + \beta_2^2)}{[4\beta_1 \beta_2 - (1 + \beta_2^2)^2]} \tag{785}$$

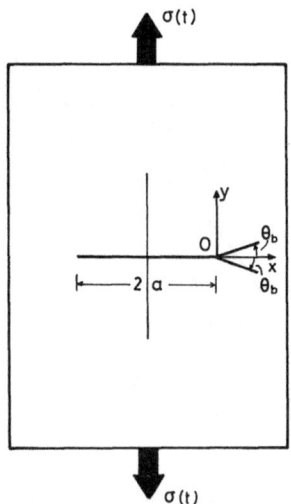

Fig. 187. Geometry of the specimen.

where F_j, G_j are given by Rels (562) and (563) and the quantity $K_I(t)$ is the time-dependent or dynamic stress intensity factor which is generally an unknown provided only by experimental methods.

By applying the Det.criterion of fracture (Det.$(\sigma_{ij}) = \sigma_{xx}\sigma_{yy} - \tau_{xy}^2$), we obtain [246]:

$$\text{Det.}(\sigma_{ij}) = \frac{B_I^2 K_I^2(t)}{2\pi r} D \tag{786}$$

where:

$$D = -(1 + \beta_2^2)(1 + 2\beta_1^2 - \beta_2^2)F_1^2 - \frac{16\beta_1^2\beta_2^2}{(1 + \beta_2^2)}F_2^2$$

$$+ \frac{8\beta_1\beta_2(1 + \beta_1^2)}{1 + \beta_2^2}F_1F_2 - 4\beta_1^2(G_2 - G_1)^2 \tag{787}$$

The variation of the values of the stress tensor determinant around the crack tip is calculated by Rel. (786). Because in dynamic brittle fracture the magnitude of the plastic zone around the tip of moving cracks decreases considerably with the speed of the crack [272, 278], the direction θ_b (Fig. 187), for which Det.(σ_{ij}) represents points lying on a circle of radius r, can be evaluated by:

$$\left.\frac{\partial D}{\partial \theta}\right|_{\theta = \theta_b} = 0, \quad \left.\frac{\partial^2 D}{\partial \theta^2}\right|_{\theta = \theta_b} < 0 \tag{788}$$

It is expected that at high crack velocities, the Det.(σ_{ij}) will present two symmetric maxima about the crack line ($\pm \theta_b$) since only symmetric configurations are considered.

The study of the prediction of an eventual bifurcation of a propagating crack under mode-I initial deformation is based on the variation of the relative crack velocity. Figures 188 and 189 present the variation of the bifurcation angle, θ_b, versus relative crack velocity, c/c_2, for values of Poisson's ratio, v, between 0.20 and 0.45 and for the plane-stress and plain-strain states, respectively. In these figures it

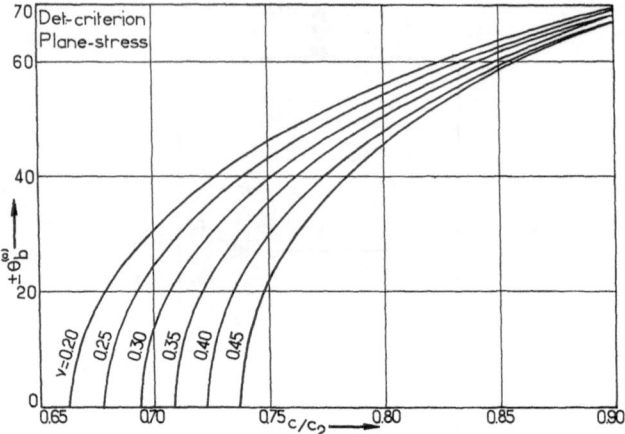

Fig. 188. Variation of the bifurcation angle, θ_b, versus relative crack velocity, c/c_2, for various values of Poisson's ratio, v and for the plane-stress state.

can be observed that there are critical relative velocities c/c_2 where the bifurcation angle θ_b is zero. For relative crack velocities greater than the critical crack velocities the bifurcation angle increases. As the relative crack velocities increase the bifurcation angle tends to a constant value .of 70°. Furthermore, for a constant relative crack velocity, the bifurcation angle decreases as the Poisson's ratio increases. Obviously then, for each Poisson's ratio (for each material) the bifurcation angle increases as the relative crack velocity increases. This means that, if the relative crack velocity is greater than the critical one, bifurcation occurs with angle θ_b greater than the angle which corresponds to critical velocity.

.   A theoretical study of bifurcation of a propagating crack was undertaken within the context of the elastodynamic theory and the Det.-criterion of fracture. Critical relative crack velocities, by which bifurcation can probably be started, are given.

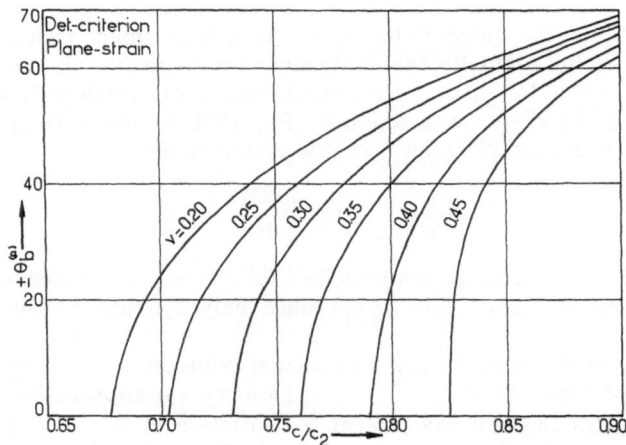

Fig. 189. As in Fig. 188, for plane-strain state.

For greater relative crack velocities than the critical velocities, crack bifurcation can occur because there are many reasons which cause the branching of a crack.

A detailed experimental study of the mode of propagation and bifurcation of a transverse crack inside a bi-phase specimen is reported in Refs [270, 271]. In these papers, the influence of the mechanical properties of each phase and the effect of the external loading rate on the crack propagation velocity and the initiation of the crack bifurcation was studied by using high-speed photography and dynamic caustics. It was shown that the propagating crack tended to bifurcate either in the hard or in the mesophase layer, and the bifurcation of the crack depended strongly on the strain rate, the elastic modulus and the Poisson's ratio of the phases, as well as on the extent of the mesophase layer, which relies upon the degree of adhesion between phases.

From this experimental study it was deduced that, as the elastic modulus of the phases is reduced, the relative crack velocities are increased. Phenomena of branching appeared at phases with high elastic modulus (brittle materials) in cases where the external loading created a low strain rate, while phenomena of branching appeared at phases with low elastic modulus (ductile materials) in the case of external loadings with high strain rates. In models with small differences in the elastic moduli of the phases and with external loading with low strain rates, only weak phenomena of bifurcation appeared.

These results indicated that, as the elastic modulus decreased [167], the relative crack velocity increased and, if the strain rate of external loading increased, phenomena of early bifurcation and microbranching appeared more frequently.

An important factor influencing the appearance of bifurcation is the state of adhesion between phases in composites. It was shown in this study that these mesophases play an important role in advancing and attracting bifurcations to themselves; this is beneficial for the structure, since the mesophases constitute poles, splitting the strain energy at the crack tip into parts and thus curbing any tendencies to failure.

An indirect effect of the development of either the main branches in the propagating cracks, or the micro-cracks in a chevron-like distribution along the main cracks, is the subsequent absorption, by them, of part of the strain energy and the creation of significant values of K_{II}^d stress intensity factors, which phenomena hinder the overall failure procedure of the cracked structure.

Figure 190 presents a photograph of a (50%–0%–50%) plasticized biphase specimen showing the crack propagation and the angles of bifurcation θ_b. Figure 191 presents a series of photographs showing the steps in crack propagation and

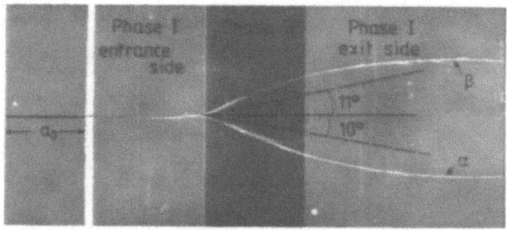

$\dot{\varepsilon} = 8s^{-1}$ (50-0-50) model

Fig. 190. Crack propagation in a specimen having soft phase I and a hard phase II. The applied strain rate was $\dot{\varepsilon} = 8/s$.

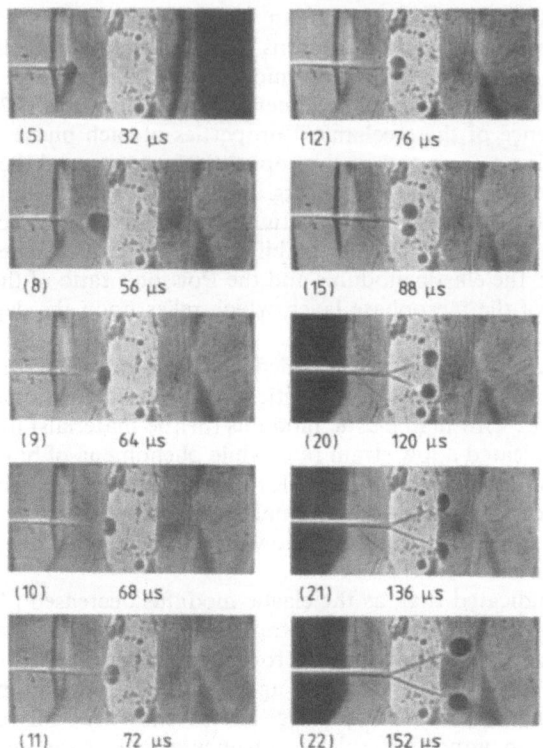

Fig. 191. Series of photographs showing the crack propagation in a (50%–0%–50%) plasticized biphase specimen subjected to dynamic loading with a strain rate $\dot{\varepsilon} = 8/s$.

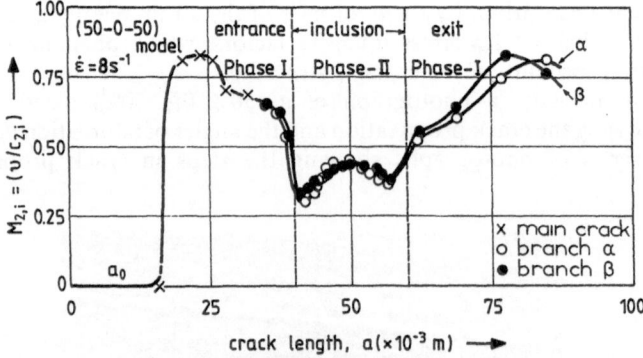

Fig. 192. Variation of the crack propagation relative velocities $M_{2,i} = c/c_{2,i}$ versus the crack length a for a (50%–0%–50%) plasticized biphase specimen subjected to dynamic loading with a strain rate $\dot{\varepsilon} = 8/s$.

bifurcation and the caustics which are formed at the propagating crack tips. Figure 192 presents the variation of the crack propagation relative velocities $M_{2,i} = c/c_{2,i}$ versus the crack length a for a (50%–0%–50%) plasticized biphase specimen subjected to a strain rate $\dot{\varepsilon} = 8/s$.

It may be deduced from these experimental curves that when a crack traverses the boundary between a region of low elastic modulus and a region of high modulus (entrance side to the inclusion) the relative velocity diminishes and if this reduction causes the crack velocity to become lower than the critical branching velocity the eventual branching of the crack is avoided. The opposite phenomenon occurs when the crack traverses the exit interface leaving the inclusion and entering the matrix. In this case the propagating crack moves from high to low modulus material and then the relative crack velocities exceed their critical values. At this interface there is always a great tendency for bifurcation of the crack, provided the strain energy at the crack tip attains the relevant limit so that the branching phenomenon can be permanently established. Otherwise the branches degenerate into micro-branches.

An exhaustive theoretical study of the bifurcation of a propagating crack in a thin elastic plate presenting abrupt variations of its elastic moduli was undertaken in Ref. [167].

It has been shown that the main cause for changing the relative velocity of a propagating crack and the subsequent eventual development of a bifurcation is the variation of the elastic modulus of the medium. Poisson's ratio variations which are in phase with any variation of the elastic modulus especially in high polymers is an important factor contributing to the development of branching.

However, as soon as the environmental conditions of a propagating crack are appropriate for an eventual branching of the crack, the amount of elastic strain energy stored at the crack tip regulates the possibility of stable evolution of branching or of degeneration of the branching phenomenon to microbranching.

The elastic strain energy is intimately related with the quality of material (brittle–ductile). Then, for brittle materials the branching phenomenon is more intense since the respective strains around the crack are reduced, whereas large strains appearing in the crack tip region of ductile materials result in high energy absorptions and therefore hinder the branching phenomenon.

The experimental results are in general agreement with the theoretical results. However, since evaluation of the instantaneous velocities and angles of inclination of the moving crack is difficult to attain with high accuracy the discrepancies between the two groups of values should be attributed equally to both of them. In general the method of caustics, when applied to polymeric plates properly selected to be free from irregularities, voids and other discontinuities, yields satisfactory experimental results whose accuracy is reflected in its agreement with the corresponding theoretical results.

Chapter 7
Experimental Det.-Criterion of Fracture

7.1 The Maximum Shear Stress, Isochromatic and Isopachic Fringe Patterns

For plane conditions the maximum shear stress τ_m is related to the Cartesian components of stress by [53, 279, 280]:

$$(2\tau_m)^2 = (\sigma_{xx} + \sigma_{yy})^2 - 4\,\text{Det.}(\sigma_{ij}) \tag{789}$$

where $\text{Det.}(\sigma_{ij}) = \sigma_{xx}\sigma_{yy} - \tau_{xy}^2$, and the stress-optic law relates the *isochromatic fringe* order N to τ_m by the relation:

$$2\tau_m = \frac{Nf_\sigma}{d} \tag{790}$$

where f_σ is the material fringe value and d is the specimen thickness.

Substituting Rel. (790) into Rel. (789), we obtain:

$$4\,\text{Det.}(\sigma_{ij}) = (\sigma_{xx} + \sigma_{yy})^2 - \left(\frac{Nf_\sigma}{d}\right)^2 \tag{791}$$

Relation (791) defines the isochromatic fringe pattern in the local field at the crack tip. An isochromatic fringe is plotted in Figs 193 and 194 for fringe order $N = 5$ and for $\beta = 90°$ and $30°$ respectively.

The first invariant $I_1 = \sigma_{xx} + \sigma_{yy}$ gives the lines of constant thickness of the specimen around the crack tip. These lines are called *isopachics* [281, 282]. The fringe order of the isopachic pattern N is related to the first stress invariant I_1 by the relation:

$$I_1 = \sigma_{xx} + \sigma_{yy} = \frac{Nf_p}{d} \tag{792}$$

where f_p is the isopachic fringe value. Relation (792) defines the isopachic fringe pattern in the local field at the crack tip. An isopachic fringe is plotted in Figs 193 and 194 for fringe order $N = 5$ and for $\beta = 90°$ and $30°$ respectively.

Substituting Rel. (792) into Rel. (789), we obtain:

$$4\,\text{Det.}(\sigma_{ij}) = \left(\frac{Nf_p}{d}\right)^2 - \left(\frac{Nf_\sigma}{d}\right)^2 \tag{793}$$

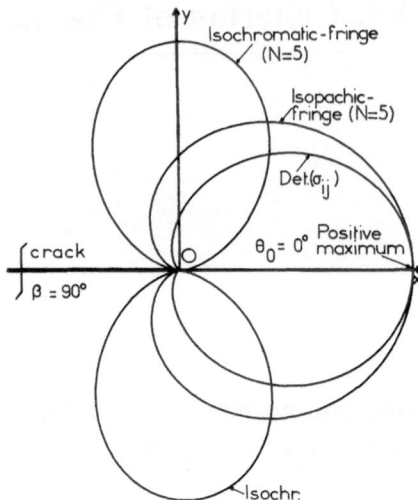

Fig. 193. Theoretical patterns of isochromatic and isopachic fringes and the distribution of $r\,\mathrm{Det.}(\sigma_{ij}) = f(\theta, \beta)$ around the crack tip for singular solution, for fringe order $N = 5$ and for $\beta = 90°$.

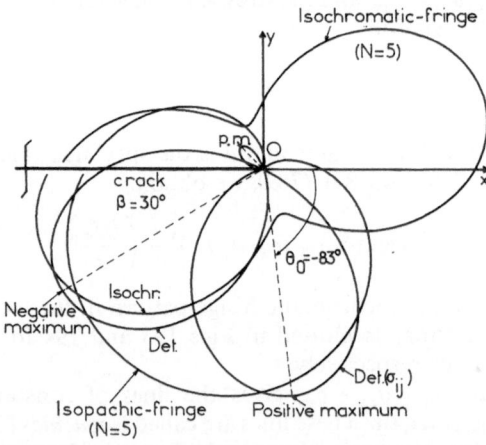

Fig. 194. Theoretical patterns of isochromatic and isopachic fringes and the distribution of $r\,\mathrm{Det.}(\sigma_{ij}) = f(\theta, \beta)$ around the crack tip for singular solution, for fringe order $N = 5$ and for $\beta = 30°$.

Relation (793) correlates the isopachic and isochromatic fringe patterns; their difference gives the values of the $\mathrm{Det.}(\sigma_{ij})$. The distribution of $r\,\mathrm{Det.}(\sigma_{ij}) = f(\theta, \beta)$ around the crack tip is also plotted in Figs 193 and 194. This distribution, which is obtained by the difference of the isopachic and isochromatic fringe patterns, is the same as those presented in Figs 142(a) and (b), respectively. Therefore, it is concluded that the distribution of $r\,\mathrm{Det.}(\sigma_{ij}) = f(\theta, \beta)$, around the crack tip, can be obtained experimentally from isopachic and isochromatic fringe patterns, which can possibly be obtained in the same photograph.

7.2 Experimental Method

Relation (793) can be experimentally proven [86]. Therefore, for a cracked specimen under loading the isopachic and isochromatic fringes can be taken in the same photograph, and hence the distribution of Det.(σ_{ij}), around the crack tip, can be experimentally calculated.

A thin birefringent plate of Lexan, width $w = 0.0391$ m, thickness $d = 0.0023$ m and length $l = 0.200$ m was used, with an edge crack of length $a = 0.003$ m.

The results of this experiment are presented as photographs in Figs 195 and 196. Figure 195 presents the experimental steps that compose the photograph of Figs 195(d) and 196. Figure 195(a) presents the unloaded cracked specimen; Fig. 195(b), the loaded specimen with a static stress at infinity of 13.1 MN/m^2, under which condition caustics are formed at the crack tip; and Fig. 195(c) the isochromatic fringes which are formed for the same load. Figure 195(d) presents the caustics, the isochromatic and the isopachic fringes. This photograph, repeated in Fig. 196 is formed by the superposition of photographs (a), (b) and (c) of Fig. 195. Numbers superimposed on the photographs indicate the order of fringes.

On the composite photograph of Fig. 196 the distribution of $r\,\text{Det}.(\sigma_{ij}) = f(\theta, \beta)$ is plotted; this was derived from the difference of isopachic and isochromatic fringes of order $N = 5$. This distribution of Det.(σ_{ij}) presents a maximum at the position $\theta = 0°$, in the crack direction. This means that the direction of crack initiation is $\theta = \theta_0 = 0°$, which is predicted by the Det.-criterion of fracture [85, 86].

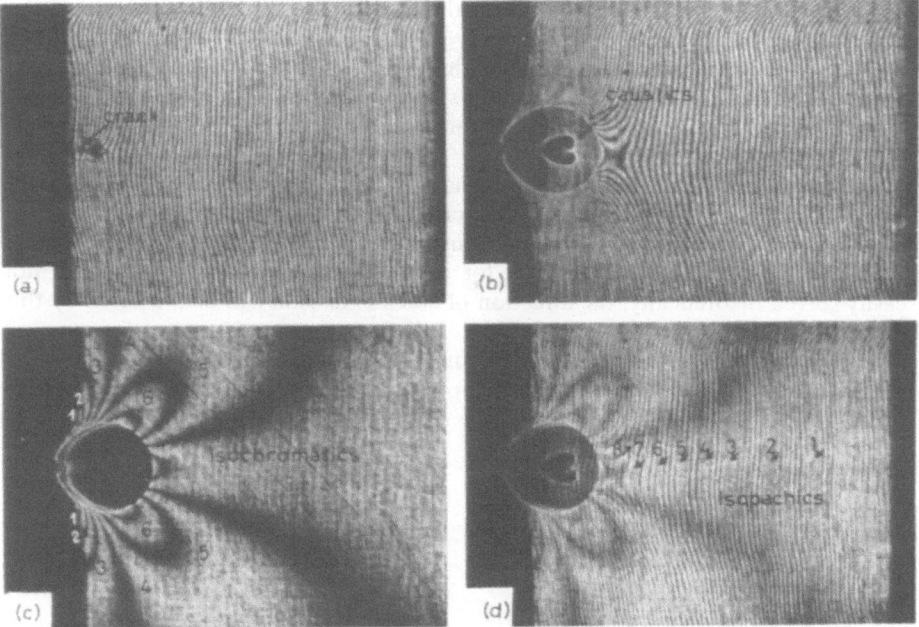

Fig. 195. Photographs of caustics, photoelastic and interferometric experimental methods. **(a)** The unloaded specimen. **(b)** The loaded specimen with $\sigma_\infty = 13.1$ MN/m^2 and the caustics which were formed at the crack tip. **(c)** The isochromatic fringe patterns which were formed around the crack tip. **(d)** The isopachic fringe patterns which were formed by the superposition of the photographs (a), (b) and (c).

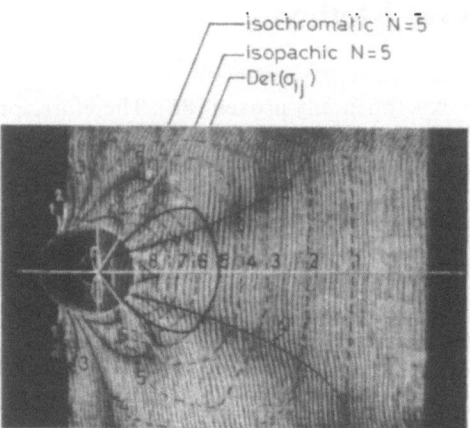

Fig. 196. The result of the superposition of the photographs (a), (b) and (c) of Fig. 195, and the calculation of the distribution of $r\,\mathrm{Det.}(\sigma_{ij}) = f(\theta, \beta)$ for fringe order $N = 5$.

In this study the stationary value of the second stress invariant as a local fracture parameter was taken and a new criterion of fracture was developed. For two-dimensional crack problems, the second stress invariant reduces to the determinant of the stress tensor and therefore, for plane stress and plane strain conditions, the Det.-criterion of fracture was developed. For various crack problems, this criterion of fracture gives theoretical results which are in agreement with the experimental results. The interest in this study is that the distribution of the determinant of the stress tensor is calculated experimentally from isopachic and isochromatic fringe patterns and therefore, prediction of the direction of crack initiation can be experimentally calculated. Experimental stress analysis near the crack tip and far from it is thus accomplished by the combination of the photoelastic, interferometric and caustics methods.

The validity of this experimental method is proved by the experiment of Fig. 196. This experiment proves that the Det.-criterion of fracture and the theoretical prediction of the direction of crack extension can be conveniently and accurately predicted experimentally. Therefore, for many cracked problems, which are not easily solved theoretically, the direction of crack extension can be predicted by this experimental method.

The experimental Det.-criterion of fracture proves that the proposed Det.-criterion of fracture method is simpler and more powerful than other fracture criteria used up to now.

References

1. Inglis CE (1913) Stresses in a plate to the presence of cracks and sharp corners. Trans Instn Nav Archit 55: 219–230
2. Muskhelishvili NI (1919) Sur l'intégration de l'equation biharmonique. Izvestiya, Ross Akad nauk: 13(6): 663–686
3. Griffith AA (1920) The phenomenon of rupture and flow in solids. Phil Trans Roy Soc London Ser A 221: 163–198
4. Griffith AA (1924) The theory of rupture. Proceedings First International Congress Appl Mech, Delft, pp 55–63
5. Wolf K (1923) Zur Bruchtheorie von A. Griffith. Zeitschr ang Math, Mech 3: 107–112
6. Obreimov IV (1930) The splitting strength of mica. Proc Roy Soc London A127: 290–297
7. Westergaard HM (1939) Bearing pressures and cracks. J Appl Mech 6: 49–53
8. Sneddon IN (1946) The distribution of stress in the neighbourhood of a crack in an elastic solid. Proc Roy Soc London A187: 229–260
9. Sneddon IN (1961) Crack problems in mathematical theory of elasticity. Report No ERD-126/1, North Carolina State College
10. Sneddon IN, Elliot HA (1946) The opening of a Griffith crack under internal pressure. Quart Appl Math 4: 262–267
11. Elliot HA (1946) An analysis of the conditions for rupture due to Griffith cracks. Proc Phys Soc 59: 208–223
12. Williams ML (1957) On the stress distribution at the base of a stationary crack. J Appl Mech 24: 109–114
13. Irwin GR (1948) Fracture dynamics, fracturing of metals. Am Soc Met Publ Cleveland: pp 147–166
14. Irwin GR (1957) Analysis of stresses and strains near the end of a crack traversing a plate. J Appl Mech 24: 361–364
15. Orowan EO (1950) Fundamental of brittle behavior of metals. In: Murray WM (ed) Fatigue and fracture of metals. Wiley, New York: pp 139–167
16. Zheltov YP, Khristianovitch SA (1955) On the mechanism of hydraulic fracture of an oil-bearing stratum. Izvestiya 5: 3–41
17. Barenblatt GI (1962) The Mathematical theory of equilibrium cracks in brittle fracture. Advances Appl Mech 7: 55–125
18. Hult JAH, McClintock FA (1957) Elastic–plastic stress and strain distribution around sharp notches under repeated shear. Proceedings Ninth International Congress Appl Mech 8: 51–58
19. McClintock FA (1958) Ductile fracture instability in shear. J Appl Mech 25: 582–588
20. McClintock FA, Sukhatme SP (1960) Travelling cracks in elastic materials under longitudinal shear. J Mech Phys Solids 8: 187–193
21. McClintock FA, Irwin GR (1965) Plasticity aspects of fracture mechanics. Symposium Fracture Toughness Testing and its Applications. ASTM STP 381: 84–113
22. Rice JR (1966) Constrained plastic deformation near cracks and notches under longitudinal shear. Int J Fract 2: 426–447
23. Rice JR (1966) Plastic yielding at a crack tip. Proceedings First Int Conf on Fracture, Sendai, Japan, 1: 283–308
24. Rice JR (1967) Stresses due to a sharp notch in a work-hardening elastic–plastic material loaded by longitudinal shear. J Appl Mech 34: 287–298

25. Smith E (1965) The spead of plastic from stress concentration. Proc Roy Soc London A285: 46–57
26. Smith E (1966) Fracture at stress concentration. Proceedings First Int. Conf. on Fracture, Sendai, Japan, 1: 133–151
27. Bilby BA, Cottrell AH, Swinden KH (1963) The speed of plastic yield from a notch. Proc Roy Soc, London A272: 304–314
28. Dugdale DS (1960) Yielding of steel sheets containing slits. J Mech Phys Solids 8: 100–104
29. Goodier JN, Field FA (1962) Plastic energy dissipation in crack propagation. In: Drucher and Gilman Eds, Fracture of Solids. pp 103–118
30. Theocaris PS, Gdoutos EE (1974) The modified Dugdale–Barenblatt model adapted to various fracture configurations in metals. Int J Fract Mech 10: 549–564
31. Manogg P (1964) Anwendung der Schattenoptik zur Untersuchung des Zerreissvorgangs von Platten. Dissertation 4/64, University of Freiburg
32. Theocaris PS (1970) Local yielding around a crack tip in plexiglas. J Appl Mech 37: 409–415
33. Theocaris PS (1970) Plastic strains at the roots of sharp notches in perspex. Proceedings Fourth Int. Conf. on Experimental Stress Analysis, Cambridge, England 1: 513–523
34. Theocaris PS (1971) Reflected shadow method for the study of constrained zones in cracked plates. Appl. Optics 10: 2240–2247
35. Theocaris PS (1973) Stress intensity factors in yielding materials by the method of caustics. Int J Fract Mech 9: 185–197
36. Theocaris PS, Gdoutos EE (1972) An optical method for determining opening-mode and edge sliding-mode stress intensity factors. J Appl Mech Ser E 39: 91–97
37. Theocaris PS (1978) The method of caustics applied to elasticity problems. In: G Holister (ed) Developments in Stress Analysis-1. Applied Sciences Publishers, London, England, Chapt 2: 27–63
38. Theocaris PS (1972) Reflected shadow method for the study of the constrained zones in cracked birefringent plates. J Strain Analysis 7: 75–83
39. Theocaris PS, Papadopoulos GA (1977) Stress corrosion crack growth in aluminum alloys. Engng Fract Mech 9: 781–794
40. Theocaris PS, Papadopoulos GA (1980) Complex stress intensity factors at cracks in birefringent plates by the method of reflected caustics. Materialprüfung 22: 246–253
41. Theocaris PS, Papadopoulos GA (1981) Stress intensity factors from reflected caustics in birefringent plates with cracks. J Strain Analysis 16: 29–36
42. Eftis J, Subramonian N, Liebowitz H (1977) Crack border stress and displacement equations revised. Engng Fract Mech 9: 189–210
43. Eftis J, Subramonian N (1978) The inclined crack under biaxial load. Engng Fract Mech 10: 43–67
44. Eftis J, Subramonian N, Liebowitz H (1977) Biaxial load effects on the crack border elastic strain energy and strain energy rate. Engng Fract Mech 9: 753–764
45. Liebowitz H, Lee JD, Eftis J (1978) Biaxial load effects in fracture mechanics. Engng Fract Mech 10: 315–335
46. Irwin GR (1958) Discussion on: The dynamic stress distribution surrounding a running crack – a photoelastic analysis. Proc Soc Exp Stress Analysis 16: 93–96
47. Smith DG, Smith CW (1972) Photoelastic determination of mixed mode stress intensity factors. Engng Fract Mech 4: 357–366
48. Theocaris PS, Gdoutos EE (1975) A photoelastic determination of K_I stress intensity factors. Engng Fract Mech 7: 331–339
49. Theocaris PS, Gdoutos EE (1976–7) Discussion on limitation of the Westergaard equation for experimental evaluations of stress intensity factors by WT Evans and AR Luxmoore. J Strain Analysis 11: 177–185 (1976): J Strain Analysis 12: 349–350 (1977)
50. Etheridge JM, Dally JW (1977) A critical review of methods for determining stress – intensity factors from isochromatic fringes. Exp Mech 17: 248–254
51. Ioakimidis N, Theocaris PS (1978) A simple method for the photoelastic determination of mode I stress intensity factors. Engng Fract Mech 10: 677–684
52. Sanford RJ, Dally JW (1979) A general method for determining mixed-mode stress intensity factors from isochromatic fringe patterns. Engng Fract Mech 11: 621–633
53. Dally JW, Sanford RJ (1978) Classification of stress-intensity factors from isochromatic-fringe patterns. Exp Mech 18: 441–448
54. Rossmanith HP (1979) Analysis of mixed mode isochromatic crack-tip fringe patterns. Acta Mechanica 34: 1–38
55. Cotterell B (1966) Notes on the paths and stability of cracks. Int J Fract Mech 2: 526–533
·56. Williams JG, Ewing PD (1972) Fracture under complex stress–angled crack problems. Int J Fract Mech 8: 441–446
57. Theocaris PS (1981) The elastic strain-energy density in cracked plates derived from caustics. Proceedings

Int. Symposium on Observed Specific Energy and Strain Energy Density Criterion, in Memory of Late Professor L Gillemot, Budapest (Sih G, Czoboly E, Gillemot, eds) 1980. M Nijhoff: pp 17–39

58. Theocaris PS, Papadopoulos G (1981) Mixed-mode elastodynamic forms of caustics for running cracks under constant velocity. Proceedings US-Greece Symposium on Mixed-Mode Crack Propagation (Sih G, Theocaris PS eds) Sijthoff and Noordhoff: pp 125–141

59. Theocaris PS, Michopoulos J (1983) The exact form of caustics in mixed-mode fracture. A comparison with approximate solutions. Acta Mechanica 46: 77–93

60. Theocaris PS, Katsamanis F (1978) Response of cracks to impact by caustics. Engng Fract Mech 10: 197–210

61. Theocaris PS (1977) Dynamic behavior of polymers studied by reflected caustics. Proceedings First Nat Symposium of Tensometry, Iasi, Roumania, 2: 207–221

62. Schirrer R (1978) The effects of a strain-rate-dependent Young's modulus upon the stress and strain fields around a running crack tip. Int J Fracture 14: 265–279

63. Broberg KB (1967) In: Recent progress in applied mechanics (Broberg KB, Hult J, Fritthorf N, eds), Almquist and Wiksell Publ, Stockholm

64. Freund LB, Clifton RJ (1974) On the uniqueness of plane elastodynamic solutions for running cracks. J of Elasticity 4: 293–299

65. Rosakis AJ (1980) Analysis of the optical method of caustics for dynamic crack propagation, Brown University. Engng Fract Mech 13: 331–347

66. Theodorescu PP (1975) Dynamic of linear elastic bodies. Editura Academiei Republicii Socialiste Romania and Abacus Press, Bucharest: Chapt 3, pp 279–296

67. Freund LB (1976) The analysis of elastodynamic crack-tip stress fields. In: S Nemat-Nasser, ed. Mechanics Today, Pergamon Press, 3: Chapt II, pp 55–91

68. Theocaris PS, Ioakimidis N (1979) The equations of caustics for crack and other dynamic plane elasticity problems. Engng Fract Mech 12: 613–615

69. Erdogan F, Sih GC (1963) On the crack extension in plates under plane loading and transverse shear. J Basic Engng Trans ASME Ser D 85: 519–527

70. Wu CH (1978) Elasticity problems of a slender z-crack. J of Elasticity 8: 183–205

71. Wu CH (1978) Maximum-energy-release-rate criterion applied to a tension–compression specimen with crack. J of Elasticity 8: 235–257

72. Sih GC (1972) Introductory chapter: A special theory of crack propagation. In: Mechanics of fracture 1. Noordhoff

73. Sih GC (1972) Application of the strain-energy-density theory to fundamental fracture problems. Institute of Fracture and Solid Mechanics, Technical Report, Lehigh University, USA

74. Sih GC (1973) Some basic problems in fracture mechanics and new concepts. Engng Fract Mech 5: 365–377

75. Sih GC (1974) Strain-energy-density factor applied to mixed-mode crack problems. Int J Fract 10: 305–321

76. Sih GC (1975) Introductory chapter: A three-dimensional strain energy density factor theory of crack propagation. In: Mechanics of fracture 2. Noordhoff

77. Sih GC (1973) A special theory of crack propagation. In: G Sih, ed. Methods of analysis and solutions of crack problems. Noordhoff Int Publ Holland: pp XXI–XLV

78. Sih GC, Macdonald B (1974) Fracture mechanics applied to engineering problems – strain energy density fracture criterion. J Engng Fract Mech 6: 361–386

79. Sih GC, Kipp ME (1974) Fracture under complex stress. The angled crack problem, discussion. Int J Fract 10: 261–265

80. Riedmüller J (1975) Experimentelle Bestimmung der Energiedichteverteilung in der Nähe von Risspitzen bei Oberlagerter normal-und scherbeanspruchung, Wissenschaftlicher. Bericht August 1975

81. Theocaris PS (1981) The causic as a means to define the core region in brittle fracture. Engng Fract Mech 14: 353–362

82. Theocaris PS (1981) Experimental determination of the core region in mixed-mode fracture Proceedings USA–Greece Symposium on Mixed Mode Fracture (Sih GC, Theocaris PS, eds) Noordhoff, Holland. pp 21–36

83. Theocaris PS, Papadopoulos G (1982) The distribution of the elastic strain-energy density at the crack-tip for fracture modes I and II. Int J Fract 18: 81–112

84. Theocaris PS, Andrianopoulos NP (1982) The Mises elastic–plastic boundary as the core region in fracture criterion. Engng Fract Mech 16: 425–432

85. Papadopoulos GA (1987) The stationary value of the third stress invariant as a local fracture parameter (Det-criterion). Engng Fract Mech 27: 643–652 (Errata, Engng Fract Mech 32: 665 (1989))

86. Papadopoulos GA (1989) New concepts on the Det-criterion. Engng Fract Mech 33: 283–293

87. Love AEH (1944) Mathematical theory of elasticity. Cambridge University Press, Cambridge, UK, pp 151–161
88. Dugdale DS (1968) Elements of elasticity. Pergamon, Oxford
89. Muskhelishvili NI (1963) Some basic problems of the mathematical theory of elasticity. 4th Edn, Noordhoff, Groningen
90. Timoshenko SP, Goodier JN (1970) Theory of elasticity. 3rd Edn, McGraw-Hill, New York
91. Irwin GR (1958) Fracture. Handbuch der Physik, Springer-Verlag, Berlin, 79: 551–590
92. Irwin GR (1962) The crack extension force for part through crack in a plate. J Appl Mech 29: 651–654
93. Born M, Wolf E (1970) Principles of optics. 4th Edn, Pergamon Press, London, pp 110–113
94. Coker EG, Filon LNG (1957) A treatise on photoelasticity. 2nd Edn, University Press, Cambridge, UK
95. Neumann FE (1841) Die Gesetze der Doppelbrechung des Lichts in Comprimiert oder ungleichförmiging erwämten unkrystallinischen Körpern. Abh.d. Kön. Acad. d. Wissenschaften zu Berlin Part II: 1–254
96. Maxwell C (1853) On the equilibrium of elastic solids. Trans Roy Soc Edin 20: 87–120
97. Favre H (1929) Sur une nouvelle méthode d'optique de détermination des tensions intérieures. Revue d'Optique Théorique et Instrumentale 8: 5–8
98. Theocaris PS, Gdoutos E (1974) An interferometric method for the direct evaluation of principal stresses in plane-stress fields. J Phys D: Appl Phys 7: 472–482
99. Raftopoulos D, Karapanos D, Theocaris PS (1976) Static and dynamic mechanical and optical behaviour of high polymers. J Phys D: Appl Phys 9: 869–877
100. Papadopoulos GA (1982) The method of reflected caustics in the study of stress and strain-energy density concentration at a crack under static and dynamic loading. Doctoral dissertation, Athens, Greece
101. Theocaris PS, Papadopoulos GA (1982) The influence of biaxiality of loading on the form of caustics in cracked plates. Acta Mechanica 44: 201–222
102. Bowie OL (1973) Solution of plane crack problems by mapping technique. In: Sih GC (ed) Methods of analysis and solutions of crack problems. Noordhoff International Publishing, Leyden, pp. 1–55
103. Paris PC, Sih GC (1965) Stress analysis of cracks. ASTM STP 381
104. Theocaris PS, Papadopoulos GA (1983) The optical method of reflected caustics in birefringent cracked plates under biaxial load. Materialprüfung 25: 238–241
105. Theocaris PS, Papadopoulos GA (1983) The experimental evaluation of the biaxiality factor k in cracked plates by the method of reflected caustics. Mechanics Research Communications 10: 297–306
106. Theocaris PS (1973) Optical stress rosette based on caustics. Applied Optics 12: 380–387
107. Theocaris PS, Ioakimidis NI (1979) On the determination of stress-optical constant by the method of reflected caustics. J Phys D: Appl Phys 12: 497–504
108. Kartalopoulos SV, Raftopoulos DD (1976) A rapid optical method for the determination of stress-optical constants of optically isotropic and anisotropic materials. J Phys D: Appl Phys 9: 2545–2553
109. Theocaris PS, Ioakimidis NI (1980) An improved method for the determination of mode I stress intensity factors by the experimental method of caustics. J of Strain Analysis 14: 111–118
110. Dally JW, Riley WF (1978) Experimental stress analysis, 2nd Edn, McGraw-Hill, New York
111. Cartwright DJ, Rook DP (1979) Green's functions in fracture mechanics. Proceedings Symposium on Fracture Mechanics – Current Status, Future Prospects. University of Cambridge
112. Bueckner HF (1970) A novel principle for the computation of stress intensity factors. Z. Angewandte Mathemat Mechan 50: 529–546
113. Rice JR (1972) Some remarks on elastic crack-tip stress fields. Int J Solids Structures 8: 751–758
114. Zienkiewicz OC (1971) The finite element method in engineering science. McGraw-Hill, New York
115. Sneddon IN, Lowengrub M (1969) Crack problems in the classical theory of elasticity. Wiley, New York, pp 17–18, 118–127
116. Sneddon IN (1973) Integral transform methods. In: Sih GC (ed) Methods of analysis and solutions of crack problems. Noordhoff Int. Leiden
117. Georgiadis HG, Papadopoulos GA (1987) Determination of SIF in a cracked plane orthotropic strip by the Wiener–Hopf technique. Int J Fract 34: 57–64
118. Georgiadis HG, Papadopoulos GA (1988) Cracked orthotropic strip with clamped boundaries. J. Appl Mathem and Phys (ZAMP) 39: 573–578
119. Georgiadis HG, Papadopoulos GA (1990) Elastostatic of the orthotropic double-cantilever-beam fracture specimen. J Appl Mathem and Phys (ZAMP) 41: 889–899
120. Papadopoulos GA, Georgiadis HG, Poniridis PI (1988) Finite length crack in a viscoelastic strip under impact – II Numerical results. Engng Fract Mech 29: 355–363
121. Theocaris PS, Papadopoulos GA (1984) The influence of geometry of edge-cracked plates on K_I and K_{II} components of the stress intensity factor, studied by caustics. J Phys D: Appl Phys 17: 2339–2349

122. Bowie OL, Neal DM (1965) Single edge crack in rectangular tensile sheet. J Appl Mech 32: 708–709
123. Andersson H (1969) Stress intensity factors at the tips of a star-shaped contour in an infinite tensile sheet. J Mech Phys Solids 17: 405–417 (Erratum, J Mech Phys Solids 18: 437, (1970))
124. Theocaris PS, Michopoulos JG (1982) Generalization of the theory of far-field caustics by the catastrophy theory. Appl Optics 21(6): 1080–1091
125. Theocaris PS, Gdoutos EE (1974) Verification of the validity of the Dugdale–Barenblatt model by the method of caustics. Engng Fract Mech 6: 523–535
126. Rosakis AJ, Ma CC, Freund LB (1983) Analysis of the optical shadow spot method for a tensile crack in a power-law hardening material. J Appl Mech 50: 777–782
127. Rosakis AJ, Freund LB (1982) Optical measurement of the plastic strain concentration at a crack tip in a ductile steel plate. ASME J Engng Mater and Technology 104: 115–120
128. Hutchinson JW (1968) Singular behavior at the end of a tensile crack. J Mech Phys Solids 16: 13–31
129. Rice JR, Rosengren GF (1968) Plane strain deformation near a crack tip in power hardening material. J Mech Phys Solids 16: 1–12
130. Rice JR (1968) Mathematical analysis in the mechanics of fracture. In: Fracture, an advanced treatise II. Liebowitz H (ed) Academic Press, pp 191–311
131. Shih CF (1973) Elastic–plastic analysis of combined mode crack problems. PhD Thesis, Harvard University
132. Prassianakis JN, Theocaris PS (1980) Stress intensity factors at V-notched elastic, symmetrically loaded, plates by the method of caustics. J Phys D: Appl Phys 13: 1043–1053
133. Theocaris PS, Papadopoulos GA (1987) The dynamic behaviour of sharp V-notches under impact loading. Int J Solids Structures 23: 1581–1600
134. Gross B, Mendelson A (1972) Plane elastostatic analysis of V-notch plates. Int J Fract 8: 267–276
135. Theocaris PS, Ioakimidis NI (1979) The V-notched elastic half-plane problem. Acta Mechanica 32: 125–140
136. Williams ML (1952) Stress singularities resulting from various boundary conditions in angular corners of plates in extension. J Appl Mech 19: 526–528
137. Ioakimidis NI, Theocaris PS (1978) A note on stress intensity factors for single edge V-notched plates in tension. Engng Fract Mech 10: 685–686
138. Theocaris PS (1987) The mesophase concept in composites. Springer-Verlag, Chap X, p 182; Chap XI, p 226
139. Theocaris PS (1975) Stress and displacement singularities near corners. J Appl Mathem Phys (ZAMP) 26: 79–98
140. Theocaris PS, Gdoutos EE (1976) Stress singularities at equal angle biwedges and two-material composite half planes with rough interfaces. J Appl Mech 98: 64–68
141. Tsamasphyros GJ, Theocaris PS (1976) Le premier probléme aux limites pour un coin infini. J de Méchanique 15: 615–630
142. Theocaris PS, Makrakis GN (1989) Caustics and quasiconformality; a new method for the evaluation of stress singularities. J Appl Mathem and Phys (ZAMP) 40: 410–424
143. Theocaris PS (1990) Evaluation of the variable order of singularities by pseudocaustics. Exper Mech 30(3): 240–246
144. Papadopoulos GA (1990) Crack initiation under stress-corrosion and stress-assisted diffusion. Int. Conference on Mechanics, Physics and Structural of Materials, Thessaloniki, August 20–25, 1990, Abstract p 152
145. Papadopoulos GA, Poniridis PI (1991) The influence of the stress-assisted diffusion on crack initiation and caustics. Engng Fract Mech 40: 265–276
146. Papadopoulos GA (1991) Influence of the orthotropy of the ductile materials and the stress-assisted diffusion on the caustics. SPIE Int Symposium on Optical Applied Science and Engineering, San Diego, July 21–26 Vol 1554A: pp 826–834
147. Aifantis EC (1976) Diffusion of a perfect fluid in a linear elastic stress field. Mech Res Comm 3: 245–250
148. Aifantis EC (1980) On the problem of diffusion in solids. Acta Mechanica 37: 265–296
149. Varotsos P, Aifantis EC (1980) Comments on the diffusion of a gas in a linear solid. Acta Mechanica 36: 129–136
150. Wilson RK, Aifantis EC (1982) On the theory of stress-assisted diffusion I. Acta Mechanica 45: 273–296
151. Unger DJ, Aifantis EC (1983) On the theory of stress-assisted diffusion II. Acta Mechanica 47: 117–151
152. Unger DJ, Gerberich WW, Aifantis EC (1982) Further remarks on the implications of steady-state stress-assisted diffusion on environmental cracking. Scripta Metallurgica 16: 1059–1064
153. Oriani RA (1969) Hydrogen in metals. Proceedings Conf. Fundamental Aspects of SCC. The Ohio State University: pp 32–50
154. Speidel MO (1971) Current understanding of stress corrosion crack growth in aluminum alloys. NATO

Science Committee, Research Evaluation Conference, Brussels: pp 289–344

155. Edeleanu C, Law TJ (1961) The propagation of corrosion pits in metals. J Inst Metals 89: 90–98
156. Pearson EC, Huff HJ, Hay RH (1952) Canadian J Technology 30: 311–319
157. Swann PR (1971) Morphological aspects of stress corrosion failure. NATO Science Committee, Research Evaluation Conference, Brussels: pp 113–126
158. Dugdale DS (1960) Yield of steel sheets containing slits. J Mech Phys Solids 8: 100–104
159. Mills NI (1974) Dugdale yielding zones in cracked sheets of glassy polymers. Engng Fract Mech 6: 537–549
160. Sih GC, Chen EP (1981) Cracks in composite materials. In: Sih GC (ed) Mechanics of fracture, Vol. 6, Noordhoff, Leyden
161. Theocaris PS (1976) Stress concentration in anisotropic plates by the method of caustics. J of Strain Analysis 11: 154–160
162. Katsamanis F, Raftopoulos D, Theocaris PS (1977) Static and dynamic stress intensity factors by the method of transmitted caustics. J Engng Mater and Technology 99: 105–109
163. Raftopoulos D, Karapanos D, Theocaris PS (1976) Static and dynamic mechanical and optical behavior of high polymers. J Phys D: Appl Phys 9: 869–877
164. Theocaris PS, Papadopoulos GA (1980) Elastodynamic forms of caustics for running cracks under constant velocity. Engng Fract Mech 13: 683–698
165. Theocaris PS, Papadopoulos G (1983) Mixed-mode dynamic stress intensity factors from caustics. ASTM STP 791: pp. 320–337
166. Theocaris PS, Papadopoulos GA (1984) Interrelation between static and dynamic stress intensity factors and their evaluation by caustics. J of Strain Analysis 19: 127–133
167. Papadopoulos GA, Theocaris PS (1987) The variation of the dynamic modulus on crack propagation modes. Int J Fract 35: 195–219
168. Papadopoulos GA (1990) Dynamic caustics and its applications. Optics and Lasers in Engng 13: 211–249
169. Sneddon IN (1952) Rend Cir Mat Palermo 2: 57–62
170. Radok JRM (1956) Quart Appl Math 14: 289–298
171. Craggs JW (1960) On the propagation of a crack in an elastic–brittle material. J Mech Phys Solids 8: 66–75
172. Kobayashi AS, Engstrom WL, Simon BR (1969) Crack opening displacements and normal strains in centrally notched plates. Exp Mech 9: 163–170
173. Beinert J, Kalthoff JF, Maier M (1978) Neuere Ergebnisse zur Anwendung des Schattenfleckverfahrens auf stehende und schnell-laufende Brüche. Proceedings 6th Exp Stress Anal, Munich
174. Theocaris PS, Papadopoulos G (1982) The minimum elastic energy density criterion in dynamic problems of propagating cracks. J Engng Mater and Technology 104: 207–214
175. Cotterell B, Rice JR (1980) Slightly curved or kinked cracks. Int J Fract 16: 155–169
176. Papanicolaou GC, Papadopoulos GA (1981) Dynamic crack propagation in particulate composite by the method of transmitted caustics. 1st USA–Greece Symposium on Mixed-Mode Crack Propagation, Sih GC, Theocaris PS, (eds), Sijthoff and Noordhoff, pp 367–384
177. Theocaris PS, Papanicolaou GC, Papadopoulos GA (1981) The effect of filler-volume fraction on crack-propagation behavior of particulate composites. J Composite Mater 15: 41–54
178. Papadopoulos GA, Papanicolaou GC (1988) Dynamic crack propagation in rubber-modified composite models. J Mater Sci 23: 3421–3434
179. Papadopoulos GA, Papanicolaou GC (1990) The effect of eccentricity on dynamic crack propagation behaviour of rubber modified PMMA models. J Mater Sci 25: 4066–4074
180. Papadopoulos GA (1991) Crack propagation in PCBA–PMMA sandwich plates. J Mater Sci 26: 569–578
181. Papadopoulos GA (1992) The effect of crack layer on dynamic crack propagation behaviour in polystyrene. J Mater Sci 27: 2154–2160.
182. Botsis J, Chudnovski A, Moet A (1987) Fatigue crack layer propagation in polystyrene – Part I Experimental observations. Int J Fract 33: 263–276
183. Botsis J, Chudnovski A, Moet A (1987) Fatigue crack layer propagation in polystyrene – Part II Analysis. Int J Fract 33: 277–284
184. Chudnovski A, Moet A (1985) Thermodynamics of translational crack layer propagation. J. Mater Sci 20: 630–635
185. Rice JR (1964) Mathematical analysis in mechanics of fracture. In: Liebowitz H (ed) Fracture, Vol II, Academic Press, NY, pp 192–308
186. Burech FE (1972) About the process zones surrounding the crack tip in ceramics. In: Fracture, 3, Pergamon Press, London, pp 929–950
187. Claussen N (1976) J Amer Ceram Soc 59: 49–60

188. Pompe HA, Bahr HA, Gille G, Kreher W (1978) Increased fracture toughness of brittle materials by microcracking in a energy dissipative zone at the crack tip. J Mater Sci 13: 2720–2723
189. Donald AM, Kramer EJ (1981) Micromechanics and kinetics of deformation zones at crack tips in polycarbonate. J Mater Sci 16: 2977–2987
190. Griffith AA (1920) The phenomenon of rupture and flow in solids. Phil Trans Roy Soc, London, Ser A 221: 163–198
191. Theocaris PS, Papadopoulos GA, Milios J (1984) Crack interaction in bending due to impact. Int J Impact Engng 2(2): 131–149
192. Kolsky H, Rader P (1969) Stress waves and fracture. In: Liebowitz H (ed), Fracture, Vol I, Academic Press, NY
193. Bodner SR (1973) Stress waves due to fracture of glass in bending. J Mech Phys and Solids 31: 1–8
194. Dao K, Hermann G (1977) An experimental study of crack propagation in beams during fracture in bending. In: Sih GC and Chow GL (eds), Fracture mechanics and technology
195. Kalthoff JF, Winkler S, Beinert J (1977) The influence of dynamic effects in impact testing. Int J Fract 13: 528–532
196. Kalthoff JF (1987) Shadow optical method of caustics. In: Kobayashi AS (ed) Handbook on experimental mechanics. Prentice-Hall Chap. 9: 430–500
197. Theocaris PS (1980) Deeply cracked strips subjected to dynamic three-point bending by caustics. Proceedings (Zbormik) 18th Czech Conf Exp Strain Analysis, Javornicky I, (ed) pp 293–303
198. Theocaris PS, Andrianopoulos NP (1981) Dynamic three-point-bending of short beams studied by caustics. Int J Solids Structures 17: 707–715
199. Bohme W, Kalthoff JF (1982) The behaviour of notched bend specimens in impact testing. Int J Fract 20: R139–143
200. Kalthoff JF, Bohme W, Winkler S (1982) Analysis of impact fracture phenomena by means of the shadow optical method of caustics. Proceedings VII Int Conf on Experimental Stress Analysis, Haifa, Israel, 23–27 August: pp 148–160
201. Theocaris PS (1978) Dynamic propagation and arrest measurements by the method of caustics on overlapping skew-parallel cracks. Int J Solids Structures 14: 639–653
202. Theocaris PS, Papadopoulos GA (1982) The propagation of skew-parallel cracks in an infinite plate studied by the method of dynamic reflected caustics. Proceedings VII Int Conf on Experimental Stress Analysis, Haifa, Israel, 23–27 August: pp 172–186
203. Theocaris PS, Milios J (1982) Dynamic interaction of collinear cracks. Acta Mechanica 43: 243–260
204. Kobayashi AS, Chan CE (1976) A dynamic photoelastic analysis of dynamic-tear-test specimens. Exp Mech 16: 176–181
205. Kobayashi AS, Ramulu M, Mall S (1982) ASME J Pressure Vessel Technol 104: 25–30
206. Melin S (1983) Why do cracks avoid each other? Int J Fract 23: 37–45
207. Theocaris PS, Papadopoulos GA (1984) The dynamic behaviour of an oblique edge crack under impact loading. J Mech Phys and Solids 32(4): 281–300
208. Kolsky H (1963) Stress waves in solids. Dover, NY
209. Sherwood JW (1958) Elastic wave propagation in a semi-infinite solid medium. Proc Phys Soc 71: 207
210. Cagniard L (1962) Reflection and refraction of progressive seismic waves. McGraw-Hill, NY
211. Ewing WM, Jardetsky WS, Press F (1957) Elastic waves in layered media. McGraw-Hill, NY
212. Davids N (1960) Some problems of transient analysis of waves in plates. Proceedings Int Symposium on Stress Wave Propagation in Materials, Davids N (ed), Interscience Publ, London, pp 271–288
213. Miklowitz J (1952) Elastic waves created during tensile fracture. J Appl Mech 20: 122
214. Thiruvenkatachar VR (1960) Recent research in stress waves in India. Proceedings Int Symposium on Stress Waves Propagation in Materials, Davids N (ed). Interscience Publ, London, pp 1–14
215. Eichelberger RJ (1960) Effects of very intense stress waves in solids. Proceedings Int Symposium on Stress Wave Propagation in Materials, Davids N (ed), Interscience Publ, London, pp 133–168
216. Broberg KB (1960) Some aspects of the mechanism of scabbing. Proceedings Int Symposium on Stress Wave Propagation in Materials, Davids N (ed), Interscience Publ, London, pp 229–246
217. Rinehart JS (1960) Role of stress waves in the communication of brittle, rocklike materials. Proceedings Int Symposium on Stress Wave Propagation in Materials, Davids N (ed), Interscience Publ, London, pp 247–270
218. Freund LB (1973) Crack propagation in an elastic solid subjected to general loading – III Stress wave loading. J Mech Phys Solids 21: 47–61
219. Freund LB (1974) Crack propagation in an elastic solid subjected to general loading – IV Obliquely incident stress pulse. J Mech Phys Solids 22: 137–146
220. Schardin H (1959) Velocity effects in fracture. Averbach BL, Felbeck DK, Hahn GT, Thomas DA (eds), Wiley, NY
221. Kalthoff JF, Shockey DA (1977) Instability of cracks under impulse loads. J Appl Phys 48: 986–993

222. Kobayashi AS, Mall S, Emery AF (1978) Fracture 1977. Advances in research on the strength and fracture of materials. 4th Int. Conf. on Fracture, 19–24 June 1977, Waterloo. Taplin DMR (ed), Pergamon Press, NY 3A: p 79

223. Homma HH, Ushiro T, Nakazawa H (1979) Dynamic crack growth under stress wave loading. J Mech Phys Solids 27: 151

224. Theocaris PS, Serafetinidis A (1984) Propagation of a slant crack under impact studied by caustics. Int J Impact Engng 2(3): 251

225. Theocaris PS, Georgiadis HG (1984) Emission of stress waves during fracture. J Sound Vibr 92(4): 517–528

226. Theocaris PS (1981) Elastic stress intensity factors evaluated by caustics. In: Sih GC (ed) Mechanics of fracture 7. Nijhoff, The Hague, Chap. 3

227. Theocaris PS (1979) Experimental study of plane elastic contact problems by the pseudocaustics method. J Mech Phys Solids 27(1): 15–32

228. Theocaris PS (1973) Stress singularities at concentrated loads. Proceedings of the 3th Int. Congress on Experimental Stress Analysis and Experimental Mechanics, Exp Mech 13(12): 511–518

229. Theocaris PS, Razem C (1977) Deformed boundaries determined by the method of caustics. J Strain Anal 12: 223–232

230. Theocaris PS, Stassinakis C, Mamalis A (1983) Roll pressure distribution and coefficient of friction in hot-rolling by caustics. Int J Mech Sci 25(11): 833–844

231. Theocaris PS, Ioakimidis N (1971) Some properties of generalized epicycloids applied to fracture mechanics. Z Angew Math und Physik 22(5): 876–890

232. Hill R (1964) The mathematical theory of plasticity. Clarendon Press, Oxford

233. Palaniswamy K, Knauss WG (1972) Propagation of a crack under general in-plane tension. Int J Fract Mech 8: 114–117

234. Hussain MA, Pu SL, Underwood J (1974) Strain energy release rate for a crack under combined mode I and II. ASTM STP 560: 2–28

235. Theocaris PS, Andrianopoulos NP (1982) A modified strain-energy density criterion applied to crack propagation. J Appl Mech 49: 81–86

236. Wang M-H (1985) A modified S theory. Engng Fract Mech 22: 579–584

237. Theocaris PS, Andrianopoulos NP (1982) The T-criterion applied to ductile fracture. Int J Fract 20: R125–130

238. Yehia NAB (1985) On the use of the T-criterion in fracture mechanics. Engng Fract Mech 22: 189–199

239. Von R Mises: (1913) Göttingernachrichten, Math Phys Klasse, 582

240. Von R Mises: (1926) Zeitschr ang Math Mech 6: 199

241. Theocaris PS (1984) A higher-order approximation for the T-criterion on fracture in biaxial fields. Engng Fract Mech 19: 975–991

242. Shah RC (1974) Fracture analysis. ASTM STP 560: pp 29–60

243. Pook LP (1971) The effect of crack angle on fracture toughness. Engng Fract Mech 3: 205–218

244. Papadopoulos GA (1987) The influence of geometry of edge-cracked plates on crack initiation. Engng Fract Mech 26: 945–954

245. Papadopoulos GA (1988) Crack initiation under biaxial loading. Engng Fract Mech 29: 585–598 Errata, Engng Fract Mech 32: 665, (1989))

246. Papadopoulos GA (1988) Dynamic crack-bifurcation by the Det.-criterion. Engng Fract Mech 31: 887–893

247. Papadopoulos GA, Poniridis PI (1988) Crack initiation from blunt notches under biaxial loading. Engng Fract Mech 31: 65–78

248. Papadopoulos GA, Poniridis PI (1988) The influence of geometry of edge-blunt-notched plates on crack initiation. Engng Fract Mech 31: 271–287

249. Papadopoulos GA, Poniridis PI (1989) Crack initiation under biaxial loading with higher-order approximation. Engng Fract Mech 32: 351–360

250. Haigh BP (1919) The strain energy function and the elastic limit. British Association of Advancement of Sciences, pp 486–495

251. Nadai A (1950) Theory of flow and fracture of solids. McGraw-Hill, NY

252. Sih GC, Cha CK (1974) A fracture criterion for three-dimensional crack problems. Engng Fract Mech 6: 699–723

253. Bradley W, Kobayashi A (1970) An investigation of propagating cracks by dynamic photoelasticity. Exp Mech 10: 106–113

254. Schroedl MA, Smith CW (1973) Local stress near deep surface flows under cylindrical bending fields. Progress in flow growth and fracture toughness testing. ASTM STP 536: pp 45–63

255. Etheridge J, Dally J, Kobayashi T (1978) A new method of determining the stress intensity factor K from isochromatic fringe loops. Engng Fract Mech 10: 81–93

256. Etheridge J, and Dally J (1978) A three-parameter method for determining stress intensity factor from isochromatic fringe loops. J Strain Anal 13: 91–94
257. Rossmanith H, Chona R (1981) A survey of recent developments in the evaluation of stress intensity factors from isochromatic crack-tip fringe patterns. Fifth Int Congress on Fracture Mechanics, Cannes, 5: 2507–2516
258. Theocaris PS, Spyropoulos CP (1983) Photoelastic determination of complex stress intensity factors for slant cracks under biaxial loading with higher-order terms effects. Acta Mech 48: 57–70
259. Francis PH, Ko WL (1976) The effect of root radius on the direction of crack extension under combined mode loading. Int J Fract 12: 243–251
260. Francis PH (1981) Crack extension trajectories from blunt notches in combined mode loading. Proceedings of the US–Greece Symposium on Mixed-Mode Crack Propagation, Sih GC, Theocaris PS, (eds). Sijthoff and Noordhoff, The Netherlands pp 323–330
261. Sih GC (1978) Strain energy density and surface layer energy for blunt cracks or notches. Introductory chapter in: Sih GC (ed), Stress analysis of notch problems. Mechanics of fracture–5. Sijthoff and Noordhoff, The Netherlands
262. Creager M (1966) The elastic stress field near the tip of a blunt crack. Master's thesis, Lehigh University, USA
263. Creager M, Paris PC (1967) Elastic field equations for blunt cracks with reference to stress corrosion cracking. Int J Fract 3: 247–252
264. Ewing PD, Swedlow JL, Williams JG (1976) Further results on the angled crack problem. Int J Fract 12: 85–93
265. Gdoutos EE (1980) The influence of specimen's geometry on the crack extension angle. Engng Fract Mech 13: 79–84
266. Yoffé EH (1951) The moving Griffith crack. Phil Mag 42: 739–750
267. Sih GC (1977) Introductory chapter. In: Mechanics of fracture–4. Noordhoff, Holland
268. Theocaris PS, Milios J (1980) Dynamic crack propagation in composites. Int J Fract 16: 31–51
269. Theocaris PS, Pazis D (1983) Crack deceleration and arrest phenomena at an oblique bimaterial interface. Int J Solids and Structures 19: 611–623
270. Theocaris PS, Siarova M, Papadopoulos GA (1986) Crack propagation and bifurcation in fiber composite models – I: Soft-hard-soft sequence of phases. J Reinforced Plastics and Composites 5: 23–50
271. Theocaris PS, Papadopoulos GA (1986) Crack propagation and bifurcation in fiber composite models – II: Hard-soft-hard sequence of phases. J Reinforced Plastics and Composites 5: 120–140
272. Theocaris PS, Georgiadis HG (1984) Rayleigh wave emitted by a propagating crack in a strain-rate dependent elastic medium. J Mech Phys Solids 32: 491–510
273. Kobayashi AS, Wade BG, Bradley WB, Chius ST (1974) Crack branching in homalite-100 sheets. Engng Fract Mech 6: 81–92
274. Theocaris PS, Georgiadis HG (1985) Bifurcation predictions for moving cracks by the T-criterion. Int J Fract 29: 181–190
275. Congleton JJ (1973) In: Sih GC (ed), Dynamic crack propagation, Noordhoff, Holland, pp 427–438
276. Ramulu M, Kobayashi AS, Kang BSJ (1982) Dynamic crack branching – A photoelastic evaluation. 15th Nat Symposium on Fracture Mechanics, University of Maryland
277. Eringen AC, Suhubi ES (1975) Elastodynamic, Vol II: Linear Theory. Academic Press, NY: pp 565–573
278. Atkinson C (1968) A simple model of a relaxed expanding crack. Arkiv för Fysik 35: 469–476
279. Frocht MM (1948) Photoelasticity, Wiley, NY
280. Post D (1954) Photoelastic stress analysis for an edge crack in a tensile field. Proc Soc Exp Stress Anal 12: 99–116
281. Post D (1955) A new photoelastic interferometer suitable for static and dynamic measurements. Proc Soc Exp Stress Anal 12: 191–202
282. Post D (1956) Photoelastic evaluation of individual principle stresses by large field absolute retardation measurements. Proc Soc Exp Stress Anal 13: 119–132

Subject Index